TEXT BOOK
OF
ATOMIC PHYSICS

DPH PHYSICS SERIES

TEXT BOOK OF ATOMIC PHYSICS

By

D.K. Jha

DISCOVERY PUBLISHING HOUSE
NEW DELHI-110002

Published by:

Tilak Wasan

DISCOVERY PUBLISHING HOUSE PVT. LTD.
4383/4B, Ansari Road, Darya Ganj
New Delhi-110 002 (India)
Phone : +91-11-23279245, 23253475, 43596065
E-mail : discoverypublishinghouse@gmail.com
sales@discoverypublishinggroup.com
web : www.discoverypublishinggroup.com

Edition: **2020**

ISBN: 978-81-7141-951-7

Text Book of Atomic Physics

Printed at:
Infinity Imaging Systems
Delhi

Preface

Atomic Physics tackled the difficult study of the nucleus of the atom. It is not only in the realm of theoretical knowledge that the discoveries of Atomic Physics have been the instrument of marvellous progress; they have been equally beneficial in their applications.

The present book covers maximum things in brief along with the theory of relativity and the quantum theory in its present form of Wave Mechanics. The study of the nucleus of the atom, in this book undertaken chiefly from the experimental point of view, and the structure and properties of the nucleus are discussed in the light of the experimental results.

This provides maximum knowledge to the B.Sc. and M.Sc. students of Atomic Physics.

I wish to express my deepest sense of gratitude to Mr. Tilak Wasan, the Prop. of Discovery Publication House, for his continued interest in bringing out this edition.

–Author

Contents

1

STRUCTURE OF AN ATOM

1.1 DALTON'S ATOMIC THEORY

In 1803 the British Chemist, John Dalton propounded the modern theory of the atomic construction of matter. Dalton proposed his atomic hypothesis which consisted of the following postulates.

- All chemical elements are made up of extremely small particles known as atoms.
- The atoms cannot be further sub divided by any chemical process.
- All atoms of the same element are identical (weights and other properties)
- When different elements combine chemically it is the atoms of these elements which combine together.

The weights and other properties of the atoms remain unchanged during the reaction. So the total weight of each element taking part in the reaction is equal to the combined weights of all the atoms present in it.

Therefore the total weight of all the elements taking part in the reaction must be equal to the total weight of all the substances produced in the reaction. This is the law of conservation of mass. By this we can understand that the elements taking part in the reaction always combine together in definite proportions by weight.

The ancient concept of atomism had involved some difficulties. First of all if the different elements were constituted with the common fundamental units, the atoms while the differences between elements were due merely to size and shape of the otherwise essentially same entities, the permanent existence of different elements in nature could not be adequately accounted for.

Another difficulty arose from the fact that, the atoms were not infinitely small since they had a size and shape proper to each element.

On account of these objections the atomic theory of matter did not gain universal acceptance for a very long period of the Christian era right up to the 19th century.

The arguments addaced in favour of atomism are as follows:

(a) *Compressibility of matter* which is clearly perceived in the gaseous state. If a gas which is evidently a material body is assumed to be continuous in structure, it is extremely difficult to understand how it can be compressed to a very small fraction of its original volume.

(b) The phenomenon of diffusion, osmosis etc. prove the existence of discrete minute particles with spaces between moving continuously, even in the smallest quantity of matter, in all the three states.

(c) The regular forms of crystals governed by definite simple laws also argue to the arrangement of discrete atoms in a well designed space-lattice.

(d) *The law of multiple proportions*, according to this law, when the same two elements combine to form different compounds, the different higher proportions of one element, which combine with the same constant amount of the other are simple integral multiples of the lower.

(e) *The periodic law among element,* the elements are arranged according to their atomic weights not only confirm the atomic structure of the elements, but also indicated that atoms must be built up according to some system.

As regards the internal structure of the atom with which we are directly concerned here progressive researches, chiefly those conducted by Rutherford on the scattering of α particles by matter showed that the atom could be split further into

(a) a central positive nucleus,

(b) configurations of electrons, developing the nucleus at distances relatively great which are responsible for the observed chemical and physical properties of the element concerned.

Number of types of units are used in Atomic physics they are as follows:

Length

$$\text{Angstrom} = 10^{-8} \text{ cm.}$$

This is used to measure the wave length of ordinary light.

In the case of X-rays and γ-rays the unit is still smaller the X.U. = 10^{-11} cm.

Mass

The electron mass or the rest mass of the electron is measured as follows:

$$m_0 = 9.028 \times 10^{-28} \text{ gm.}$$

Electric Charge

The fundamental unit is the charge of the electron, that is as follows:

$$4.767 \times 10^{-10} \text{ e.s.u.}$$

or

$$1.59 \times 10^{-20} \text{ e.m.u.}$$

Energy

In spectroscopy the natural unit is the Rydberg which is the ionisation energy of the hydrogen atom.

The energy in ergs can be evaluated as follows.

$$1\text{eV} = 4.767 \times 10^{-10} \times 1/300$$

$$= 1.589 \times 10^{-12} \text{ erg.}$$

therefore, One electron volt $= \dfrac{e(\text{e.s.u})}{300}$ erg

1.2 THE THOMSON ATOM MODEL

The Thomson atom model not only satisfied the requirements of the stability of the atom, but was also able to explain to a certain extent the origin of spectral lines.

Thomson considered the following points.

(a) Electrons enter into the constitution of all atoms.

(b) Since the atom as a whole is electrically neutral the quantity of positive and negative charges in it must be the same.

The important points, in which Thomson had to tackle were,

(i) The total number of electrons in an atom,

(ii) The way in which they were distributed, along with the positive charges, in the atom.

Thomson used X-rays scattering Phenomenon.

When a beam of X-rays passes through matter, it should be scattered.

The scattering coefficient 6, according to the classical theory being given by,

$$\sigma = 8\pi e^2 n/3m^2c^4.$$

where e and m are the charge and mass of the electron, n the number of electrons per unit volume and c the velocity of light. Now σ can be experimentally determined.

From the value of n, the number of electrons per atom can be readily computed. Thomson found that this number was proportional to the atomic weight of the element.

Thomson made the assumption that the positive charges were uniformly distributed in a sphere of atomic dimensions. This concept seemed to him most suited to mathematical treatment, while electrons were so arranged inside the positive sphere that their mutual repulsions were exactly balanced by the force of attraction towards the centre of the sphere.

Later Thomson showed that, in an atom with a single electron like the hydrogen atom, the electron must be situated at the centre of the positive sphere.

In the three electron system, the electron should be at the corners of a symmetrically placed equilateral triangle, the side of which was equal to the radius of the sphere. Proceeding in this manner, Thomson detailed the arrangement of electrons ranging from 1 to 100 inside the positive sphere.

According to Thomson hydrogen can give rise only to a single spectral line. Evidently Thomson's model was defective somewhere.

Lord Rutherford's α-particles proved that the assumption of Thomson about the uniform distribution of positive charges in a sphere of atomic dimensions, was wrong. This gave rise to a very different conception known as the nuclear atom model.

1.3 THE RUTHERFORD NUCLEAR ATOM MODEL

Earnest Rutherford proposed that, the electrons revolve in orbits round the nucleus much like the planets revolving round the sum, in their respective orbits.

In the case of the planets the gravitational attraction of the sun provides the centripetal force mv^2/r of the rotational motion of the planets in their orbits.

Experiments on the scattering of α-particles by thin foils of matter showed that although most of the α-particles suffered only a small deflection due to multiple scattering. Yet there were a certain number that were scattered through much larger angles.

When an α-particle enters the positive sphere the charge in the shell outside the path of the α-particle will exert no deflecting force on it. This means that there would be only a small deflection due to a single encounter.

Calculating the probable angle of scattering of α-particles as a function of the atomic charges it is found that,

$$N_Q = N_0 \; e^{-} \; (Q/Qm)^2$$

where N_Q = Number of particles scattered at an angle Q.

N_0 = The total number of incident particles.

Q_m = The most probable angle of scattering.

It is evident from this relation that the probability of large scattering is necessarily very small, Since as Q increases N_Q will decrease very rapidly.

In 1911, Rutherford proposed his atomic model capable of giving to an α-particle a large deflection due to a single encounter.

Rutherford atomic model had some drawbacks.

According to Rutherford, the electron revolving in its orbit is acted upon by the centripetal force its motion is accelerated. Thus it will lose energy by the emission of electromagnetic radiation. Due e attractive force of the nucleus it will move in orbits of continually decreasing radii and its path will be a spiral. Finally, it will fall upon the nucleus, itself and will disappear.

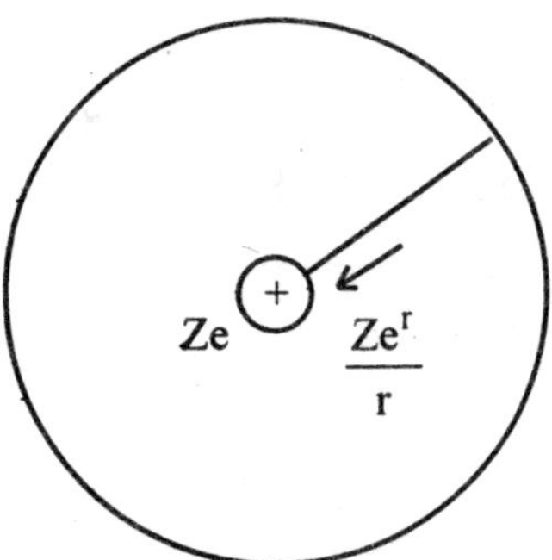

Fig. 1.1 : The Rutherford Nuclear Atom Model.

According to electromagnetic theory, revolving electron should radiate energy continuously.

All these drawbacks were solved in 1913 by Niels Bohr a Danish Physicist. He put forward his proposals based on the quantum idea which had been proposed by Max Planck a few years earlier.

1.4 BOHR'S ATOMIC MODEL

Bohr Postulated his theory as follows:

(i) The electrons cannot revolve in all possible orbits, (as suggested by the classical theory) but only in certain definite orbits, satisfying quantum conditions. The privileged quantum orbits are the non radiating paths of the electron.

(ii) Radiation of energy takes place only when an electron jumps from one permitted orbit to another. The energy thus radiated which is equal to the difference in the energies of the two orbits involved, must be a quantum of energy *hv*.

Bohr considered the simplest of all atoms *viz.,* the hydrogen atom. According to Rutherford model, there is only one electron, revolving in the orbit of this atom round the nucleus which is a single proton.

Bohr assumed that the electron revolves in a circular orbit.

In general if an atom has Z electron in its orbits and (Z – 1) of these are removed from it then the (Z – 1) fold positively charged ion which is left behind resembles a hydrogen atom with one electron revolving in the orbit round the nucleus carrying Z units of positive electricity.

1.5 ELECTRONIC STRUCTURE

When we consider a linear simple harmonic oscillator its displacement x at any instant t is given by

$$x = A \sin 2\pi vt$$

where, A = Amplitude

V = Frequency.

The kinetic energy of the oscillator at the instant t

$$= \frac{1}{2} m(dx/dt)^2$$

where 'm' is the mass of the oscillator and dx/dt its linear velocity at the instant considered.

As we known,

$$\text{The total energy} = \frac{1}{2} m\left[(dx/dt)m_2 x\right]^2$$

$$= \frac{1}{2} m(2\pi vA)^2$$

$$= 2\pi^2 V^2 A^2 m.$$

According to the quantum theory this energy should be an integral multiple of hv

Therefore, $nh\nu = 2\pi^2V^2A^2m$ (n = an integer)

$$nh = 2\pi^2A^2\nu m.$$

The momentum P_x of the Oscillator at the instant 't' is given by

$$Px = m\,(dx/dt)$$
$$= m \times 2V\pi A \cos 2\pi\,\nu t$$

$$\because \left.\begin{array}{r} m\cdot 2\pi\nu A = B \\ Px = B\cos 2\pi\nu t \end{array}\right\} \therefore Px/B = \cos 2\pi\nu t$$

From equation (1)

$$x/A = \sin 2\pi\nu t$$

$$\therefore \quad x^2/A^2 + Px^2/B^2 = 1.$$

By this we can understand that the relation between P_x and x is given by an ellipse.

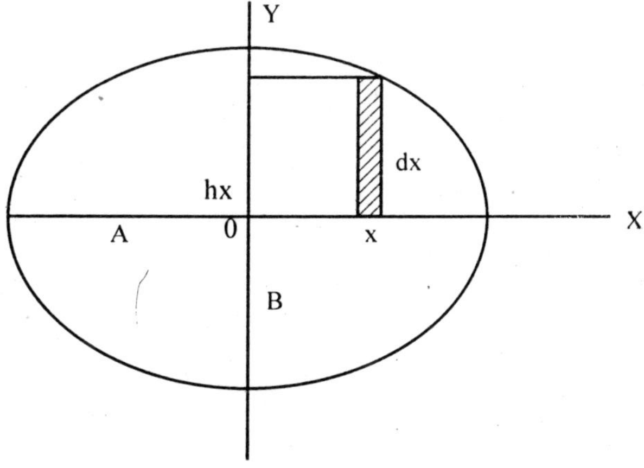

Fig. 1.2 : The Ellipse Showing the Relation Between P_x and x.

1.6 SPECTRAL LINES

The spectra of elements, become possible only when the wavelengths of the spectral lines could be measured.

The measurement of wavelength is based on the phenomenon of interference.

In 1801 Young was the first to make such estimates of wavelengths from data obtained with Newton's rings. After twenty years Fraunhofer developed the diffraction grating as a method of measuring wavelengths.

In 1868 Angstrom Published an elaborate table of wavelengths in the solar spectrum. Modern accuracy were made by Rowland, and he improved the technique of grating and using them.

Most of the atomic spectra are very complex, and the lines appear at first sight to be distributed at random.

The features of spectral lines can be noticed as follows :

1. Physical appearance of the lines for instance 'sharp' and 'diffuse'.
2. The behaviour of the lines, when the emitting atom is subjected to a magnetic field etc., it was shown that, the spectral lines of an element could be grouped into several series. Of such series for any one element the most intense are four called the principal, sharp, diffuse and fundamental.

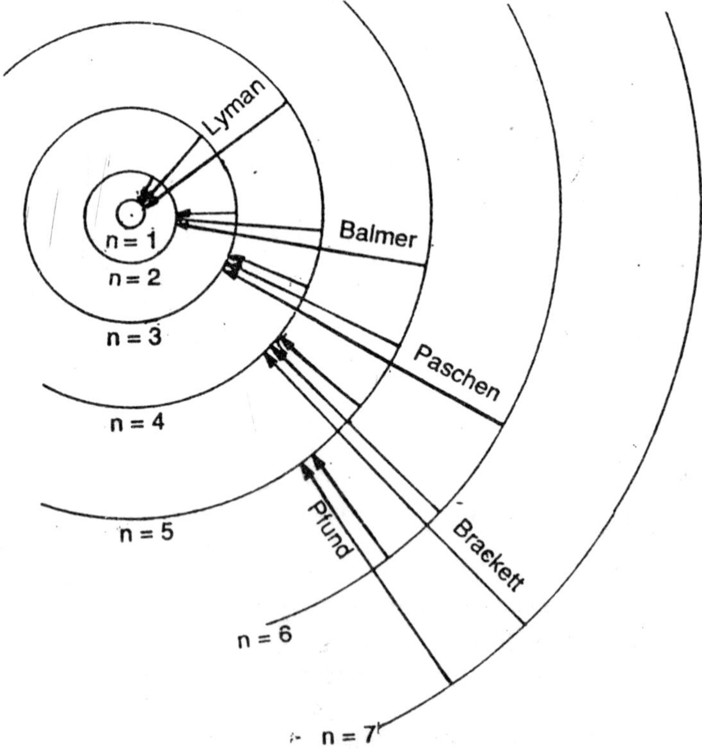

Fig. 1.3 : Origin of the Different Spectral Series of Hydrogen According to Bohr.

Physical similarities in the spectra of elements having thus been noted, several investigators sought for numerical relations in the lines of a series as well as between the different series of a given element and of different elements.

In 1883 Balmer had shown that the experimentally determined wave number of the spectral lines of hydrogen in the visible region could be represented by a formula.

$$\tilde{V} = A\left[\frac{1}{4} - \frac{1}{n^2}\right] = A\left[\frac{1}{2^2} - \frac{1}{n^2}\right]$$

A is a constant and n takes on different integral values for the different lines of the spectrum:

$$n = 3, 4, 5...$$

The value of A found from the measurement of the wavelength

$\lambda = \frac{1}{\tilde{V}}$ of the spectral lines was $1.09678 \times 10^7\ m^{-1}$.

The waves numbers of the different spectral lines of hydrogen can be expressed by a single mathematical formula as above.

The origin of the different spectral series of hydrogen due to the transition between the different orbits can be explained with the help of a simple fig.

The transitions are indicated by drawing arrows from the initial to the final orbits.

RYDBERG'S FORMULA

Rydberg in 1889 established that all the series in optical spectra could be arranged according to a general relation of the form

$$\overline{V} = \overline{V}_\infty - \frac{R}{(n + \mu)^2}$$

where R was found to be a universal constant for all series now called the Rydberg constant. n is an integer, μ a fraction less than unity which is practically constant for all the lines of a series; V_∞ the limiting or convergent wave number in the series, corresponding to $n = \infty$.

The great merit of Rydberg's formula consists in the fact that wave number of any line can be expressed as the difference of two terms, one fixed represented by $\overline{V}_\infty$ and the other variable which is obtained by giving different integral values to n. Rydberg's formula can be written as follows:

$$\overline{V} = R\left[\frac{1}{m^2} - \frac{1}{n^2}\right]$$

'm' is fixed and 'n' is variable.

1.7 THE MOTION OF THE NUCLEUS

It was assumed that the mass M of the nucleus was so great, compared with the mass m of the electron that the nucleus remained fixed at the centre of the circular orbits.

In the hydrogen atom the nuclear mass is only about 2000 times that of the electron. On account of its finite mass the nucleus will also revolve.

But the electron and the nucleus will evidently rotate about their common centre of gravity, O, the former in the larger circle of radius r_1, while the latter in the smaller one of radius r_2.

A simple theorem on centre of gravity gives the relation

$$Mr_2 = Mr_1.$$

As before if 'a' represents the distance between the nucleus and the electron, we get

$$\begin{aligned} r_1 &= a - r_2 \\ &= a - mr_1/M \\ (1 + m/M)\, r_1 &= a_i \\ \therefore \qquad r_1 &= M/(M + m) \qquad \text{...(1)} \\ r_2 &= Mr_1/M = a\,(M + m) \qquad \text{...(2)} \end{aligned}$$

The total Kinetic energy is made of two parts that of the electron and that of the nucleus.

$$\begin{aligned} \therefore \qquad \text{Total K.E.} &= \frac{1}{2}mv^2 + \frac{1}{2}MV^2 \\ &= \frac{1}{2}mw^2r_1^2 + \frac{1}{2}Mw^2r_2^2 \end{aligned}$$

Substituting the values of r_1 and r_2 from equations (1) and (2) and simplifying we get

$$\begin{aligned} \text{Total K.E.} &= \frac{1}{2}\{mM/(M+m)\}w^2a^2 \\ &= \frac{1}{2}\mu w^2a^2 \end{aligned}$$

$$\mu = \frac{mM}{M+m} = \frac{m}{\left\{\frac{1+m}{m}\right\}} \qquad ...(3)$$

We can write the final equation for the wave numbers of the spectral lines of the atom, in which the nuclear mass is not infinite.

$$\overline{V} = \frac{2\pi^2 mE^2 e^2}{Ch^3} \cdot \frac{M}{M+m}\left(\frac{1}{n_1^2} - \frac{1}{n_2^2}\right) \qquad ...(4)$$

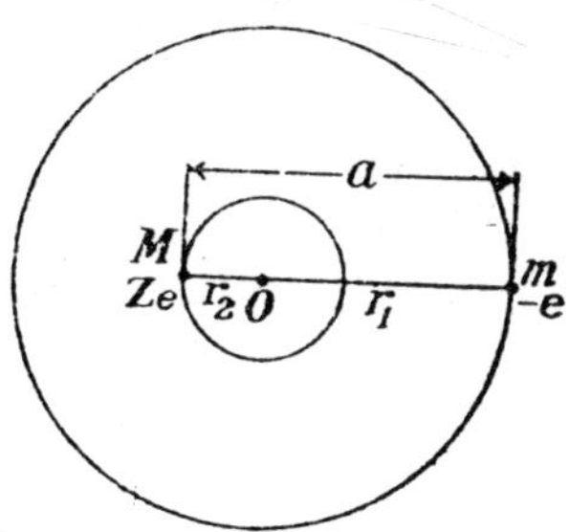

Fig. 1.4

Including the atomic number Z by the relation, E = Ze equation (4) becomes

$$\overline{V} = \frac{2\pi^2 me^4}{Ch^3} \cdot Z^2 \cdot \frac{M}{(M+m)} \cdot \left(\frac{1}{n_1^2} - \frac{1}{n_2^2}\right) \qquad ...(5)$$

The Rydberg constant for any element is given by

$$R = \frac{2\pi^2 me^4}{Ch^3} \cdot \frac{M}{M+m}$$

$$= R_\infty \cdot \frac{1}{1 + m/M}$$

where $R_\infty = 2n^2 me^4/Ch^3$, and M_8 the mass of the nucleus of the element of atomic number Z.

From the general expression for the Rydberg constant we get,

$$R_H = R_\infty \{M_H/(M_H + m)\}$$

$$R_{H_e} = R_{\infty} \cdot \left\{ \frac{M_{H_e}}{(M_{H_e} + m)} \right\}$$

For corresponding lines in the two spectra

$$\frac{R_{He}}{R_H} = \frac{4M_H (M_H + m)}{M_H (4M_H + m)}$$

$$= \frac{4(M_H + m)}{4M_H + m}.$$

The values of the Rydberg constant estimated from the Hydrogen.

$$R_H = 109{,}677.7 \text{ em}^{-1}.$$

1.8 EXCITATION OF ATOMS

Atoms can be excited and even ionised by the following methods.

1. Electronic Bombardment

The common method of excitation of atoms by can be carried by bombarding then with suitably accelerated electrons.

Slow moving electrons, whose velocity is below a certain critical value, cannot excite the atom on which they collide. The energy transferred from the electron to the atom in an elastic collision, even at its maximum value is inappreciable.

When the velocity of the colliding electron exceeds the critical value inelastic collision occurs, in which the electron loses a much larger amount of energy than that computed above and the struck atom suffers internal change of energy leading to its excitation.

We can conclude that an electron accelerated by a P.D. of 5.12 volts colliding with a Na atom imparts to it sufficient energy to raise an electron in it to infinity.

When the atom returns to the normal state, the electron might drop into any of the intervening quantum levels and thereby generate the entire are spectrum of sodium.

2. Thermal Excitation

The excitation of the atom and the subsequent emission of radiation are due to rise in temperature. At high temperatures with the increase in transitional energy of the atoms, the proportion of atoms, which have sufficient energy to excite one another by collision, is large enough to produce a perceptible luminescence.

The further increase of temperature, the transitional motion becomes increasingly violent and is able to produce higher violent and is able to produce higher excitation so that new spectral lines appear. In the electric spark discharge the lines of the ionised atom are seen.

When an excited atom subjected to collide with another unexcited atom it may lose its excitation energy. The kinetic energy of translation is converted into excitation energy.

Sensitized fluorescence when mercury and thallium were subjected to temperature, about 800° C, the wavelengths 5351, 3776, 3519, 2918 and 2218 Å were seen. The anomalous thallium line 2238° A, can not evidently be excited by the transfer of the excitation energy of Hg atom to the thallium atom.

3. Irradiation of Atoms by Light Energy

Atoms can be excited also by energy supplied in the form of light. Irradiating hydrogen with ultra violet light and supposing that the incident light contains one of the wavelengths of the Lyman series, the atoms which absorb this light will be raised to the second excited state. The electron will move from the first to the third orbit.

The phenomenon of emission of light by which have been excited by absorption of light is called fluorescence.

When an atom is irradiated with light of frequency corresponding to one of its absorption lines, the return of the atom from the excited to the normal state may occur in stages.

The critical potential of excited atoms can be determined by, the electrical method, or by the spectral method.

The spectral method is an indirect one, and is capable of giving more accurate results than the electrical method, since the measurement of wavelengths can be done with greater precision than that of accelerating potential.

1.9 THE SOMMERFELD ATOMIC MODEL

In 1915 sommerfeld modified Bohr's theory. He argued that the electron is moving around and under the influence of a massive nucleus like a planet round the central massive sun.

Considering the electron (–e) moving in an elliptical orbit its position at any instant can be fixed interms of polar coordinates r and Q.

Where r is the radius vector, that is the distance of the electron from the nucleus (+E) at one of the foci of the ellipse and Q the vectorial angle.

The tangential velocity V of the electron at the instant considered can be resolved into two components, one radial that is along the radius vector equal to dr/dt and the other transverse that is at the right angles to the radius vector equal to r (dQ/dt).

Sommerfeld assumed that since the elliptical orbits should satisfy the quantum conditions just as the circular orbits, the circle being the special case of the ellipse to, each of these two momenta, the phase integral of the quantum theory might be applied.

$$\oint Pr \cdot dr = n_r \cdot h$$

and

$$\oint P_Q \cdot d_Q = n_Q \cdot h. \qquad ...(1)$$

The three quantum numbers are related as follows

$$n = n_r + n_Q \qquad ...(2)$$

where,

n = total quantum number

n_r = the radial quantum number

n_Q = azimuthal quantum number.

The total energy W of the system is partly potential and partly kinetic

So,

$$W = -\frac{E_e}{r} + \frac{1}{2}m\left(\frac{dr}{dt}\right)^2 + \frac{1}{2}mr^2\left(\frac{dQ}{dt}\right)^2 \qquad ...(3)$$

The electron in the elliptical orbit satisfying the quantum conditions, can be shown as follows.

$$1 - E_2 = \frac{b^2}{a^2} = \frac{n_Q^2}{(n_Q + n_r)^2} \qquad ...(4)$$

where,

E = essentricity of the ellipse

a = semi-major axes

b = semi-minor axes.

$$W = \frac{2\pi^2 mE^2e^2}{h^2}\left(\frac{1}{n_Q + n_r}\right)^2 \qquad ...(5)$$

from equation (4) we get

$$(1 - E^2)^{1/3} = \frac{b}{a} = \frac{n_Q}{n_Q + n_r} = \frac{n_Q}{n}$$

By this we can understand that, for a given value of the total quantum number there can be only a limited number of elliptical orbits that are permitted.

The extension to elliptical orbits is not however, without effect since it enabled sommerfeld to proceed further and find a solution to the problem of the fine structure of spectral lines on the basis of the variation of the mass of the electron with velocity.

1.10 THE VECTOR ATOM MODEL

The Vector Atom Model is an extension of the Rutherford-Bohr-sommerfeld atom model. The original simple theory of Bohr was absolutely incapable of explaining the fine structure of spectral lines even in the simplest one electron systems, such as H and He^+ atoms.

New ideas were introduced partly by analogy, partly by empirical methods in the interpretation of more complex spectral phenomena and their relation to atomic structure which finally resulted in what is known as the vector atom model.

The two characteristic features of vector atomic model are as follows:

(i) quantisation of direction

(ii) the spinning electron hypothesis.

To quantise spatially we need of course a certain preferred direction with respect to which the orbits may receive their orientation. Such a favoured direction may be obtained by an external field of force.

To determine the permitted orientation relative to the field direction, we are guided by the fact that the projections of the quantised orbits on the field direction must themselves be quantised.

It has been known for many years that the yellow sodium line is actually a doublet. With a spectroscope of sufficiently high resolving power it is found that the yellow D line splits into two components.

In 1925 two Dutch physicists, Uhlenbeck and Goldsmith put forward the hypothesis of the spinning electron according to which the electron revolves not only in an orbit round the nucleus but also about an axis of its own some what like a planet in the solar system. In other words the electron is endowed with a spin motion over and above the orbital motion. Thus the 'spinning electron' introduced brought in profound modifications in the atom model.

Accordingly to 'spinning electron' idea the electron will be endowed with two angular momenta, and two magnetic movements one due to orbital motion of the electron, and also due to the spin of the electron. Likewise the total magnetic moment of the atom is due to both the orbital and spin magnetic moments.

The atomic magnet is the resultant of two magnets, one arising from the orbital motion and the other from the spin motion.

There are several ways in which the different vectors of the electrons may combine to give the vectors representing the atom as a whole. The method of combination depends on the interation or "coupling" between the component vectors. Since the orbital and spin motions of the electron producing magnetic fields result in mutual perturbation.

Under certain circumstances the interation between he spin and orbital vectors in each electron may be stronger than that between either the spin vectors or the orbital vectors of the different electrons.

Each electron may be considered separately and its contribution to the total angular momentum of the atom may be obtained by combining first its individual spin and orbital vectors by the relation $j = L + S$. The vector sum of all the individual, J vectors of the electrons gives the total angular momentum J of the atom.

1.11 PAULI'S PRINCIPLE

Pauli's principle states that every completely defined quantum state in an atom can be occupied by only one electron.

In otherwords it is impossible for two electrons in an atom to be identical as regards all their quantum numbers, that is one of the two in such a case will be excluded from entering into the constitution of the atom. Hence the name exclusion principle".

The principle may be stated in another way.

Two systems of quantum numbers which are deducible from each other by interchange of two electrons represent only one state. The principle enunciates the indistinguishability of electrons which have identical quantum numbers. Hence the name equivalence principle."

The principle defines a certain minimum individuality of the electron in the atom.

The principle was first introduced by Pauli to explain certain experimental facts.

The principle finds its chief use in the elucidation of electronic structure and atomic spectra. It is also helpful in defining the special quantum properly of closed shells, as well as accounting for certain limitations of terms multiplicity actually observed.

It has been realised later that the principle is much more universal and fundamental, holding good for the totality of electrons in any molecule, may even for the more comprehensive system of conduction electrons, that belong to a metal in the bulk state on the one hand and the constituent particles of the nucleus in the ultramicroscopic state on the other.

The Selection Rules

All the possible combinations of permitted energy states of an atom do not actually appear as spectral lines. For the vector atom model three selection rules have been devised, on for L another for J and a third for S.

The intensity rules have been devised to supplement the selection rules in order to predict also the intensity of the lines that occur.

Lande discovered a rule regarding the interval in frequency between he different levels, constituting a multiple. It is called the Lande interval rule.

1.12 FRANCK AND HERTZ EXPERIMENT

J. Franck and G. Hertz conducted series of experiments after the formulation of Bohr's theory.

In the Fig. T is a sealed tube within which a metal filament 'F' can be heated by passing an electric current through it so that there is electron emission from F. 'P' is a collector plate at the other end of the tube and 'G' is a wire-grid placed just infront of 'P'. Thus the electrons from F have to pass through the grid 'G' to reach 'P'.

The pressure within T which contains a small amount of the experimental gas, is kept quite low.

The grid 'G' is kept at a positive potential with respect to 'F' so that the electrons gain kinetic energy in travelling from F to G which becomes maximum when they reach G. The collector 'P' is kept at a slightly negative potential with respect to 'G'. The collector 'P' is kept at a slightly negative potential with respect to 'G'. This retarding potential V_0 is usually about 0.5 volt, so that if the electron energy is less than eV_0 while passing through G, they are unable to reach P.

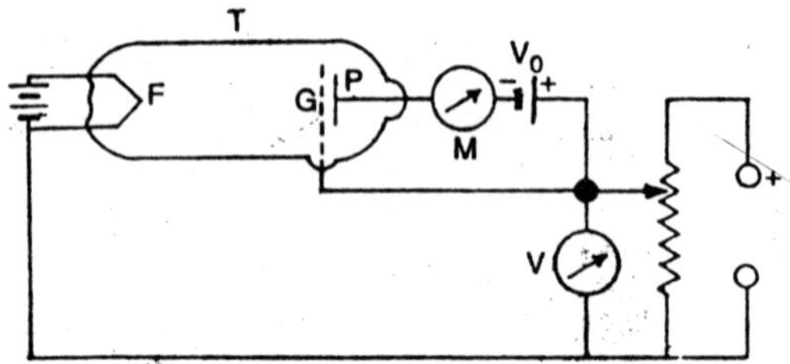

Fig. 1.5 : Franck and Hertz Experiment.

If an electron travelling from F to G collides with a gaseous atom then the atom may receive some energy from the colliding electron. If the collision is elastic, the n the energy transfer is negligibly small since the atoms are much heavier than the electrons. If the collision is with an electron bound in the atom, then there may be considerable energy transfer from the impinging electron to the atomic electron.

If the energy of the impinging electron is not sufficient to raise the electron bound in the atom from the ground level to the first excited energy level by collision, then there will not be any energy transfer between the two inspite of the collision.

Thus, the electrons coming from F will reach the grid G with the full energy gained by them and will be able to overcome the retarding potential V and fall on 'P'. The meter μ will then record an electric current. As the P.D. between F and G is increased the energy gained by electrons, as they reach G becomes equal to the energy difference between ground level and the first excited level of atomic electrons, then the latter may be transferred to the higher-level by gaining sufficient energy from the incident electrons. When this happens, the entire energy of an incident electron is transferred to the atomic electron.

The corresponding potential difference between, F and G is known as the resonance potential.

As the potential difference between F and G is further increased the current recorded by M begins to rise again. The collisions between the on-coming electrons and the atomic electrons now take place some distance behind G towards F.

Thus even after losing their entire energies by collision, the on coming electrons again acquire sufficient, energy when they reach G and are able to overcome the retarding potential V_0 to full on ‘P’ so that M again record, some current.

The different maxima of the current occur at potentials which are integral multiples of the resonance potential so that the difference in the potentials between two consecutive peaks is equal to the resonance potential.

2

X-RAYS

2.1 INTRODUCTION

Early investigations of the general properties of some invisible rays indicated that, they might be of the same nature as visible and ultraviolet light but of very short wavelength.

X-rays were discovered by W.C. Rontgen in Germany in 1895, almost accidentally while investigating the properties of cathode rays.

Roentgen thought that X-rays might be longitudinal, ether waves, unlike the transverse waves of ordinary light.

In 1912 Prof. Laue got an ingenious idea, which solved the problem to a very great extent. He expressed his opinion that, ultra violet light was not suited for these rays on account of their very short wavelength.

Later the identity of X-rays with light in all respects except that they occupy a narrow region of the spectrum beyond the ultraviolet was established.

Rontgen investigated the properties of the radiation in detail. He found that many substances which were opaque to visible or ultraviolet rays were transparent to the X-rays.

Rontgen noticed that X-rays could penetrate through the skin and flesh of the body of an animal. If some flesh of the body of an animal. If some organ of the body of an animal such as leg or hand of a human being was placed on a fluorescent screen and X-rays were allowed to fall on the organ the shadow of the bone could be observed on the screen.

Later different hospitals in Europe began to use X-rays as a valuable new aid in the field of surgery.

Using the basic principle of generating X-ray by making fast moving electrons strike a solid target, X-ray sources of increasing efficiency have been devised.

2.2 PRODUCTION OF X-RAYS

The gas tube, the coolidge tube and the recently developed betatron mark the different important stages of perfection of technique in the production of X-rays.

The gas tube is simply the cathode ray discharge tube with certain modifications.

The cathode ray discharge tube with certain modifications, consists cathode, C; which is made up of aluminium while the anode A is usually put in a side tube and connected to the target T whose front surface facing the cathode is sloped to have an inclination of 45° to the axis of the cathode ray stream and is arranged to be at the focus of the concave cathode.

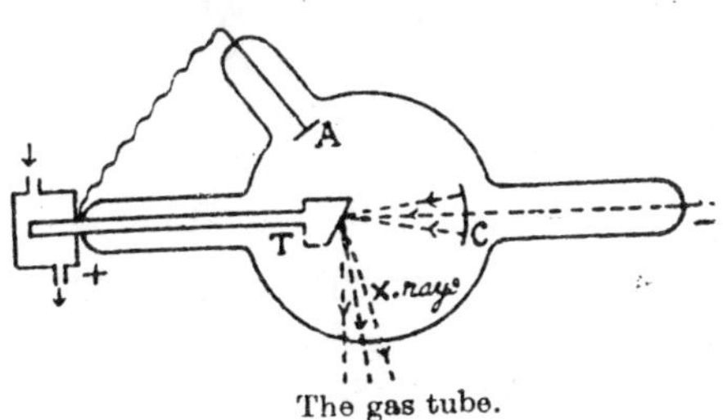

Fig. 2.1 : The Gas Tube.

The target is usually made of a metal of high melting point like platinum, molybdenum or tungsten.

With the specially shaped cathode and target the rays leaving the cathode very nearly at right angles to its concave surface are brought to a focus at the centre of curvature, a point on the target which in consequence becomes a source of X-rays.

The gas pressure in the tube is of the order of 0.001 mm.

The required high tension of about 30,000 to 50,000 volts is obtain from an induction coil.

When the tube is to be operated for a long time, special arrangements are made to cool the target by a current of water.

In 1913 Dr. Coolidge introduced a new type of tube. This tube overcame most of the drawbacks of the gas tube.

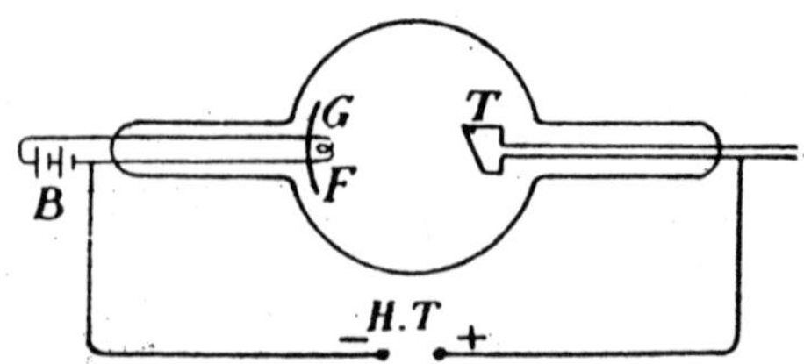

Fig. 2.2 : The Coolidge Tube.

In the coolidge tube the cathode is made up of a filament of tungsten 'F'. This filament is heated by a small battery 'B' and made to emit thermkionic electrons which constitute the cathode ray steam.

The tube is exhaust to the highest attainable vacuum of the order of 0.0001 mm. and more.

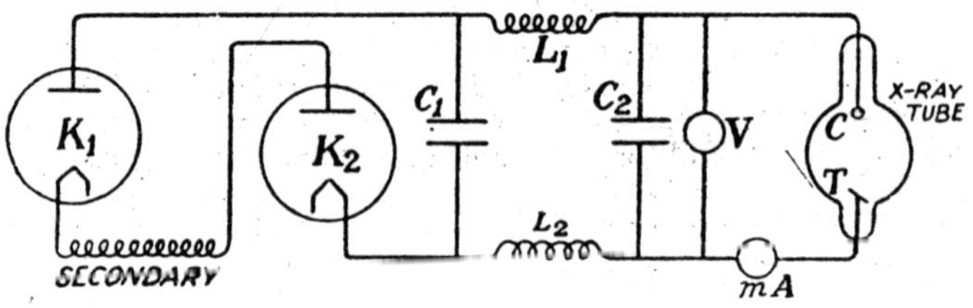

Fig. 2.3 : Use of Kenotrons for X-ray Tube.

The fast moving electrons striking the target produce hard X-rays. The very high voltage of the order of 50,000 volts and more developed in the secondary, with about 200 volts in the primary is evidently alternating which when applied to the X-ray tube will produce intermittent X-rays, the tube operating only during that half cycle, when the filament is negative with respect to the target. The life of a tube is lengthened by supplying it with a high tension, which has been already rectified. The great advantage of the coolidge tube over the gas type is the possibility of controlling the current through the tube independently of the voltage applied. The intensity of the X-rays produced can be easily controlled by the filament temperature. The quality of the X-ray will depend practically on the high voltage applied and can therefore be readily controlled.

Coolidge tubes have been designed to operate at voltages ranging from a few hundred to about one million volts. The coolidge tube is more stable in operation than the gas tube.

Detachable anode tubes are also in use when different substances are to be employed as targets.

The modern hot cathode coolidge tube is of great service in researches demanding high precision and accuracy.

2.3. PROPERTIES OF X-RAYS

The following are the some of characteristic properties of X-rays.

1. X-rays can penetrate through most substances. (but lead glass is almost completely opaque to X-rays.
2. X-rays can produce fluorescence in different substances (*e.g.*, rock salt, uranium glass etc.)
3. X-rays can affect photographic plates. (the intensity of X-rays can be measured with the help of photographic plates.)
4. X-rays ionize the gas through which they travel. (So X-ray intensity can also be measured by its ionizing power.)
5. X-rays are not deflected by electric or magnetic fields.
6. X-rays travel in straight lines like ordinary light.
7. X-rays can reflect and also they can refract.
8. X-rays can polerized.
9. X-rays can be diffracted with the help of crystaline substances.

The ionising power of X-rays has been used from the first as a quantitative means of measuring the intensity of an X-ray beam. X-rays do not ionise gases directly by themselves but by indirect means. Fast moving electrons acting on the molecules and atoms of the gas, that break them up into positive and negative ions, which by their motion render the gas, conducting and give rise to an ionisation current.

The measurement of ionisation current offers a quantitative means of measuring the intensity of on X-rays beam. X-rays are electromagnetic radiations, like ordinary light, or ultraviolet radiations. Their wavelengths can be measured by diffraction experiments using crystals.

$$\text{Wavelength of X-rays} \sim 10^{-2} \text{ to } 100\text{Å}$$

$$\text{Wavelength of u.v radiation} \sim 100 \text{ to } 1000 \text{ Å}$$

$$\text{Wavelength of visible radiation} \sim 4000 \text{ to } 8000 \text{ Å}.$$

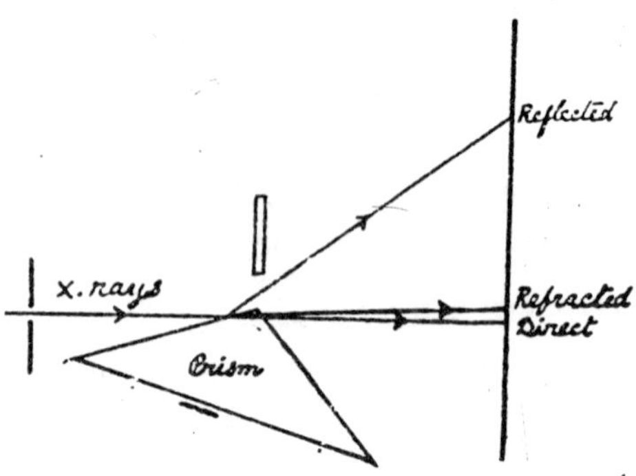

Fig. 2.4 : Prismatic Refraction Method.

By this we can understand that the wavelength of X-rays are much shorter than visible radiation.

2.4 INTENSITY OF THE X-RAYS

The intensity of the X-rays increases with increasing potential difference between the cathode and the anode. It is found that, X-rays of different wavelengths are emitted from the X-ray tube.

If the intensity I of the X-ray is measured as a function of the wavelength λ and the variation I is plotted graphically;

I varies continuously with λ upto potential differences of the order of 50,000 volts between the anode and the cathode.

For a given potential difference there is a minimum wavelength λm below which no X-rays are emitted. With increasing wavelength the intensity at first increases and after attaining a maximum it begins to decrease again.

The variation of I with λ for several different values of the p.d. It is found that the minimum wavelength λm decreases with increasing p.d. For a given wavelength the intensity is higher when the p.d. is higher.

For the same potential difference the intensity increases with the increase of the atomic number Z of the target. The is a minimum wavelength lm of the emitted X-rays whose values depend on the potential difference applied.

With a low Z target such as molydenum the variation of I with λ has the appearance, and there is continuous variation of I with λ for VL 20,000 volts.

For targets with still lower Z *e.g.*, Cu, such peaks begin to appear at still lower p.d.

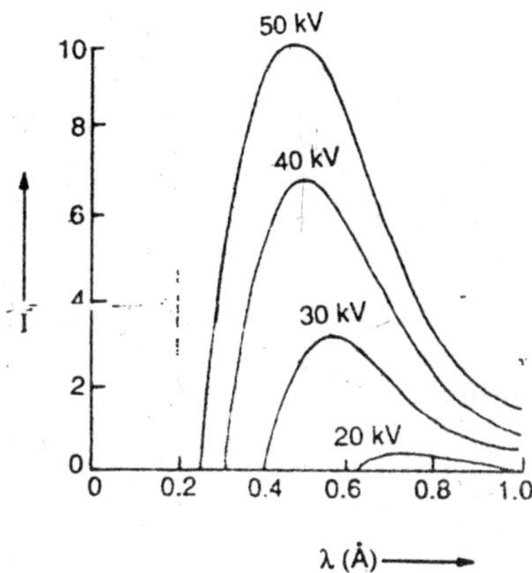

Fig. 2.5 : Intensity vs. Wavelength of X-rays Emitted from Tungusten at Different Voltages.

The positions of the discrete peak is, in the intensity distribution curve are characteristic of the nature of the target. They do not depend on the potential difference applied.

The heights of the peaks increase with increasing potential difference.

2.5 SCATTERING OF X-RAYS PRAGMATIC REFRACTION

In 1896 Imbert and Bertin Sans, accidentally discovered scattering of X-rays. Their experimental arrangement is shown diagrammatically in the figure.

Between the source (S) of X-rays and the photographic plate (P) was placed a thick copper strip (C). A plane mirror (M) was so arranged that a beam of X-rays if reflected would pass to the plate 'P' on which would appear an image or shadow of an obstacle O.

A plate of paraffin was just as effective as the mirror. The conclusion was that X-rays were diffused or scattered from the mirror or paraffin surface.

Careful analysis of the ordinary scattered light reveals the existence of two types of scattering.

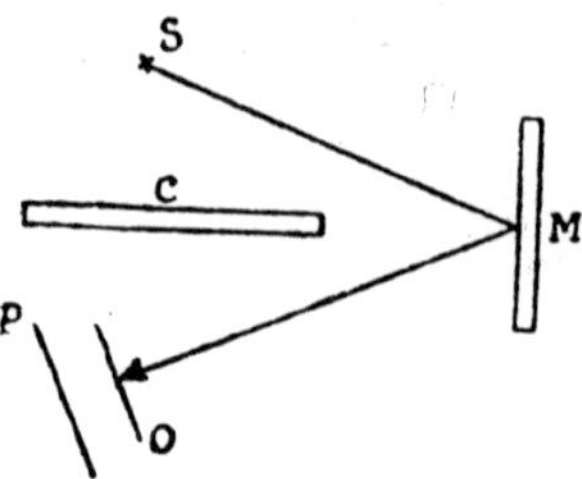

Fig. 2.6 : Arrangement of Demonstrating the Scattering of X-rays.

1. Coherent Scattering

The scattered light has the same wavelength and frequency as the incident light.

2. Incoherent Scattering

The scattered light has a wavelength different from that of the incident light. These two types of scattering occur with X-rays also.

Coherent scattering of X-rays fully understood on classical principles, which leads to diffraction effects with crystal gratings. In coherent scattering of X-rays usually called the Compton effect.

On account of the much shorter wavelength of X-rays the information obtained by their scattering regarding molecules, atoms and electrons is much more precise and detailed than that given by ordinary scattered light.

Investigations of the scattering of X-rays by liquids are giving us new information about the arrangement of atoms in molecules and from the scattering by gases we are learning the distribution of the electrons in the atoms themselves. Very often along with the scattering of light another phenomenon known as *'fluorescence'* first discovered by stokes also occurs.

X-ray scattering is also accompanied by the fluorescent X-rays, ordinarily known as characteristic X-rays.

POLARISATION OF X-RAYS

According to the more recent electromagnetic theory, light is due to transverse vibration in ether; the electric vector is perpendicular to the direction of propagation.

In 1906 Barkla demonstrated on experiment that, X-rays can be polarised like light, and hence they are also transverse waves. The difficulty in his experiment was that the intensity of the beam obtained from the second scattered was very small and consequently wide apertures had to be used for the chambers with the result that the chambers were receiving rays which might have been scattered at angles different from 90°.

Barkla was able to show that the secondary beam was at least 75% polarised. In 1924 Compton and Hagenow repeated the experiment with improved technique.

A very intense primary beam obtained from a tungsten target X-ray tube operated about 1,30,000 volts was used which permitted first better collimation and greater definition of the scattering angle and the use of very thin scatterers like paper aluminium etc.

Barkla, analysing directly the primary beam from the X-ray tube found that it was only partially polarised, about 20%. This was confirmed later by other workers, and it was found that the polarisation was more complete if the beam was filtered to remove the softer components as also if thin targets were used.

Further an increase in the speed of the parent cathode beam reduced the amount of polarisation.

All the above facts found a ready interpretation on the basis of the classical electromagnetic theory of the production of X-rays.

2.6 BRAGG'S X-RAY SPECTROMETER SEEMAN'S SPECTROMETER

Bragg's X-ray spectrometer is similar in form to the ordinary optical spectrometer consisting of three parts, *viz.,* a source of X-rays a crystal held on a circular table which is graduated and provided with vernier and a detecting device.

X-rays from an X-ray tube are made to pass through two narrow lead slits g_1 and g_2 and thereby collinated into a narrow beam.

This fine pencil is allowed to fall on a cleavage face of a crystal mounted on the circular table 'T' which can rotate about a vertical axis, its angular position being determined accurately by the vernier V.

The choice of the crystal depends upon several factors such a the range of wavelengths to be used, the reflecting power of the crystal and the case with which a good surface can be obtained.

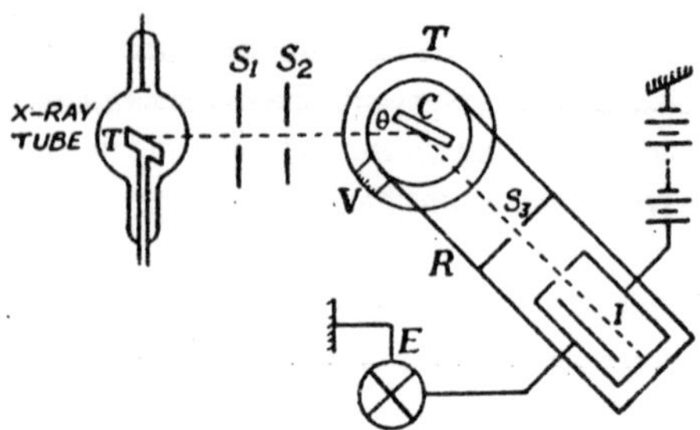

Fig. 2.7 : Bragg X-ray Spectrometer.

The crystals used are rocksalt, calcite mica and quartz.

An arm 'R' rotating about the same axis as the crystal table carries an ionisation chamber I;

The degree of ionisation in the chamber and consequently the intensity of the X-ray beam that enters the chamber is measured by an electrometer 'E'.

If the beam of X-rays is incident on the face of the crystal at a glancing angle θ, the reflected beam will make an angle 2θ, with the direction of the incident beam.

Hence, if the arm containing the ionisation chamber is set in this direction it will receive the reflected beam whose intensity can then be accurately measured by the rate of deflection of the needle of the electrometer. The ionisation chamber is replaced by a camera if photographic recording is desired.

Considering a homogeneous incident beam of X-rays, and using a given face of the crystal, it is evident that if Bragg's relation is true intense reflections will occur for certain specified glancing angles.

for *e.g.*, $\lambda = 2d \sin \theta_1$, for $n = 1$

Seeman's crystal spectrometer reduces the time of exposure very much and at the same time produces well-defined and intense traces. To obtain these effects, Seeman made use of the appreciably large focal spot on the target of a coolidege tube and a wedge of lead as a slit.

Seeman's spectrometer is useful for obtaining the X-ray spectrum of an unknown substance, which is placed on the target and measuring the wavelengths by comparison with the traces of a known substance recorded on the same film.

The first reliable value of the wavelength, of X-rays was obtained by Bragg with his instrument. It is a direct and simple method in so far as all measurements are made directly on the angular scale of the crystal table; high precision in the horizontal adjustment is not necessary and no linear distances are to be measured.

2.7 DIFFRACTION OF X-RAYS

The German physicist Von Laue was the first to point out that diffraction of X-rays could be produced more easily with the help of crystals.

The wavelength of ordinary light can be measured with the help of a diffraction grating in which the grating elements must be comparable to the wavelength to be measured.

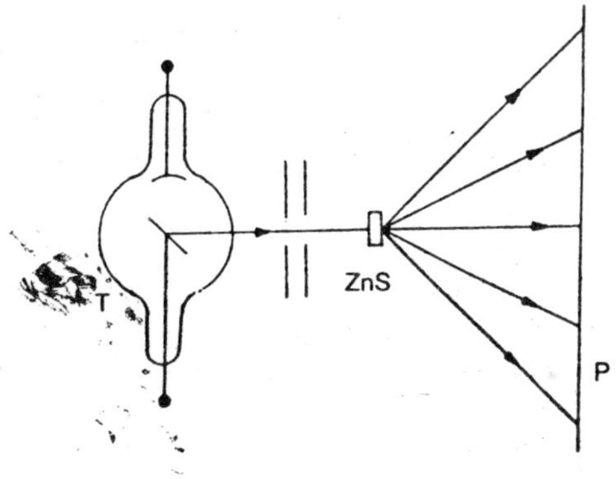

Fig. 2.8 : Friedrich and Knipping's Experiment.

To produce X-ray diffraction Friedrich and Knipping performed an experiment.

A collimated beam of continuous X-rays from an X-ray tube was passed through a zinc Sulphide Crystal, and the radiation, coming out on the otherside was allowed to fall on a photographic plate P. When the plate was developed, it was found that, there was an intense black spot at the centre, that is at the point where the incident beam hit the plate normally.

Surrounding this central spot there were a number of additional black spots arranged in a regular pattern.

The British physicist W.L. Bragg suggested a simple explanation for the production of Laue diffraction pattern.

According to Bragg there are certain special planes known as cleavage planes within a crystal along with the atoms in a crystal are more densely concentrated. If the crystal is hit with a sharp pointed end it splits along these cleavage planes.

The grater concentration of atoms along some planes can be studied thoroughly. This may be regarded as the two dimensional representation of the section of an ideal three dimensional crystal.

According to Bragg the incident X-rays are preferentially reflected from the cleavage planes which are rich in atoms. For this reason the rays are reflected preferentially in certain directions giving rise to the production of the black spots on the photographic plate in these directions.

The German physicist Von Laue estimated the distance between the atoms in a crystal from a knowledge about the number of atoms per unit volume in the crystal.

The regularity in the arrangement of the atomic layers inside the crystalled Laue to conclude that crystals are most suitable for producing diffraction of X-rays.

2.8 X-RAY CRYSTALLOGRAPHY

The analysis of crystal structure by means of X-rays consists in the demonstration of the actual existence of the space patterns and the determination of their arrangement by the measurement of the size and shape of the unit cell.

The space lattice can be determined by examining a large number of X-ray spectra and noticing systematic absence of spectra of a general type.

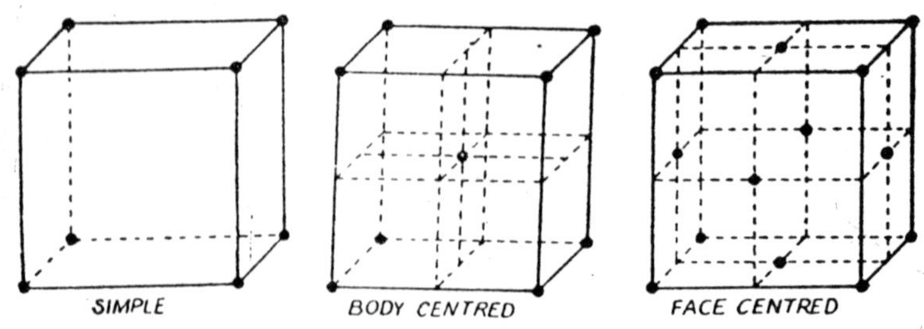

Fig. 2.9 : Three Types of Cubic Crystal.

If the unit cell contains many atoms in general positions then the real problem of structure determination begins.

The 'parameters' or co-ordinates of the atoms in the unit cells must be found by checking against the observed intensities. Here, ionic radius, grouping of the atoms into radicals etc. are used.

The method of procedure in the analysis of crystals of more complicated structure is similar to that used in the simple cubic type.

There are three sub-types in cubic crystals with different concentrations of atoms in special planes, and consequent different spacing of these planes. They are as follows:

1. The simple cube in which an atom or a molecule lies at each corner of the cube.
2. The body centred cube, in which there is an additional atom lies at the centre of the cube.
3. Face centred cube, in which an additional atom lies at the centre of each of the six faces of the cube.

X-ray analysis of crystal structure, can be carried by the following methods:

(i) The Laue Spots Method

The method can be carried by diffracting X-rays with crystals. It is to be borne in mind that a crystal behaves not as a plane grating but as a three dimensional grating.

The diffraction of X-rays produced by such a cubical system of lattice points can be easily understood by first considering the diffraction due to the lattice-points in one of the three principal directions; that is due to a line lattice or a single one dimensional point grating.

The diffracted rays will have appreciable intensity only if the rays due to the action of all the lattice points agree in phase

The following equation gives the value of the particular wavelength that can have maximum intensity in the direction, α, β, γ, for a given direction of incidence α_0, β_0, γ_0,

$$\lambda = -2a\frac{(\alpha_0 n_1 + \beta_0 n_2 + \gamma_0 n_3)}{n_1^2 + n_2^2 + n_3^2}$$

With a plane grating light of any arbitrarily chosen value of λ will have maximum intensity after diffraction in a particular direction, where as with a crystal grating only for certain definite values of λ diffraction will take place the other wavelengths being lost either by partial or complete destructive interference.

It should be borne in mind that the Laue spots are obtained only in the case of single large crystals using a heterogeneous X-ray beam consisting of a continuous band of wavelengths.

The symmetry of the Laue spots affords also valuable information as to the crystal symmetry. The Pattern is formed by different wavelengths renders the method more difficult to apply and less powerful than methods employing monochromatic radiation.

(ii) The Bragg's X-ray Spectrometer Method

We have already studied Bragg's spectrometer, and the method of using it. (in 2.6) An X-ray tube with a molybdenum target gives the characteristic k radiation of the metal consisting of two principle wavelengths. With only a comparatively faint continuous spectrum.

The wavelengths of the two chief lines k_α and k_β, differ enough to be reflected at different angles so that effectively a monochromatic X-ray beam is used.

The ionisation spectrometer method has the great advantage of giving accurate quantitative estimate of the intensities of the reflected beams, which is of the greatest importance in determining crystal structure. Since the crystal is dealt with face by face the origin of the spectra is in all cases certain.

But this method is laborious and slow requiring a great deal of time and labour. In practice a fairly large crystal with definite faces is required to get good results.

(iii) The Rotating Crystal Method

This method is devised by schiebold and polanyi. If the crystal is rotated slowly about a fixed axis, a large number of planes will come successively into the reflecting positions and the corresponding diffracted beams will flash out for a moment as the crystal passes through the correct setting, producing upon the photographic plate, arranged to receive them, a pattern of spots known as rotation photographs. Hence the method is also called the rotation photograph method.

With complete rotations of the crystal too many spots are obtained on the film; hence it is usual to rock the crystal back and forth through a range of only 80; in order to limit the spots on any photograph to those of certain indices.

The angular rate of rocking must be kept constant, if one desires to compare intensities and a clockwork motor is used to secure this condition. If rotation photograph are taken with rotation of the crystal about all the three axes separately the size of the unit cell of the structure can be estimated.

(iv) The Powdered Crystal Method

In many truly crystalline solids individual crystals large enough even for the rotation photograph method never occur. Such substances are readily examined by this method.

If a monochromatic X-ray beam is allowed to fall to a fine powder, the orientation of the minute crystal grains being completely at random, a certain number of then will lie with a given set of lattice planes making the correct angle with the incident beam for reflection of the grains will have another set of planes in the correct position for reflection and so on.

For each set of planes and for each order there will be such a cone of diffracted rays. Their intersection with a photographic plate with its plane normal to the incident beam form a series of concentric circular halos, from the radii of which the glancing angle and hence the spacings of the planes can be deduced.

2.9 APPLICATION OF X-RAYS

X-ray applications in Industry Cover a very wide range. They can be marked as follows.

X-rays are used to detect and photographic defects with in a body such as a metal plate a machine part or tubes intended to withstand high pressure etc.

X-rays are used to analyse and control alloys and other composite bodies by determining the crystal form in an ingot and by following the effect of composition, heat and mechanical treatment on a given specimen.

X-rays are also used to study the structure of cellulose, rubber fibers, etc.

X-rays found their most important practical application in the branch of science, dedicated to the physical welfare of the human body.

X-rays have a wide range of use, in radiography and in X-ray therapy.

The effect of powerful X-rays on living tissue is destructive if exposed for a long time.

This is due to the fact that X-rays on absorption by the tissue give rise to photoelectrons of high speed which collide with the molecules of surrounding matter and knock out electrons. This changes the structure of the molecules and the tissue dies away.

Hence, if X-rays are allowed to act upon abnormal tissues. Which are the cause of a disease the affected tissues will die away leading to the cure of the disease.

3

SOLIDS AND THEIR PROPERTIES

3.1 INTRODUCTION

We know well that matter can exist in three states of aggregation, solid liquid and gaseous. Solids can be grouped into two classes the amorphous and the crystalline.

In the amorphous substances the atoms or the molecules are strongly bounded to one another.

In crystalline solids a definite regularity is observed in the arrangement of the atoms or molecules. It is easier to explain the various properties of crystalline solids than those of amorphous solids.

From the study of bulk properties in the macroscopic state of slide it has been shown that solids in general are characterised by their elasticity of volume and shape. Many metals and alloys in the solid state are microcrystalline or polycrystalline.

It is usual to distinguish the polycrystalline bodies from single crystals which may be defined as solid objects of uniform chemical composition, which as they occur in nature, or are formed in the laboratory have a very regular structure, that is bounded by plane faces, the interrelation of which exhibits a typical symmetry *e.g.*, rock salt.

All solids occupy finite volume, the distances between their atoms cannot exceed a certain limiting value. This can happen only if an attractive force acts between them.

In a crystalline solid the atoms are arranged in a regular pattern, each atom occupying the positions of stable equilibrium. When the atoms are displaced from these positions forces come into play which tend to restore their equilibrium positions. As a result atoms vibrate about these positions.

In solids the force acting upon the atom is electrostatic in nature. The nature of the force depends upon the electron distribution, in the atoms.

The properties of different solids, and their nature differ from each other.

3.2 CRYSTALLINE SOLIDS

1. Ionic Crystals.
2. Covalent Crystals.
3. Metallic Crystals.
4. Molecular Crystals.
5. Hydrogen-bonded Crystals.

1. Ionic Crystals

In the atoms of the alkaline elements just after the inert gas, the outermost electron is very loosely bound in S-shell on the other hand, in the atoms of the halogens just before the inert gases, there is a deficit of one electron in the outer-most 'P' orbit.

The halogen atom on acquiring the extra electron in its outermost p-orbit becomes transformed to a negative ion, and also achieves the inert gas electron configuration. These two oppositely charged ions attract each other. This type of binding is known as the ionic bond and the type of crystal formed due to such bonds is known as an ionic crystal.

In ionic crystals due to the long range nature of the Coulomb force the ions located at the different lattice points ineract not only with their immediate neighbours but with all other ions throughout the crystal.

Ionic crystals have usually low electrical conductivity at low temperatures. They exhibit ionic conductivity at higher temperatures. Their absorption coefficients are high for infra-red radiation. They can be cleaved easily. Ionic crystals have high melting points.

2. Covalent Crystal

In covalent crystals the neighbouring atoms share their electrons, which results covalent bonds.

The molecules of hydrogen nitrogen and oxygen, consist covalent bonds.

As shown in the figure the region between the atoms A and B is usually much greater than he probability of their location outside this region as shown by the curve 'C'.

It can be shown on the basis of the quantum mechanical theory that the exchange interaction between the two hydrogen atoms becomes attractive when the spins of the two electrons are antiparallel. Such state is known as symmetric state.

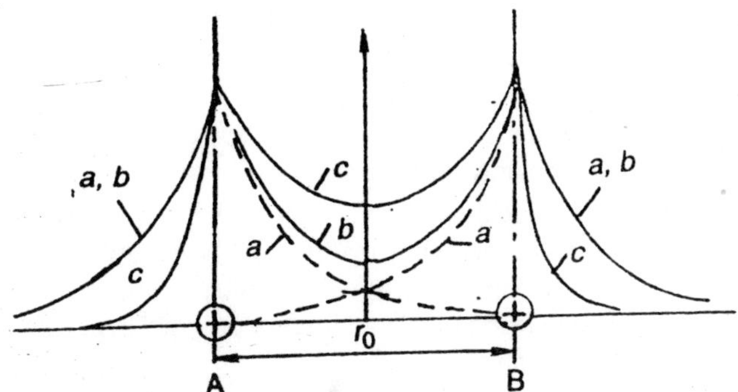

Fig. 3.1 : Probability of Locating Two Electrons in H_2(c). Curve a is for Two Isolated H Atoms. Curve b is for the Two H Atoms Brought Nearer.

If the spins of the two electrons are parallel, then the exchange force becomes repulsive. The corresponding state is called antisymmetric.

Covalent Crystals have great hardness and cannot be cleaved easily. Their electrical and heat conductivities are low. Their melting points are usually quite high. Covalent crystals of diamond and other group four elements have tetrahedral lattice structure.

3. Metallic Crystals

In metallic crystals the outermost orbits of the atoms are generally very loosely bound. Because of this they have high mobility and can move about freely from one place to another inside the metallic crystal lattice much like the molecules in a gas. These are called free electrons.

The electrons gas is uniformly distributed throughout the metal lattice. In metallic crystals, three types of lattice structure can be seen; face-centred cubic, hexagonal close-packed, and body centred cubic.

4. Molecular Crystals

In the molecular crystals, the bond between the neighbouring atoms associated with dipole moments.

The electrostatic interaction between these dipoles in the condensed states produce weak binding between the neighbouring atoms in these crystals. This type of force is known as the dispersion force.

In many organic crystals molecular bonds can be observed. The molecules of some of then possess permanent dipole moments.

The dipole interaction between the neighbouring molecules produce attraction between them. These permanent electric dipole them. These permanent electric dipole moments induce dipole moments in the neighbouring molecules. The interaction between the permanent and induced moments also produces an attractive force.

All these interactions along with the dispersion force mentioned earlier produce the binding in such crystals.

The combination of these interactions is known as Vander Wall force. Molecular crystals have low melting and boiling points. They are very compressible. They have usually face centred cubic structure. The interaction energy due to Vander Wall force is proportional to r^{-6}. Where r = distance between the molecules.

5. Hydrogen Bonded Crystals

Hydrogen bond can be observed in the crystal of ice. Such bonds are usually produced between a strongly electro-negative atom *e.g.*, oxygen and the hydrogen atom.

In the ice crystal each molecule is surrounded by four neighbouring molecules. They are located at the vertices of a regular tetrahedron and are bound together by hydrogen bond.

Hydrogen-bonded molecules have a tendency to polymerize. One can understand that, the strength of the oxygen hydrogen bond in the molecules of ice crystal is much weaker than the strength of covalent bond in an water molecule.

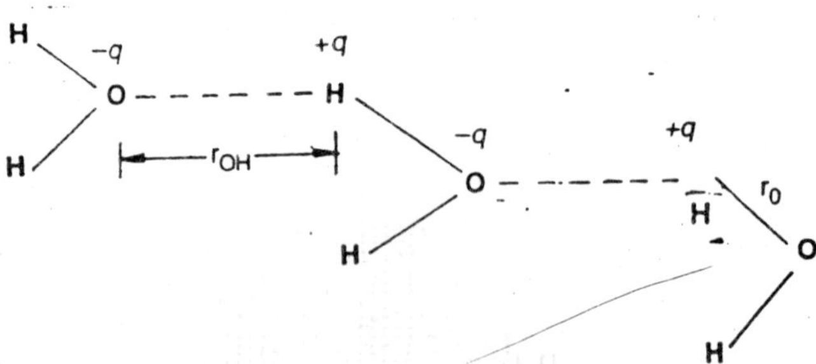

Fig. 3.2 : Hydrogen Bond.

In an actual crystal none of the above bonds is found to exist exclusively. In general a mixture of two or more bonds is found to be present in a crystal.

3.3 MILLER INDICES

In a crystal lattice one can find number of planes, which have larger concentration of atoms at the lattice sites. The orientations of such parallel planes are usually expressed interms of a set of three numbers, known as miller indices.

To get the miller indices of a crystal line one has to determine the intercepts of the line between two lattice points along the two crystal axes in units of the sides of the unit cell.

The miller indices of all planes parallel to PQR are the same. The set of parallel crystal planes with given values must be drawn so as to embrace all the lattice points in the crystal. It is then possible to find the distances between the point of intersection of the adjacent crystal planes with the crystal axes.

Let us consider the set of parallel crystal lines passing through the lattice points. The consecutive lines have intercepts separated by 2a and b along the crystal axes. One can draw another set of lines, parallel to the former, showed by the dashed lines in the figure.

These pass through the lattice points, which do not fall on the previous set of lines. Since the crystal lines with given h and k values must cover all the lattice points.

It can be proved that in the case the distances between the points of intersection of the adjacent crystal planes with a given set of miller indices h, k, s, with the crystal axes are equal to a/h, b/k and c/*l*, respectively.

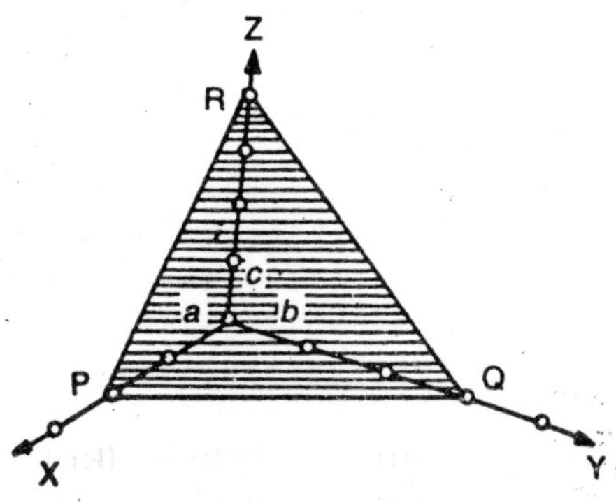

Fig. 3.3 : Miller Indices for a Three Dimensional Lattice.

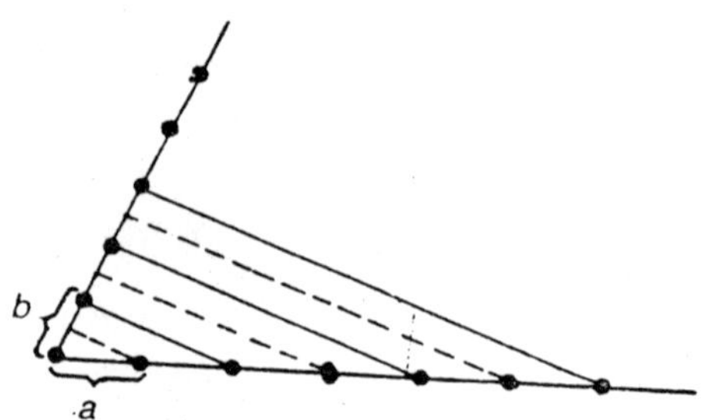

Fig. 3.4 : Lattice Points in a Two Dimensional Lattice.

Miller indices can be both positive and negative. To get the Miller indices of the plane, we have to take the reciprocals of the intercepts. Since the reciprocals of the infinite intercepts are zero, the Miller indices of these planes perpendicular to the edge.

3.4 SPECIFIC HEAT OF SOLIDS

In a crystalline solid the atoms are arranged in a regular three dimensional lattice. But the atoms are not stationary but vibrate about their mean positions of rest at the lattice sites.

In a three dimensional arrangement of ~ atoms the total energy is equal to that of 3~ linear harmonic oscillators. When heat energy is applied to the substance, this total energy is changed which can be calculated theoretically assuming suitable models.

Albert Einstein was the first to explain the decrease of the specific heat of solids at low temperature on the basis of the then newly discovered quantum theory.

Paul Debye in 1912 pointed out that this assumption cannot (all the atoms in the crystal vibrate with the same frequency, independently of each other) be correct, because there are strong interatomic forces between the atoms at the different lattice sites so that the vibrations of any one of the atoms strongly influence the vibrations of its neighbouring atoms.

So, Debye proposed that instead of considering the independent vibrations of the individual atoms, it was necessary to consider the vibrational modes of the entire crystal. The strong interactions between the neighbouring atoms in the crystal transmit the vibrations of any one atom to the others throughout the crystal and a collective motion in the nature of an elastic wave involving all the atoms is excited with in the crystal such collective motion is known as the normal node of a lattice. The number of normal modes is equal to the number of degrees of freedom 3~ where N is the number of atoms.

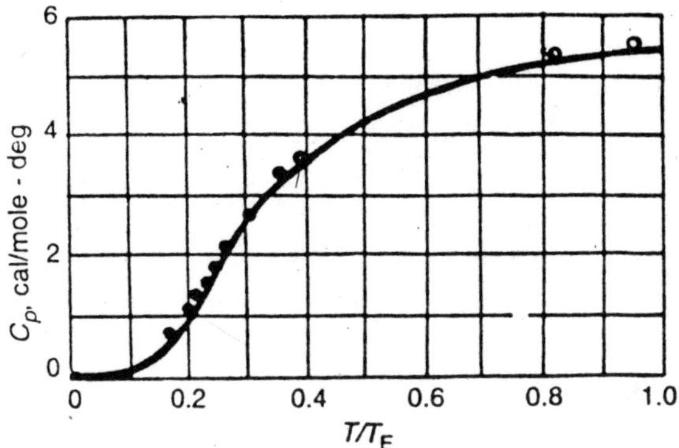

Fig. 3.5 : Cυ vs. T for Diamond After Einstein.

Through in an actual crystal the atoms are arranged in a regular pattern as discrete mass point with finite distances in between, in Debye's theory, the crystal is assumed to be, a continuous medium through which stationary waves of different frequencies which constitute the normal modes of the crystal are propagated with the velocity of elastic waves.

A modification in Debye's theory was introduced by Max Born who proposed a different procedure. According to Born this should be based on a common minimum cut-off wave length for the longitudinal and transverse waves.

In Debye's theory a solid is regarded as isotropic. However a crystalline solid shows marked anisotropy in many of its physical properties. Thus the velocity of the elastic waves is usually different in different directions.

According to quantum mechanics, a linear harmonic oscillator can exist in a discrete energy level of energy

$$E_n = \left[n + \frac{1}{2}\right] \hbar\,\omega$$

$$n = 0,\ 1,\ 2,\ \ldots \text{ etc.}$$

Spacings being equal to, $\hbar\omega = h\nu$.

When a crystal is heated each of the linear harmonic oscillators can absorb a minimum amount of energy hv. This minimum amount of energy by which the vibrational energy of an oscillator can change at a time may be regarded as the quantum of energy, associated with the elastic vibrations of the oscillators. They are called, phonons. They are analogous to the quantum of energy associated with the electromagnetic oscillations which are known as photons like to photons, the phonons also carry on energy hv each. 'λ' be the wave length of the elastic waves and 'v' is their velocity in the crystal, then the momentum of the phonos is,

$$p = hv/v = h/\lambda.$$

The phonon concept has been found useful in explaining many other properties of solids apart from the specific heat. For example the thermal conductivity of solids can be understood by considering the motion of the phonons in the solid much like the motion of the molecules in a gas.

The thermal conductivity of a gas can be explained on the basis of the kinetic theory of the transport of heat energy due to the motion of the gas molecules.

Similar considerations can be applied to the motion of the phonons. It is necessary to known, the mean free path against phonon-phonon interaction to carry out such calculations.

3.5 LAUE'S METHOD BRAGG METHOD

Von Loue in Germany was the first to point out that, since the X-ray wavelengths were of the same order of magnitude as the spacing between the atoms in the crystal the crystal could be used for diffracting an X-ray beam. Observations of the diffraction maxima and minima would therefore give an idea about the structure of the crystal. In Laue's method a single crystal is exposed to a beam of continuous X-rays usually in the range 0.2 to 2A. The beam is collimated by a pinhole, and the diffracted beam is received on a flat film placed on the other side of the crystal. A small crystal may be used.

Laue pattern is specially suitable for checking the orientation of the crystals used in different types of experiments. This is due to the fact that, the pattern reprocessed the symmetry of the crystal in the orientation used for the experiment.

This method is very easy for obtaining qualitative data on the reflections from a large number of planes simultaneously and hence is much less time, than any other method.

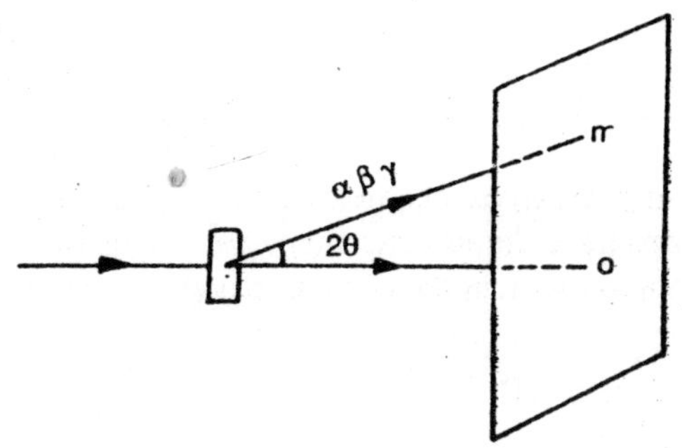

Fig. 3.6 : Laue Method.

Bragg Method

In Bragg method monochromatic X-rays are reflected from a crystal and maxima of the reflected beam are observed. With the help an ionization chamber. As the crystal plane is rotated in small steps the ionization chamber is turned through twice the angle at each step and the ionization current is measured as a function of the angle of reflection.

The method developed by the Bragg is of importance mainly for the accuracy of intensity measurements. It is also possible to obtain very accurate value of sin q by this method, and hence of the dimensions of the unit cell.

A.W. Hull on American physicist developed the powder photograph method for determining the crystal structure.

A monochromatic beam of X-rays is allowed to fall on the specimen, which is in the form of finely ground powder, kept inside a small tube. The powder is actually made up of a large number of microcrystals oriented at random. As a result there is always a substantial number of crystals which reflect the incident beam from a given family of important planes in a particular direction.

The reflected rays from the different crystals with appropriate orientations will proceed along straight lines lying on the curved surface of a cone of semivertical angle 2q. Due to reflections from a given set of planes there will be a number of such cones corresponding to different orders of reflection.

These reflected rays given rise to a series of concentric circular lines on a photographic plate kept at right angles to the incident beam.

For crystals with high symmetry and with smaller number of atoms per unit cell the diffraction patterns are relatively simple. Powder photograph method is thus specially suitable for the analysis of the structures in the case of many metals and alloys where all the conditions are full filled.

The rotation photograph method permits great accuracy in crystal structure determination. Several improvements of the method were later introduced. Instead of repeatedly rotating through a full circle the crystal is sometime rocked back and forth through a limited range of angles. This eliminates the possibility of the, overlapping of the spots produced by reflections.

4

THE ELECTRON

4.1 INTRODUCTION

Thomson's experiments established that the cathode rays are made up of a beam of negatively charged particles. Due to the very large potential difference applied between the two electrodes in the discharge tube they acquire very high kinetic energy in travelling from the cathode to the anode. As a result they travel with very high velocity inside the discharge tube. It has been found that irrespective of the nature of the gas present in the tube, these negatively charged particles have always the same properties. This shows that they are universal constituents of all matter. These have since been named electrons, a name originally coined by Johnstone Stoney, during his studies on electrolytic conduction.

Prof. J.J. Thomson at the cavendish laboratory, cambridge was the pioneer worker in the determination of the charge and mass of the electron.

In 1897 using cathode rays and subjecting them to electric and magnetic fields Thomson, first determined the values of, electric charge (e) and mass of the electron (m). He made it by using, Wilson's Cloud Chamber, he measured the electronic charge 'e' and combining the two results, obtained the value of the electronic mass.

The ratio e/m could be determined without any difficulty but the measurement of 'e' was more trouble some with the apparatus then available he could not make very accurate measurements, but subsequent technical improvements, made by H. A. Wilson, Millikan and others led to much better values for both charge and mass.

Today there exist general entirely different experimental methods of high precision.

Since all substances are made up of atoms and since the electrons are present in all substances we may conclude that, they must be present within the atoms of all substances. Since the atoms are electrically neutral the atoms must therefore be made up of two parts one of which carries positive electricity while the other an equal amount of negative electricity.

4.2 THE SPECIFIC CHARGE OF THE CATHODE RAYS

Cathode rays are deflected by electric and magnetic fields. By subjecting a narrow pencil of cathode rays to electric and magnetic fields acting at right angles of the pencil caused by the two fields, e/m can be determined.

The cathode rays produced in a highly evacuated discharged tube and shot out normal to the surface of the cathode 'C' are reduced to a narrow pencil by marking them pass through fine slits in the anode A, and in the metal plug B which is electrically connected with A. This pencil travelling in a straight line with a velocity that is accelerated between 'C' and 'A' but remains uniform beyond 'A' strikes a fluorescent screen SS and produces a small patch at 'P'. On the path of the pencil electric and magnetic fields are set up;

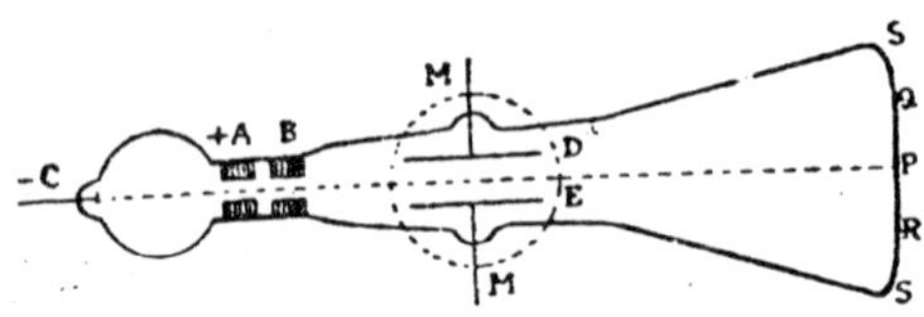

Fig. 4.1 : Thomson's Apparatus.

The electric field is produced by maintaining two horizontal plates D and E at a high difference of potential; this field acts in the plane of the diagram, and at right angles to the direction of motion of the cathode ray particles.

The cathode ray passing through such a field will be deflected in the vertical plane towards that plate which is positively charged. The magnetic field is produced by an electromagnet MM with its direction perpendicular to the plane of the diagram and hence at right angles both to the electric field and the direction of motion of the cathode ray. Such a field will deflect the cathode ray also in the vertical plane, downwards or upwards, according as the field is directed from front to back or back to front. Thus the pencil of cathode rays can be subjected to crossed electric and magnetic fields that act in directions perpendicular to its own.

The lines of force of the two fields are arranged to act over the same region so that the cathode ray will be under their influence over the same length of path.

The cathode ray beam after emergence from the electric and magnetic field region falls upon screen, on which it produces a luminous spot.

Let e and m be the charge and mass of the cathode rays and 'V' their velocity. Let 'X' be the electric field and B the magnetic flux density through which the cathode rays pass. Then the force exerted on them by the electric field is

$$1eV = 4.767 \times 10^{-10} \times 1/300$$

$$F_e = X_e$$

The magnetic force is,

$$F_m = B_e V.$$

If F_e and F_m act in opposite directions, then the fields can be so adjusted that. the deflection of the luminous spot on P produced by the cathode rays due to one field is exactly compensated by the action of the other. Under this condition the two forces are equal and opposite. If X_0 be the electric in this case, the we can write

$$X_0 e = B_e V$$

$$V = X/B.$$

Thus the velocity 'V' of the cathode rays can be determined by noting either the electric deflection or the magnetic deflection.

Thomson showed that, irrespective of the nature of the residual gas in the discharge tube, the value will be the same.

4.3 THE RADIUS OF THE ELECTRON: THE MASS OF THE ELECTRON

The dimension of the electron is extremely small compared with the atom of any element which is of order of 10^{-8} cm. It must be noted however that the classical conception of the electron as a spherical charge with a fixed radius is not warranted by the modern scientific trend of though, according to which, the classical electromagnetic theory is no longer applicable to microscopic atomic system, and therefore much less to the ultramicroscopic electron.

The radius of the electron can be estimated as follows:

$$m = \frac{2\mu e^2}{3a}$$

or $$a = \frac{2\mu e^2}{3m}$$

Taking $$\mu = 1 \text{ for air,}$$

$$e = 1.6 \times 10^{-20} \text{ e.m.u.}$$

$$m = 9.1 \times 10^{-28} \text{ gram}$$

$$a = \frac{2 \times 1 \times (1.6 \times 10^{-20})^2}{3 \times 9.1 \times 10^{-28}}$$

$$= 1.875 \times 10^{-23} \text{ cm.}$$

The mass m of the electron is found by combining the measured values of e and e/m.

$$\frac{e}{m} = 1.76 \times 10^7 \text{ e.m.u.}$$

$$e = 1.6 \times 10^{-20} \text{ e.m.u.}$$

$$m = \frac{e}{\frac{e}{m}} = \frac{1.6 \times 10^{-20}}{1.76 \times 10^7}$$

$$= 9.1 \times 10^{-28} \text{ gm.}$$

The mass of the hydrogen atom 'M' can be obtained from the atomic weight of hydrogen and Avogadro's number 'N'

$$M = \frac{1.008}{6.03 \times 10^{23}} = 1.67 \times 10^{-24} \text{ gm.}$$

Comparing the masses of the electron and hydrogen atom

$$\frac{M}{m} = \frac{1.67 \times 10^{-24}}{9.1 \times 10^{-28}} = 1835.$$

This indicates that the electron is 1845 times lighter than the lightest hydrogen atom. It may be noted that the mass of the electron, according to the theory of relativity is a variable quantity dependent on its velocity, increasing with the velocity, expressed by the relation,

$$m = \frac{m_0}{\left(1 - \frac{V^2}{C^2}\right)^{\frac{1}{2}}}$$

where,

m_0 = mass of the electron when it is at rest

m = Increased mass

V = Velocity

C = Velocity of light.

From this relation it is readily seen that the increase in mass is inappreciable except for high velocities comparable with that of light.

The variation of the mass of the electron with its velocity at very high values of the latter, has been remarkably confirmed by experiments, conducted on high-speed electrons.

J.J Thomson considering the electron as a uniformly charged sphere moving with a uniform velocity in a straight line was able to show on the basis of the classical electromagnetic theory that an electro-possesses mass solely by virtue of its electric charge.

The electron, the ultimate atom of electricity has no mass in the material sense or that its mass is purely electrical. In other words a charge is equivalent to a mass eventhough it may not be connected with any material. Electrical mass has been derived on the assumption that the velocity is not large and forces due to electromagnetic induction are negligible.

4.4 PROPERTIES OF THE CATHODE RAYS

Following are some of the important properties of the cathode rays:

1. The cathode rays travel in straight lines. The cathode rays cast shadows of obstacles put in their path, they must be moving in straight lines. This is further confirmed by the fact that, if the cathode is concave in shape, then the cathode rays are focused at a point. If a small piece of platinum is kept at this point it becomes heated by the impact of the cathode rays and ultimately begins to glow.
2. The cathode rays have very high kinetic energy.
3. The cathode rays are found to be deflected by electric and magnetic fields. (This proves that, cathode rays are composed of, electrically charged particles.)
4. Cathode rays are composed of negatively charged particles.
5. Cathode rays produce fluorescence when they are incident on certain substances like barium platino cyanide, zinc sulphide etc.
6. When the cathode rays fall on some substance, they exert a pressure on it.

According to Thomson it is not merely the mechanical pressure, due to the impact of the cathode rays. (on the mica vane or a wheel kept infront of it)

Thomson's experiments established that the cathode rays are made up of a beam of negatively charged particles.

Cathode rays acquire very high kinetic energy in travelling from the cathode to the anode. As a result they travel with very high velocity inside the discharge tube.

4.5 SOURCES OF ELECTRONS

Any kind of matter can be made to emit electrons under suitable conditions. For instance, in the discharge tube the electrons, constituting the cathode rays are produced by the process of ionisation by collision under the action of a strong electric field.

Substances can be made to give up their electrons by other means also, such as raising their temperature or allowing rays of light to act upon them. The former is known as thermionic emission and the latter photoelectric emission.

These two sources of electrons are closely related to one another as regards their mechanism according to accepted ideas and are of great practical importance as they are used in the wonderful modern inventions of thermionic value and the photoelectric cell.

Thermionic emission is the phenomenon in which electric charges are emitted from hot bodies. It has been known for a long time that air becomes conducting in the neighbourhood of incandescent metals.

Guthrine in 1873 found that a red hot ball retained a negative but not a positive charge. At higher temperatures charges of both signs were emitted.

4.6 THERMIONIC EMISSION

Thermionic emission is the phenomenon in which electric charges are emitted from hot bodies.

Elster and Geital conducted a series of researches between 1882 and 1889 on the phenomenon by arranging a metallic filament and a plate in a glass bulb connected to an exhaustion pump heating the filament red hot by means of battery and measuring the charge received by the plate.

They found that the charge on the plate increased with rise of temperature of the filament, until the latter became white hot, after which the amount of charge became constant.

At that stage the pressure in the bulb was reduced, and it was found that after some decrease the charge on the plate became negative.

Further reduction of pressure increased the amount of the negative charge. It was also found that, the amount of charge depended on the chemical state of the emitter and the nature of the surrounding gas.

In 1885 by using an incandescent filament, Edison was able to demonstrate an experiment in a precise manner.

The heated filament emitted mainly negative electric charges under the conditions of low pressure and high temperature obtained in an ordinary electric lamp.

A galvanometer included in the plate circuit registered a deflection only when the plate was positive with respect to the filament, but no deflection when the plate was made negative. This clearly proved that, a current flowed in the plate circuit due to the emission of negative charges from the incandescent filament and their subsequent attraction towards the plate when it was positive, while there was no such current, when the plate was negative, since the charges coming from the filament were repelled. On account of this important discovery by Edision the phenomenon was for some time known as the Edision effect.

Later prof. O.W. Richardson conducting thorough researches on chemically pure incandescent emitters in high vacuum, obtained sure experimental results and gave a satisfactory explanation of the observed facts connected with the phenomenon. He gave the name of thermionics to the subject, and thermions to the ions emitted from but metals.

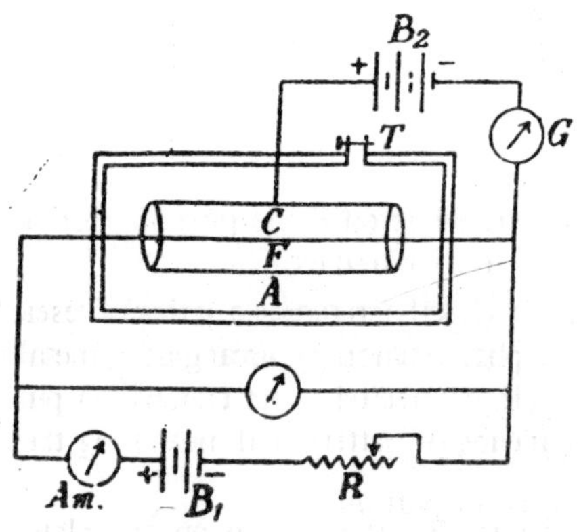

Fig. 4.2 : Apparatus for the Study of Thermionic Emission.

The apparatus used in the detailed study of the thermionic emission consists, a wire 'F' of platinum, tungsten or any other highly refractory metal stretched along the axis of a hollow metallic cylinder C.

The wire and the cylinder are enclosed in a glass bulb 'A' in which a good vacuum can be maintained by connecting to a high vacuum pump through a side tube T, which is also used to supply the bulb with any gas at any desired pressure.

The wire F is heated by means of an electric current from the battery B, connected in series with it through an ammeter A_m, and on adjustable resistance R.

A voltmeter is put in parallel with the filament. The resistance of the heated wire can be determined from the readings of the voltmeter r and ammeter.

The cylinder 'C' is connected to another battery B_2 which enables a P.D. to be applied between the cylinder and the wire. A sensitive galvanometer 'G is included in the cylinder circuit.

It is sure that the thermions experimented upon are negatively charged.

Thermionic Emission and Temperature

Thermionic current does not obey Ohm's Law. At first it increases very slowly, then more rapidly and finally becomes a constant which is known as saturation current. The value of the voltage beyond which any increase or voltage produces no increase in current is called the saturation voltage.

To investigate the relation between thermionic current and temperature the applied voltage is given such a value as to produce saturation for any temperature of the filament that is used, in the experiment.

At low temperatures there is practically no thermionic emission. The thermionic current begins at a temperature of about 1000°C and then rapidly increases with increase of temperature. It is experimentally established that the thermionic current is directly proportional to the area of the filament.

It is a matter of general observation that thermionic emission takes place in the case of pure metals only at very high temperatures above 1000°C and that alkali metals are much more active than others. Traces of impurities in the material of the filament affect the emission.

In 1908 Wehnelt made a discovery of great practical importance in the construction of thermionic valves and oscillographs, *viz.*, when a metal surface is coated with certain oxides chiefly of alkaline earths, it emits a considerably greater number of electrons at comparatively low temperatures and low applied voltages.

The oxides of calcium, strontium and barium are found to before effective, while those of magnesium, zinc, cadmium produce less pronounced effect.

The negative filament coated with an oxide of the above metals is known as the Wehnelt cathode. It is prepared by immersing the filament wire in a solution of the nitrate of calcium strontium or barium and then heating when a layer of the oxide is formed on the filament. The wire is next activated by introducing a trace of carbon in the nitrate layer, which reduces a very small percentage of the oxide coating to the pure metal state.

The thermionic emission obtained from Wehnelt cathode is much greater and more lasting than in the case of the non-coated filament.

The thermal ionisation of the small amount of metal produced by activation and absorbed in the oxide layer is responsible for the increased emission.

4.7 PHOTOELECTRIC EMISSION

Photoelectric emission was first discovered in 1873 by a telegraph operator W. Smith. He observed that when sunlight fell upon resistors, the current in the circuit varied considerably.

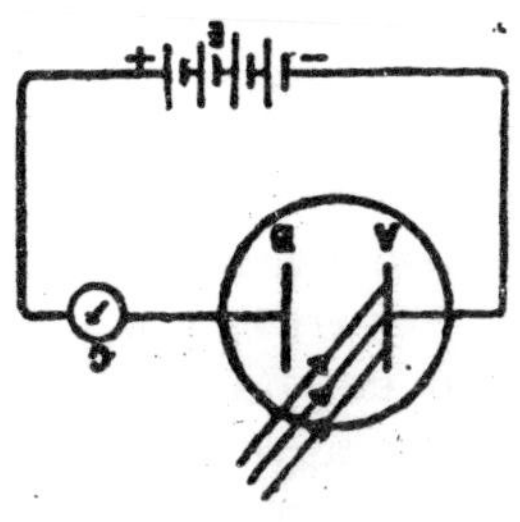

Fig. 4.3 : Apparatus for the Study of Photoelectricity.

In 1887 Hertz observed the same phenomenon, while working with resonance electric, while working with resonance electric circuits in connection with electromagnetic waves.

Two zinc plates A and B were placed in an evacuated quartz bulb and connected to a battery E and galvanometer G. When ultraviolet light fell on the plate A which was connected to the negative terminal of the battery, a current was found to flow as indicated by the galvanometer. But when the light fell on the positive plate B there was no flow of current. This proves that not only a negatively charged body loses its charge when irradiated with ultraviolet light, while no such effect occurs with a positively charged body, but also the ions emitted from bodies which lose charge under the action of light must be negatively charged, for, the observed current can be explained only by the fact that the negative ions emitted by the negative plate are attracted towards the positive plate and thus cause a flow of electricity. Since the effect is produced under the influence of light, it a called *photo-electric effect*, (photo = light) and for the same reason the negative ions and the resulting current are known as *photoelectrons* and *photoelectric current* respectively.

When plates of different materials were used it was found that zinc, magnesium, lithium, sodium, potassium and rubidium responded photoelectrically to ultraviolet light with increasing order of sensitivity. The alkali metals, sodium, potassium and rubidium, were sensitive even to ordinary visible light. This shows that the photo electric emission depends upon both the nature of the emitter and the quality of light used. It was also established in these initial experiments that the photoelectric effect takes place even in the highest vacuum, which shows that the gas surrounding the emitter plays no essential role in the effect.

Experimental Study

After the existence of the photoelectric phenomenon and some of its peculiar features had been thus established by the simple experiments of Hallwachs, Eisterand Geitel, workers such as J. J. Thomson, Lenard, Richardson, Compton and others undertook a wide series of detailed researches to fix up many other important factors connected with the phenomenon, such as the precise nature of the negative ions emitted, the relation between the photoelectric current and the intensity of light used, the velocity and energy photoelectrons and their dependence on the wavelength of the light used etc.

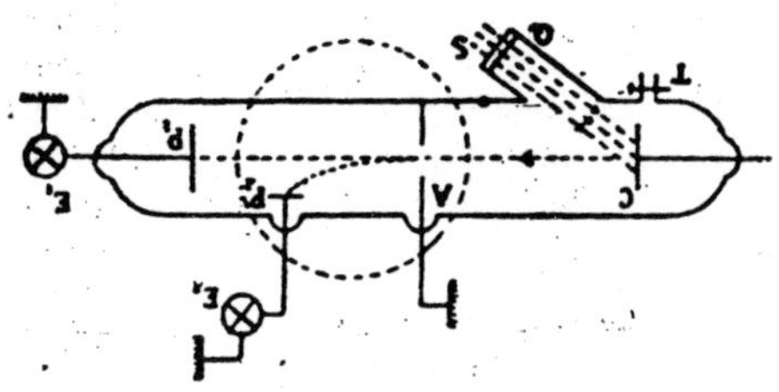

Fig. 4.4 : Lenard's Apparatus.

4.8 ENERGY OF PHOTO ELECTRONS

Prof. Lenard made an important discovery in the course of his experiments. He found that the emission velocity and kinetic energy of the photoelectrons could vary from zero upto certain maximum value.

The experiments conducted by Richardson and Compton in 1912 may be considered as the most successful.

Richardson and Compton Experiments. The apparatus used is shown in Fig. 4.5. The emitter of photoelectrons, C, a very small metal under study, is placed at the centre of a spherical glass bulb B; about 10 ohms in diameter and silvered on its inner surface. C can be raised to any desired potential by a potentiometer arrangement and the value of the applied potential read by a voltmeter V. The bulb is evacuated through the side tube T. Monochromatic light L, whose intensity can be varied in a known manner, is made to pass through a quartz window W and fallen C. The silver coating on the inside of the bulb serves as the anode A and is connected to an electrometer E by means of which the photo-electric current is measured.

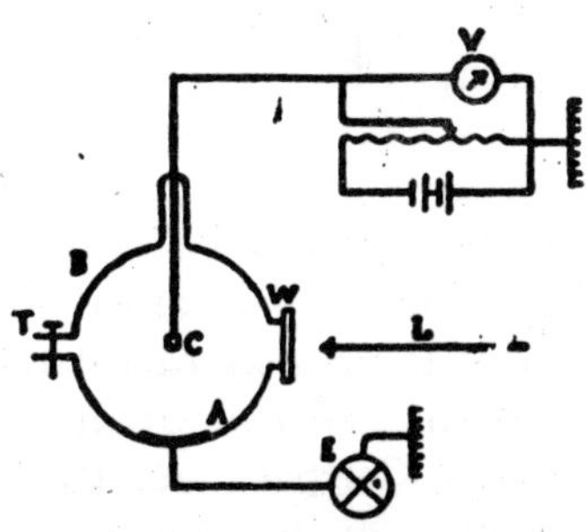

Fig. 4.5 : Richardson and Compton Apparatus.

The large spherical anode with the small emitter at its centre serves two important purposes. First, since the electric field around C is nearly radial, it is possible to measure the energy distribution of the photoelectrons irrespective of the direction of emission. This enables the "total energy" to be determined as against the "normal energy" measured in the case of parallel plates being used for electrodes. Secondly, the impact of the photoelectrons on the anodes causes a diffuse emission from the latter of a certain number of photoelectrons. In the case of parallel electrodes a considerable portion of these returns to the cathode and the observed current is not therefore the true photoelectric current.

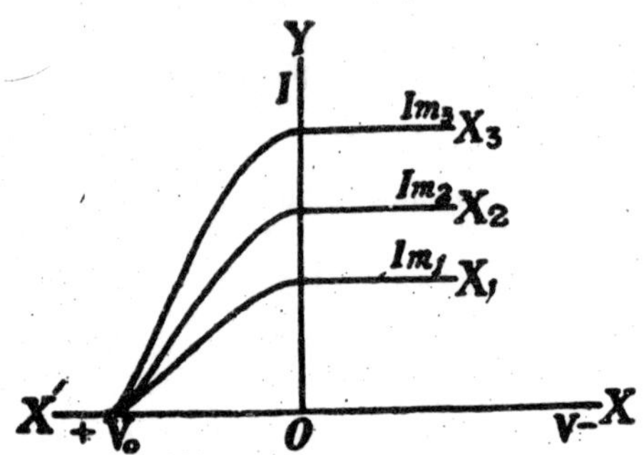

Fig. 4.6 : Retarding Potential Curve.

Rechardson and Compton next studied the more important relation between the velocities of photoelectrons and the frequency of light as follows:

Using several different monochromatic radiations of wavelengths λ_1, λ_2, λ_3, etc. to irradiate the emitter successively, in each case the photoelectric current I for different values of voltage applied to the emitter was determined. When the results were plotted, I vs. V, curves of the type shown in Fig. 4.7, were obtained, provided the intensities of illumination had been so adjusted as to give the same value for saturation current in all cases.

It is seen from the curves that if $\lambda_1 > \lambda_2 > \lambda_3$, the corresponding critical voltage values at which the current starts in the three cases are found to be $V_0^1 < V_0^2 < V_0^3$. Hence as the wavelength of light increases the critical retarding potential decreases. This means that the maximum kinetic energy of the photoelectrons given by

$$\frac{1}{2} mv_m^2 = eV_0,$$

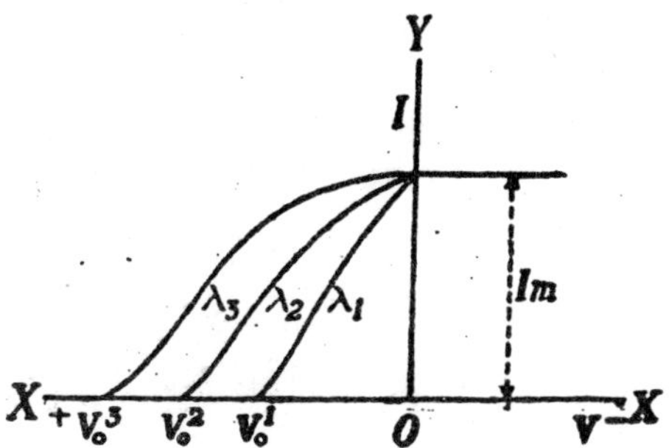

Fig. 4.7 : Frequency Variation Curves.

increases with increasing frequency of the light which causes the emission. Since it has already been shown that V_0 is independent of the intensity of illumination, we arrive at the important conclusion that the velocity and kinetic energy of photoelectrons are independent of the intensity of illumination, but dependent on the frequency of the incident light.

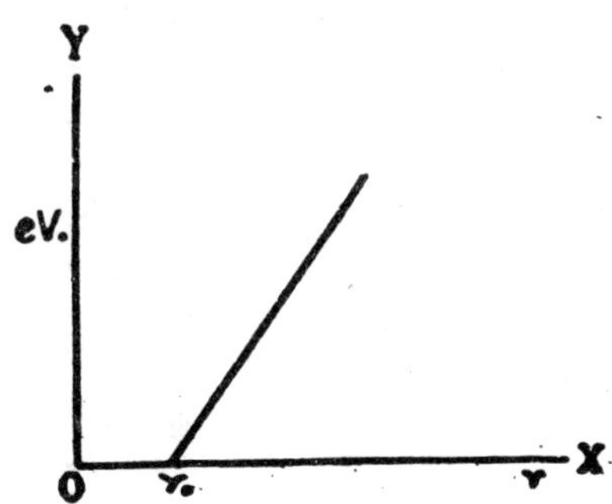

Fig. 4.8 : E_{max-v} Curve of Photoelectrons.

A very simple relation has been found to exist between the maximum energy of emission and the frequency of the light. If the value of the maximum energy given by eV_0, is plotted against the corresponding frequency ν of the light, (Fig. 4.8), a straight line is obtained which has an intercept ν_0 on the frequency axis. The meaning of this intercept is that light of frequency less than ν_0 cannot cause photoelectric emission from the emitter concerned. The quantity ν_0 is characteristic of the emitter varying from substance to substance and is known as the "*threshold frequency*" , as it represents the beginning of the photoelectric activity of the emitter.

The equation of the straight line carry may be written as

$$eVo = h(\nu - \nu_0)$$

where h is the slope of the curve, a constant.

But $$eV_0 = \frac{1}{2}mv_m^2$$

$$\therefore \quad \frac{1}{2}mv_m^2 = h\nu = w_0$$

where $w_0 = h\nu_0$, which is known as the *photoelectric work function.* This equation has had a very interesting history and is one of the most fundamental equations of Modern Physics. It is known as *Einstein's photoelectric equation*, since it was first proposed by Einstein on purely theoretical grounds as a result of the application of Planck's Quantum Theory to the photoelectric process. We shall study this equation more in detail when we deal with the Quantum Theory of radiation. It is enough to note that it has been derived here from strictly empirical data.

4.9 PHOTO ELECTRIC EFFECT

The photo electric effect was first observed in 1910 by Pohl and Pringsheim. During that time they were working with polished mirror surfaces of the alkali metals.

In general we may say that every metal shows a photoelectric effect from a long-wave threshold of 3000 to 2000 Å downwards. In the case of alkalis the threshold lies in the infra-red. The surface condition exerts a very great influence on the photoelectric effect. The presence of layers of gases on the surface or absorbed in the interior is of great importance, as it not only increases the yield as seen above (selective effect) but also shifts the threshold very considerably towards the long wavelength side. Some metals can be made photosensitive in the infra-red by preheating them in hydrogen, the resulting adsorbed surface layer producing a large shift in the threshold value.

It is found that the photoelectric sensitivity of a pure metal diminishes with time. This was attributed to some sort of fatigue, at a time when the action of surface contains nation was not fully explored. Now it has become evident that the "fatigue" effects are due either to the *slow oxidation of the surface* which results in an increase of the photoelectric work-function with the corresponding decrease in the yield of the photoelectrons, or to the ionic bombardment of the surface by the positive ions created in the surrounding gas by the photoelectrons, which produces changes in temperature at the surface and thereby alters the nature of the surface. This second cause of "fatigue" arises not with pure metals but with metals coated with a surface film.

The photoelectric effect is produced not only from metallic surfaces but also in non-metals, liquids, vapours and gases. Some metallic oxides and sulphides are photosensitive even when dry while other metallic compounds must be moistened to become so. A good number of organic compounds emit photoelectrons either in the solid state or in solution. Radiation of high frequencies such as X-rays is required to produce the effect in gases. The photo-activity of gases is known as *photo-ionisation*, as the activity results from the ionisation of the atoms of the gas. When the incident radiation has sufficient energy to remove an electron from the atom, a free electron and a positive ion appear. Experimental work on photo-ionisation in gases is beset with many difficulties and so far only some reliable results have been obtained.

Photoelectric reactions are due to photoelectric emission. The photographic plate is a practical application of such a reaction. The silver salts which are found suspended in a gelatine solution on the photographic plate are strongly photosensitive. Light falling on the plate produces ionisation which forms an invisible image of the object photographed. The latent image is not permanent; hence it has to be developed and fixed with suitable reagents. Otherwise, it will disappear slowly when the photoelectrons produced in the irradiation process and trapped in the emulsion combine again with the parent atoms.

The *act of seeing* appears to be a simple photoelectric effect as there is some evidence of electrical changes in the retina when the eye receives light.

The photoelectric effect so far considered many "surface" effect, but an inner "volume" emission of electrons may be produced if the incident radiation is capable of penetrating into the interior of the body. This liberation of electrons inside a dielectric (insulator) or a body with a high resistance (semi-conductor) results in an increase of specific inductive capacity or of conductivity. This effect, known also as photoelectric conductivity, was first noticed in selenium, as we have already seen. It is also found in rock-salt, fluorite, quartz, paraffin, etc., when they are irradiated by X-rays. The electrical conductivity of sulphur illuminated by X-rays may be increased hundred times. Thallium sulphide changes in specific conductivity under the action of visible as well as infra-red light. It should be noted that there is no electron emission in this inner effect, unlike in the case of normal outer surface effect. Both the inner and outer photoelectric effects have been made use of in the construction of photoelectric cells.

The important conclusions arrived at by fob experimental study of the photoelectric phenomenon may be resumed as follows:

(a) *The strength of the photoelectric current is directly proportional to the intensity of the incident radiation.* The more intense the light used, the greater is the number of electrons ejected and hence the stronger the current that results. The efficiency of emission of photoelectrons, although, dependent to a certain extent on the nature of the emitting surface and the wavelength of the radiation used, yet is generally low. Rough computation shows that even in the case of a good photosensitive metal, like sodium, only one in a hundred million atoms emits a photoelectron at any particular instant.

(b) *The velocity and hence the kinetic energy of photoelectrons are independent of the intensity of the incident radiation, but strictly dependent, with a direct proportionality, on the frequency of the, incident radiation.* Whatever be the intensity of the light used, the velocity and kinetic energy remain the same provided the frequency of light remains unaltered. But greater the frequency greater is the velocity of the photoelectrons and likewise lower the frequency smaller is the velocity. There exists, however, a minimum frequency, the threshold frequency, ν_0, which varies with the nature of the emitter used. Light of frequency lower than the threshold value can never free photoelectrons from the emitter, no matter how long it falls upon the surface or how great its intensity.

(c) *Photoelectric emission is an instantaneous phenomenon.* There has never been observed any time-lag between the beginning of irradiation and the starting of the photoelectric current. Indeed, precise measurements have shown that, if there be any time-lag, it is not more than 3×10^{-9} second.

The classical electromagnetic wave theory of radiation is found altogether incapable of explaining the above-mentioned sure experimental facts, for, in the first place according to it, light which is constituted of electromagnetic waves, passes over all the atoms of the emitter and radiant energy is distributed continuously over the whole wave-front and hence there is no reason why one atom rather than another should be affected and made to emit an electron. The singling out of one atom in a hundred million from the surface of sodium for instance, is inexplicable on the basis of the classical theory. Secondly, the expulsion of electrons, according to this same theory, is due to the strength of the electric field which exists in the light waves. Now, since the intensity is proportional to the field strength, on increasing the intensity, the field strength should be increased, which would, in consequence, eject electrons with greater velocity and energy. This is, however, contrary to observed facts proving the independence of the velocity of photoelectrons from the intensity of light. Thirdly, considering the source of energy of photoelectrons, experimental facts strongly suggest that the photoelectrons derive their energy from the incident light. If that is the case, it must be that the parent atom absorbs the energy from the incident light and hands it over to the photoelectron. But on the basis of the classical theory, it can be shown that it would take hours and even days for an atom to absorb enough energy from an incident wave-train of visible light of moderate intensity to supply the energy with which the photoelectron is actually ejected; yet experiment shows that the emission of the energised photoelectron takes place almost instantaneously. Finally, the existence of a threshold frequency in every case of photoelectric emission cannot be understood on the basis of the classical theory, for a satisfactory explanation of these facts, we have to resort to an altogether new idea known as the "Quantum Theory" of radiation which forms one of the corner stones of Modem Physics. As a matter of fact, the photoelectric effect with one or two other phenomena supplied the first experimental basis for the new theory, which we shall take up for study in a later section, dealing with the new revolutionary fundamental concepts of Modern Physics.

PRACTICAL APPLIANCES OF ELECTRON BEAMS

Among the many practical appliances of electronic beams there are five very interesting instruments, which have proved themselves to be of great importance from different points o view, scientific, aesthetic and utilitarian. They are:

(1) *The thermionic valve*, which forms an essential component of the modern radio or wireless communication and other techniques such as the radar, microwave spectroscopy and noser;

(2) *The cathode ray oscillograph*, which has become, over and above its many uses in the study of periodic phenomena, an indispensable part of modern television;

(3) *The photoelectric cell*, the so-called "magic eye", which is used in almost every walk of life and in particular in talkies and television;

(4) The photomultiplier, based on the phenomenon of secondary emission of electrons, and employed wherever small electronic currents are to be amplified;

(5) The electron microscope, a very important practical application, which has superseded even the most powerful of optical microscopes in the exploration of ultra-microscopic entities.

In this section we shall describe briefly the essential features of these modem and very useful instruments, insisting chiefly on the scientific principles involved in their construction and operation.

4.10 THERMIONIC VALVES

The first thermionic valve was designed in 1904 by Flaming and was known as the flaming valve. Today it is called 'diode rectifier' valve. These valves are of very great practical importance, chiefly on account of their constant use in the modern radio. These valves transmit signals and speeches from one place to another.

It consists of a filament F and a plate P enclosed in a glass bulb that is highly evacuated. When the filament is heated to incandescence by means of battery B_1 it emits electrons by the thermionic process. Now, if the plate B_1 is maintained at a *positive* potential with respect to the filament by means of another battery B_2, the electrons emitted by the filament are attracted towards the plate and a thermionic current flows from the filament to the plate, as indicated by the milliammeter A, but no current flows if P is made negative relative to F. Hence the flow of current between F and P is *unidirectional* and in consequence the arrangement is called a "valve".

Uses of the Diode Valve—Rectifiers

On account of this unidirectional flow of current, the diode valve can be used as a *rectifier*. Supposing that the battery B_2 is replaced by an alternating *e.m.f.* the current will flow only in that half of the cycle

Fig. 4.9 : Sir Ambroee Fleming.

when the potential of P is positive with respect to F and not during the other half of the cycle when P is negative. Thus, an alternating current is rectified into a direct current. This is known as "half wave rectification", in which the direct current is not steady but pulsatory, since only the half waves on one side of the zero line of the alternating current are transmitted. With the aid of a condenser and a resistance or inductance as filters it is possible to obtain a rectified current which is nearly steady. For a full wave rectification a *duo-diode* valve is usually employed, *i.e.,* two plates enclosed in one and the same bulb, which results in economy of both space and cost, the same filament being used for both. Contrivances of this kind are widely used in radio sets both in transmitters and in receivers, and are known as power packs. The diodes which are used for rectifying high voltage A.C. are known as kenotrons. A *full-wave* rectification can be obtained by a hook-up system of four kenotrons, in which one pair is idle during a half cycle while the other pair functions, so that the energy content of the whole wave is obtained.

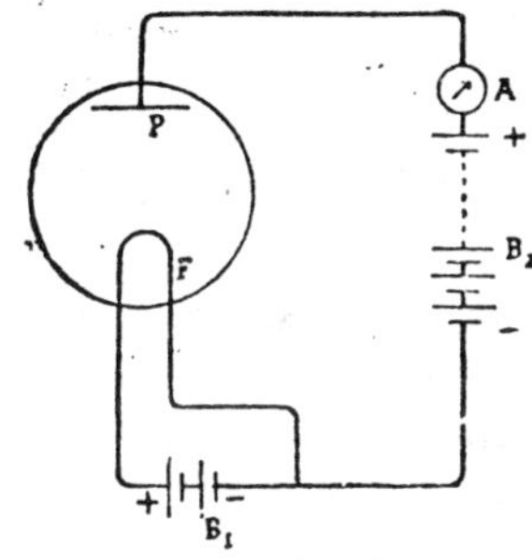

Fig. 4.10 : Diode.

Rectifiers

Gas-filled diodes are used for rectifying fairly heavy currents required for charging accumulators from the A.C. mains. Ionisation by collision of the gas molecules produces positive ions which neutralise the space charge effect and thus prevent saturation of current. This type of rectifiers needs less power for filament heating, works on a much lower plate potential and is more efficient than the vacuum tube. The "Tungar rectifier" sold in the market is a gas filled diode, which has a thick tungsten filament close to a graphite plate mounted in a tube containing argon at a pressure of about 1/10th of an atmosphere.

Mercury arc rectifiers are essentially of the type of the gas-filled diodes. The source of electrons instead of being a filament is a pool of mercury C in a highly evacuated bulb (Fig. 4.11). There are two metal or graphite anodes AA sealed inside the bulb. Once the arc has been started by tilting the tube so that the mercury makes a momentary contact with the proper anode, the discharge takes place between the anode and a spot on the pool. This hot cathode spot acts as the source of electrons, and the atoms of mercury vapour supply the positive ions to annul the space charge.

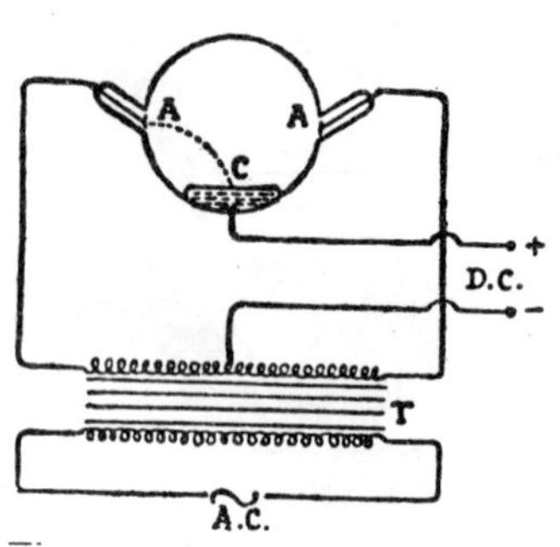

Fig. 4.11 : Mercury Arc Rectifier.

The mercury arc rectifier can be used to get a full wave rectification by designing a circuit as shown in the figure. The A.C. supply is connected to the primary of the transformer T, the secondary winding of which is provided with a mid-point tapping, and this centre-tapping point is connected to the negative terminal of the D.C. output. A connection from the pool of mercury forms the positive terminal of the D.C. output. The rectifying action is briefly as follows. As the A.C. supply pressure passes through successive cycles, the two anodes connected to the ends of the secondary will in turn become positive with respect to the mid-point of the secondary of the transformer. The arc, however, can be established only between the pool of mercury and that anode which is positive and consequently the arc will pass to the mercury pool from each anode in turn. The mercury pool will therefore be continuously maintained at a positive potential with respect to the mid-point of the secondary winding and in this way a rectified current will be obtained at the D.C, output terminals.

The principle of the mercury arc rectifier was first discovered in 1903 by P. Copper-Hewitt. The earlier types were all of the glass-bulb type and this limited the output of a single unit to a relatively small power. Later, the glass-bulb has been replaced by a steel tank, which gives a much larger output, about 5000 kilo-watts at 6000 amps. The mercury arc rectifiers have the practical advantage of being static, silent in action and efficient.

It may be noted that for low pressures and heavy currents, another type of rectifier, known as the metal plate rectifier has supplanted the mercury are rectifier. It consists of a copper disc which has been oxidised into cuprous oxide on one side. If this disc whose one face is pure metal while the other the oxide of the metal is connected to an A.C. circuit, it is found that the current flows only in one direction, *viz.,* from oxide to copper, which shows that a high resistance comes into play preventing the flow of electrons from oxide to metal. Very probably the oxygen of the cuprous oxide acts as impurity in the insulating oxide and converts it into a semi- conductor of high resistance. Whatever be the mechanism, the arrangement offers a means of converting A.C. into D.C. In practice the cuprous oxide film is produced on a copper disc by a secret process and is caused to grow out of the copper so that there is no boundary space and consequently no boundary resistance between the oxide film and the copper disc. Contact is made with the oxide film by means of a lead disc which is pressed close to the film by means of a clamping screw. One such element can deal only with about 6 volts, but a sufficient number can be arranged in series to build up the requisite voltage which may go up to about 4000 volts even. The permissible current depends upon the type of cooling. By immersion in oil current densities of 3-5 amps. per square inch may be obtained. Undue rise of temperature, however, must be avoided; otherwise 'the appliance will allow current to flow in both directions. There now exist other kinds of plate rectifiers—the selenium rectifiers, for instance.

The diode valves serve also the purpose of *defector* or *demodulator* in modern radio receivers. If the high frequency modulated wave, picked up by the receiver, is directly applied to the telephone, the original signal or sound will not be reconstructed since the average effect of the inducing *e.m.f.* over a complete cycle of the modulating signal is zero. But if the current received be first rectified by a diode valve, all the impulses in the telephone are in the same direction, which can therefore operate the telephone and make it respond to the modulating signal. The diode detector is frequently used in receivers where the detector follows stages of amplification and extreme sensitivity is not required.

The Triode Valve

The first type of valves was soon replaced by a more efficient one known as the *triode* valve which has a third electrode called the *grid* between the filament and the plate. It was first devised by Lieben in Germany (1906) and by Lee de Forest in America (1907). The modern triode valve (Fig. 4.12) consists of a highly evacuated glass bulb containing (i) a filament F usually made of thoriated tungsten or platinum coated with oxides in order to obtain a good emission of thermionic electrons at low temperatures, (ii) a grid G which may be a flat metal gauze arranged above the filament or a spiral of wire mounted with the filament as axis and (iii) a *plate* P or *anode* which is in the form of a cylinder surrounding the grid if the latter is spiral or in the form of a flat plate if the grid is flat.

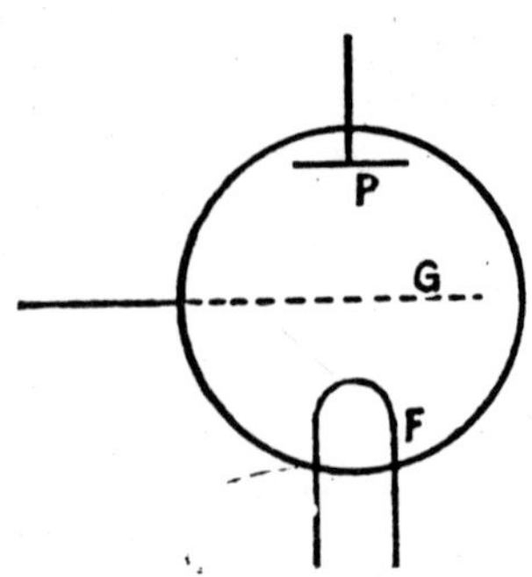

Fig. 4.12. Triode Valve.

Although, in an actual triode the form and distribution of the three electrodes are so different, the triode is ordinarily represented diagrammatically as shown in the figure. The filament is provided with two external leads for connecting it to a low tension battery which supplies the heating current, while the grid and plate are each provided with a single lead so that they may be maintained at any desired potential by the use of batteries. The three parts of the valve are well insulated from one another and the leads from them are connected to separate terminals fixed on a suitable holder of the bulb.

In the absence of the grid a thermionic current flows in the valve when the plate is at a positive potential with respect to the filament, as in the diode valve. This is known as the anode or plate current, usually of the order of milliamperes. It increases in strength as the plate potential is increased until saturation is reached due to the action of the space charge around the filament. The grid may be considered as a *controller* of this space charge. If the grid potential is *negative* with respect to the filament it will repel the electrons emitted by the filament, which means the space charge effect will increase and as a result the plate current will be reduced. If, on the other hand, the grid is made positive, it will aid the flow of electrons towards the plate and thus, diminish the space charge, which means the plate current will increase. Thus the plate current is controlled at will by the variation of the grid potential. Further, it is found that for certain values of the grid voltage, a very small change in it produces a large change in the plate current. If the grid were absent, a change of potential ten times as great would have to be applied to the plate in order to cause the same change in the plate current. Since charging the grid to an electrostatic potential requires practically no expenditure of power, except for the small loss in ohmic resistance of the grid circuit, the grid is very economic and at the same time a very efficient controller of the plate current, in a manner similar to valves which regulate the flow of steam in a pipe. This is an additional reason why the arrangement is rightly termed a thermionic valve. Due to the open construction of the grid, spiral or mesh, the grid does not act to any appreciable extent under ordinary Working conditions as a collecting electrode so that the current in the grid circuit normally remains inappreciable;

4.11 USES OF THE TRIODE

The triode can be used as a relay. It can be used as rectifier. The triode can be used as an amplifier and also as an oscillator.

The Triode as a Relay

A relay is a device by which a small amount of energy can be used to turn on and off or control a much larger source of energy. The triode acts as a *perfect relay*, since the application of a few volts to the grid allows to control, establish or cut off the plate current belonging to a source of several hundreds of volts; it is much superior to mechanical relays since it has no appreciable inertia, working, as it does, on instantaneous thermionic emission, consumes minimum energy as there exists very little grid current, and is capable of delicate regulations, unlike mechanical relays which either act wholesale or not at all.

In this connection, the principle of the *thyratron*, a gas filled triode relay, may be mentioned. A triode valve, in which a small quantity of helium, argon, neon or mercury vapour is introduced, constitutes the thyratron. Suppose we consider an ordinary vacuum triode with its grid potential sufficiently negative to prevent any plate current from flowing. If now we raise the plate potential, the relative effect of the grid will be reduced and a stage will be reached at which the plate current begins to flow. First the current strength would remain small, becoming gradually larger as the plate potential is increased. But with a gas filled valve, like the thyratron, the electrons moving towards the plate, collide with the molecules of the gas which being at a low pressure is readily ionised.

At each collision more electrons are produced and these, in their turn, collide with more molecules releasing more electrons, so that a decent current strength is rapidly built up and a discharge takes place through the gas. Also some of the positive ions formed in the ionisation process are attracted to the negative grid and form a sheath round the grid which prevents the grid from having any further effect on the flow of the plate current. Thus the grid loses its control and cannot stop the discharge when once it is started. Until the plate potential is removed or at least reduced to a value below the ionisation potential of the gas which is 10 to 30 volts in most cases, the discharge will continue. The valve becomes practically a short circuit during the discharge. When the voltage tells below the ionisation potential, the valve resumes its non-conducting capacity and the process will be repeated when the voltage builds up again in a very short time of 0.1 to 1 millisecond. This fact endows the thyratron with a high resolving power and the frequent use of it in electrical counting of individual events that occur in rapid succession. As an inertia-less, relay, it finds another important use in the construction of *ideal time-base* circuit employed in the functioning cathode ray oscillographs, as we shall see later.

The Triode as a Rectifier

The principle of the rectifying action of the triode is easily understood by the following consideration. Taking a mutual characteristic curve, (Fig. 4.13), let us suppose that the grid is maintained at a negative potential represented by a point A at the lower bend of the curve. Let an alternating *e.m.f.* be applied to the grid, so that DC and DE represent the variation in V_g, about A. From the figure we see that during the one half DE of each oscillation of V, the plate current changes much more than during the other half DC. The resulting change in the plate current, though still alternating, has for the upper half of the wave a larger amplitude than for the lower. This means a larger plate current in one direction followed by a smaller current in the opposite direction, so that the mean plate current over a complete cycle is not zero, but has a certain value in one direction. Thus the valve acts as a *rectifier*. Supposing the alternating *e.m.f.* applied to the grid is a signal received in the form of damped waves, the signal will produce small impulses of unidirectional plate current following each other at the frequency of the damped waves; the cumulative effect of these pulses will enable the telephone-receiver to respond. Thus the valve by its rectifying action becomes also a *detector*.

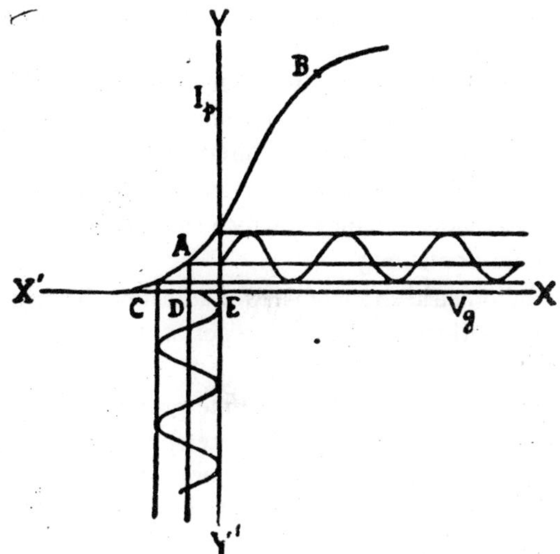

Fig. 4.13 : The Rectifying Action of a Triode.

There are two methods of using a triode as a rectifying detector :

(a) Anode Bend Tectification, and

(b) Grid-leak or Grid Current Rectification.

(a) Anode Bend Rectification

In this method the grid is given a potential corresponding either to a point A at the lower bend or to a point B at the upper bend of the mutual characteristic curve. The former is the one more commonly used since it has the following advantages:

(a) The working point of the grid being negative to the filament, there is no flow of grid current, and hence no absorption of energy from the input circuit, which means less damping,

(b) Very the current is taken from the H.T. supply of the plate circuit, while in the other case, a much larger current is drawn.

It can be easily shown that no rectification takes place if the grid potential is such that the working point falls on the straight portion of the curve. But owing to the steepness of the slope in this region, the oscillations of the plate current will have greater amplitude than the input oscillation; hence under such conditions the valve can act as an *amplifier*, but not as a *rectifier*.

Some device must be used for maintaining the grid potential at the working point A, as it tends to collect electrons and thus fall in potential. This is usually dono by what is known as a *grid-leak* consisting of a high resistance of the order of 1 to 10 megohms, suitably introduced in the grid circuit, which enables the accumulated electrons to leak away to the negative end of the low tension battery, earthed in an actual circuit.

The circuit details of the simple anode bend rectifier are shown in Fig. 4.14. The oscillating circuit LC which can input signal is connected to the grid circuit containing

(1) A potentiometric arrangement for adjusting the grid potential so that the working point may be get at the middle of the lower or upper anode bend, and

(2) A grid-leak R_1 for maintaining the grid potential steady at the working point. A high resistance telephone T is put in the plate circuit. A large condenser C_2 is shunted to the plate circuit. With this arrangement when the input signal is applied to the erid, a rectified current consisting of three components, a direct current (D.C.), a high frequency (R.F.) and a low frequency (A.F.) flow through the plate circuit. The high resistance of the telephone and the relatively large capacity condenser C_2 form what is known as the output impedance; the A. F. component carrying the modulated signal passes through the telephone, while the B.F. component consisting of the carrier wave is "by-passed" through C_2.

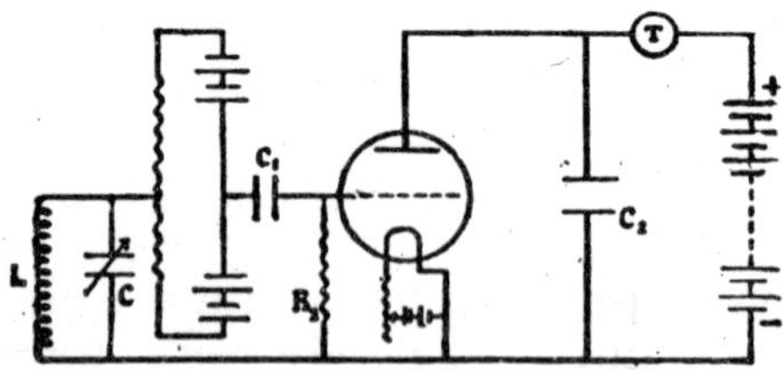

Fig. 4.14 : Circuit Diagram of Anode Bend Rectification.

(b) Grid-leak or Grid Current Rectification

This method, first devised by Laede Forest in 1907, is a more sensitive one than the anode bend rectification, but its principle is more complicated, Briefly stated in this method, the grid-filament circuit is used as if it were a diode rectifier, *i.e.*, the high frequency modulated input signal is first rectified in the grid circuit and then the low frequency modulated spinal obtained by rectification is amplified in the plate circuit. This method, therefore, makes use of the valve first as a diode rectifier and then as a triode amplifier in contrast with the previous method in which the valve may be considered first as an amplifier, then as a rectifier. In this method the grid characteristics must be used. Portlier, die flow of the grid current and the fixing up of the grid potential at the non-linear curved part of the grid characteristic curve are essential conditions which are secured by the introduction of a suitable grid-leak in the grid circuit.

The circuit details of a grid leak rectifier are shown in Fig. 4.15. In this method no auxiliary grid potential is required; the grid is insulated from other parts of the circuit by a condenser C_1 and is connected: to the negative of the filament through a high grid leak resistance R_1. Before the arrival of any oscillating voltage to the grid there is a steady flow of the plate current. When a signal is received, it passes through the condenser C_1 and is applied to the grid. As the line end of C_1 goes positive, R_1 becomes shorted by the relatively low impedance of the grid-

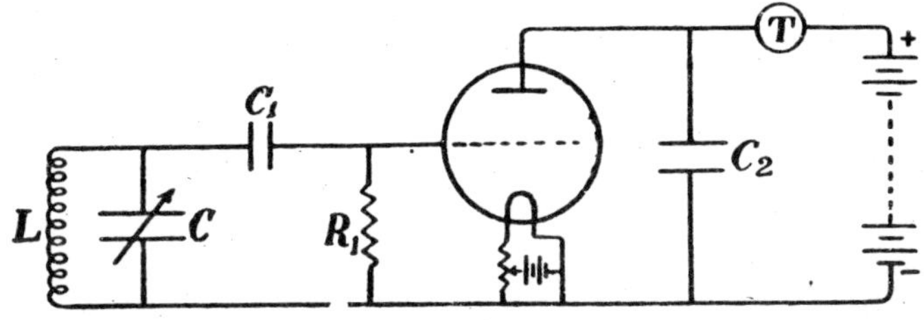

Fig. 4.15 : Circuit Diagram of Grid-leak Rectification.

filament diode section. Through this low impedance C_1 go to charged to very nearly the peak value V of the signal voltage, so as to make the grid end of C_1 negative. This charging current flowing to the grid is the grid current through the valve. When the signal goes negative, the valve becomes an open circuit and C_1 begins to discharge through R_1. If C_1R_1 is large compared to the period of the signal, the original voltage across C_1 remains substantially unchanged when the next positive peak comes in. Any charge lost by C_1 during the negative half cycle of the signal is made good during the next positive half cycle. As a result, C_1 remains always charged to V and the discharge current V/R_1 flowing through

R_1 keeps the grid at a steady voltage Y negative with reapeot to the filament. This is the case when the carrier wave is unmodulated. With a modulated carrier, the bias Y also becomes modulated, C_1R_1 being small compared to the period of the modulating A.F. signal. Thus due to detection at grid, a steady grid bias for the amplifying triode section is developed by the B.F. carrier upon which is superimposed the A.F. signal contained in the modulation.

The important point to be noted here is that the voltage developed across R_1 is primarily due to the discharge of the condenser C_1 with a high time constant and-not due to the leakage of electrons to the filament. This latter is present only daring very short pulses of grid conduction at each positive peak and even then is of no practical importance. It may be added that since C_1 is a "by-pass" for the R.F. component, the fall R.F. voltage across C_1 is also present across K, superimposed on the steady bias developed by the R.F. carrier and the A.F. component developed by the modulation. Whereas this causes some additional damping of the input tuned circuit, it plays no part in detection proper.

The Triode as an Amplifier

The main principle of the amplifying action of the triode has already been pointed out *viz.,* that changes in the grid potential produce a much greater effect on the plate current than the same changes in the plate potential. To understand this amplifying action let us consider the straight steep part ACB of the mutual characteristic curve (Fig. 4.16). Let the mid-point C of that straight portion be chosen as the working point of the valve, *i.e.,* the grid is given a negative potential represented by the point E. If an alternating potential is now applied to the grid, its potential will vary correspondingly. The positive half of the wave increases the grid potential from OE to OD (*i.e.,* the grid becomes less negative than at atari;), while the negative half decreases it from OE to OF (*i.e.,* the grid becomes more negative). When the grid potential is OD, the plate current is OK and when it is OF the other is OM. Thus, as the grid potential varies between OD and OF the plate current varies between OK and OM. By comparing the two sinusoidal curves G and P, where G represents the potential variations applied to the grid and P the current variations in the plate circuit, we see that, the variations of current in the plate circuit are greater, the steeper the curve. To obtain therefore large amplification the steepest part of the curve should be used and the grid potential should be such that the working point is in the middle of that steepest part. Under these conditions if the incoming weak signal be passed through the grid of the valve, it can be strengthened or amplified in the plate circuit.

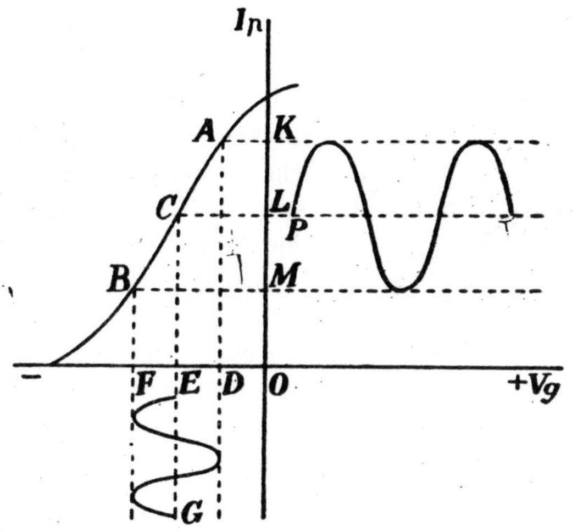

Fig. 4.16 : Amplifying Action of the Triode.

In order to utilise the valve as an amplifier it is essential that changes in the plate current should produce corresponding changes in the P.D. across an external impedance in the plate circuit, for if the plate is directly connected to the positive terminal of the high tension battery in order to maintain it at a steady potential, changes in the plate current cannot obviously produce any change in the potential of the plate, but if an external impedance, say a high resistance B is included in the plate circuit between the plate and the high tension battery or between the L.T. and the H.T. batteries, then a change in the plate current dI (due to a small change dV_g in the grid potential) will produce a change dE = RdI in the P.D. at the ends of the resistance B. The amplification produced then is

$$\frac{dE}{dV_g} = \frac{RdI}{dV_g} \qquad ...(1)$$

If μ be the amplification factor of the valve, a change dV_g in the grid potential causes a change μdV_g in the plate potential, which acts across both the valve of internal resistance B, and the external resistance R. Hence the change in current dl in the plate circuit is given by

$$dI = \frac{\mu.dV}{(R + R_i)}$$

$$\text{Amplification obtained} = \frac{dE}{dV_g} = R\frac{\mu.dV_g}{(R + R_i).dV_g}$$

$$= \mu.\frac{R}{R + R_i}$$

From this relation we see that the greater the value of R, the more nearly will the amplification actually obtained approach the amplification factor μ of the valve, since the value of $R/(R + R_i)$ approaches unity.

The external impedance in the plate circuit, which we have taken above as a non-inductive resistance R_i may be also an inductance coil or choke, a combination of these, or even a turned circuit containing inditoctance and capacity in parallel. Further, in order to get the maximum power, the external resistance R should be more or less the same as the internal resistance R_i, of the valve. In practice, the external impedance must be at least as large as the internal impedance and generally must be considerably larger, in order that small variations in the plate current may produce large alterations in the P.D. across the external impedance.

Conditions Far Distortionless Amplification : It is essential especially in radio-tolephony that the process of amplification does not produce distortion, which arises from two causes:

(a) If the mean grill-potential is such that the oscillatory input voltage gives instantaneous values of the grid potential corresponding to points on the curved portion of the mutual characteristic curve, rectification takes place, and hence the plate current waveform, as also the output voltage waveform, will not be a replica of the input voltage waveform, *i.e.,* distortion has been produced;

(b) If the grid current flows during any part of the input voltage cycle, there arises damping on the input circuit which causes the variation of the oscillatory grid-potential, which would distort the waveform of the signal already at the input stage.

To produce therefore distortionless amplification these two causes of distortion must be removed. The first is got rid off by confining the range within which the grid potential varies under the influence of the oscillatory input voltage to the straight portion of the mutual characteristic curve. The second is eliminated by making the grid potential sufficiently negative so that the peak grid potential during the positive half cycle of the input voltage does not extend into the region where the grid current flows. Thus a *relatively large negative grid bias and operation on the straight portion of the mutual characteristics* are the two conditions required to obtain distortionless amplification. In practice, the best results are obtained if the grid is biased negatively to a point half-way between the voltage at which the grid current starts to flow and the voltage at which the mutual characteristic curve starts to bend. This is known as *mid-point biasing.*

The circuit details of a simple anode resistance amplifier are shown in Fig. 4.17. The oscillating circuit LC tuned to the input signal is connected to the grid circuit containing :

(i) A potentiometric arrangement for obtaining the midpoint biasing of the grid, and

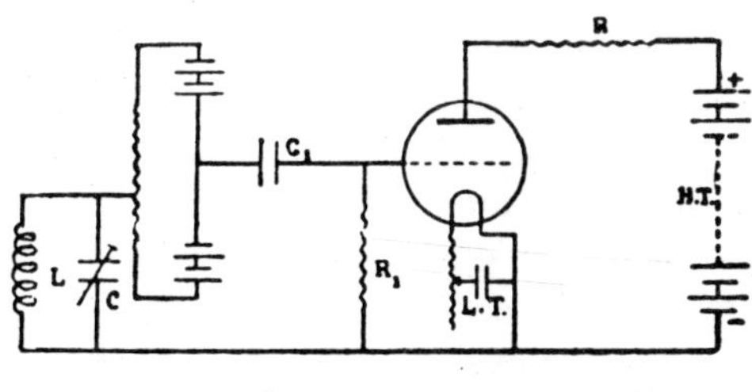

Fig. 4.17 : Circuit Diagram of a Triode as an Amplifier.

(ii) A condenser C_1 and a grid leak resistance R_1, the former to insulate the grid from the direct H.T., which becomes necessary chiefly in multi-valve amplification and the latter to serve the usual purpose of preventing an accumulation of negative charge on the grid and the consequent increase of the negative grid bias to such an extent that the plate current may become insignificant. It must be noted here that since for distortionless amplification the first valve should pass on to the second one a waveform of voltage, which is the same as that received by the first, it is necessary to minimise the typical effect of a condenser and leak, which is to introduce asymmetry in the waveform, if the grid current flows. In amplifiers, therefore, the condenser has a large value than in rectifiers, while the leak resistance is generally somewhat lower. The plate is connected to a high tension battery through the external impedance B whose value is somewhat larger than the internal resistance of the valve. With such an arrangement the oscillations of the plate current will have a Much greater amplitude than the input oscillations, but its shape will replica of the input signal. Thus, distortionless amplification is produced in the plate circuit, which can be applied through a suitable coupling circuit to the grid of the next valve for further amplification.

4.12 CLASSIFICATION OF AMPLIFIERS

Amplifiers are classified according to the operating conditions under which the valve works.

The few types of amplifiers can be classified as follows:

1. Class A Amplifier
2. Class B Amplifier
3. Class C Amplifier.

Class A Amplifier is one in which the grid bias and the input signal voltage are such that the, operation is confirmed to the, straight portion.

This type of amplification is most useful when maximum voltage amplification of the input signal is desired.

Class B Amplifier is one in which the grid bias and the input signal voltage are such, that the plate current flows only during approximately one half of the cycle of the alternating input signal voltage. The grid bias is adjusted so that the plate current is approximately zero when no input signal is present. This type operates at a much higher plate efficiency than class A and is usually employed to obtain linear amplification of audio frequency or modulated radio frequency signals at high efficiency and tolerable distortion.

Class C Amplifier is one in which the grid bias and the input signal voltage are such that the plate current flows during substantially less than half of the cycle of the input signal voltage. The grid bias is appreciably higher than necessary to cut off the plate current in the absence of the input signal voltage, usually twice as great or more. This type produces the highest plate efficiency of any type and the greatest amount of radio frequency output from a valve. It is hence widely used in the r.f. stages of transmitters. The high efficiency is obtained as a result of the high bias and large driving voltage. Grid current and plate current are caused to flow during a shorter time interval than the full positive half cycle of the driving voltage, A very large pulse of plate current flowing for a relatively short time reduces the tube heating. This increased efficiency makes it possible to raise the operating power level before the rated tube heat dissipation is exceeded.

Multivalve Amplifiers : In order to receive effectively signals coming from either a vary great distance or a transmitting station of low power, it is usually necessary to employ several valves for amplifying purposes before or after rectification. They may be connected in series (technically called "in cascade") in *parallel* or in what is known as a *push-pull system*.

In the *cascade amplification*, different inter-valve coupling circuits are employed, according to which the amplifiers are classified as resistance, inductance, tuned anode and transformer amplifiers, the general principle involved being that the amplified voltage across the impedance of the plate circuit of the first valve is applied to the grid of a second valve where it is again amplified and so on.

In the *resistance-capacity coupling* the voltage across the load resistance R of the first valve is applied to the grid of the second valve through the intermediary of a capacity which provides a low reactance path for the audio-frequency voltage from the plate of the first valve. The advantage of using this arrangement is that it gives equal amplification ovora wide range of frequencies, while the disadvantage is a lowering of the plate voltage across the load resistance. It is often used in A.F., amplifying stages, but less frequently in R.F. amplifiers, owing to the fact that the coupling resistances having self-capacities in parallel with themselves form shunt impedances, whose value is lowered the higher the frequency

In the *inductance capacity coupling*, the external impedance is an inductance coil or choke with a low resistance but with a high reactance. Across the impedance of the first valve will appear the amplified A.C. voltage which is applied to the grid of the second valve through the intermediary of a capacity. Due to the low resistance of the choke, the voltage drop across it is small. The great disadvantage of this method is that the amplification obtained from a given choke varies with the frequency amplified with the resulting "peak effect", an undesirable feature in an amplifier which may be avoided by using high resistance wire for the choke.

The *tuned anode coupling* is an improvement over the previous method, where the peak effect for different frequencies are taken advantage of. If a condenser, whose value can be varied, is shunted across the choke in the plate circuit, it becomes possible to make this parallel circuit exactly resonant to any desired frequency and thus obtain maximum amplification of a small selected range of frequency.

The last two methods can be used both for A.F. and R.F. frequencies. But on account of the high cost of choke as compared with resistance, these types of amplifiers are seldom used, except for purposes of high selectivity obtained with the tuned anode coupling.

In the transformer coupling, the output voltage in the plate circuit of the first valve is passed on to the grid of the second valve through the intermediary of a transformer (step up), which enables the passed on voltage to be increased by the transformation ratio and thus secure a much greater voltage amplification than is possible in other types of coupling. In the transformer intervalve coupling there is no need of an intermediate capacity since there is no direct connection from the plate of one valve to the grid of the next. If a condenser and leak are found in the secondary circuit of a transformer coupling it is a definite indication that the next valve is a rectifier valve using grid leak rectification. Hence the primary of the transformer is included in the plate circuit of the first valve and the voltage variations induced across the secondary are applied directly to the grid of the next valve.

The transformer amplifier can be used both for A.F. and R.F. amplifications, but the transformers used in the two cases differ in certain respects.

In A.F. a*mplification*, the transformer is wound on an iron core, which gives the result that the magnetic leakage is negligible or the coupling is approximately perfect. In addition, the ratio of the secondary to primary turns may be as much as 5 to 1 or 7 to 1 giving a bigger voltage across the secondary than the primary circuit, and hence a high voltage amplification.

In A.F. *amplification, the transformer* must have no iron core, on account of the excessive eddy currents and hysteresis losses which occur at the high frequencies involved. They may be wound on a frame of insulating material with as little self-capacity as possible, This however introduces considerable leakage and the coupling is very much looser than if iron were used. Satisfactory results have been obtained by the use of powdered iron-cores, where particles of iron, often colloidal in size and nature are embedded in an insulating material by the application of high pressure. Such a composition produces a core with the properties of iron, but in which eddy currents and hysteresis losses are minimised on account of the insulating surfaces separating the particles.

In the transformer as in the case of inductance coupling there are self-capacities between the turns of the primary and also of the secondary and similar effects as regards "peaky" amplification may be expected. The same steps are taken either to eliminate it by winding the transformers with high resistance wire or to utilise the exceptionally good amplification it gives by tuning either the primary or secondary inductances or both so that they form resonant circuits to the received frequency with a consequent increase in selectivity.

We have considered above a simple type of anode-resistance, choke or transformer amplifiers. There is a more involved type, chiefly used in R.F. work and known as reaction amplifier. As it is based on the same principle as that of a valve oscillator, we shall deal with it below after studying the triode as an oscillator.

In the *push-pull* arrangement, two valves are used, fed from a centre-tapped intervalve transformer, so that equal voltages are applied to each valve in the opposite direction. In consequence, as the voltage on one grid increases, that on the other decreases by the same amount. Corresponding increases and decreases take place in the plate current and these are passed through an output transformer with a centre tapped primary. It will be seen that any variations in voltage combine in the output stage to produce an additive effect in the secondary, but that the steady plate currents from each valve cancel each other out as far as the output transformer is concerned, so that the iron circuit of this transformer is not subject to saturation and the consequent undesirable reduction of inductance of the core. The advantage of the push-pull system is that distortion, arising from uneven amplification at widely differing frequencies ia greatly reduced. This form of distortion, due to the curvature of the valve characteristic, produces a doubling action, giving second harmonics as well as the original note. It is found that with a push-pull arrangement the curvature of one valve is counteracted by the other giving appreciably less "even harmonic" distortion.

4.13 THE RADAR TECHNIQUE

The term 'radar' is applied not a single particular instrument, but to a group of varied and flexible, but intimately related techniques which employ the same basic principle, and have more or less the same end in view.

It is the abbreviation for 'radio detection and ranging. By this device radiowaves are fixed in a particular position.

Radar, in the ultimate analysis, is a kind of *spectometer*, using radio-waves instead of sound waves; hence its main principle is best understood by a consideration of the location of an object by means of sound echoes. A sharp hand-clap is sent back as an echo by a reflecting surface such as a cliff, and the distance of the cliff from the person making the hand-clap can be measured by simply counting the seconds that elapse between the hand-clap and the hearing of the echo. If the clap is loud the echo will be proportionately loud; if the clap is continuous the weak echo may not be heard due to the loud and sustained local-noise; if there are several cliffs each sending back an echo, it becomes difficult to pick out the echo from one of them. These are the main drawbacks which limit the acoustic system of location of objects. The measurement on a particular cliff is best done by making a very loud and a very brief clap and then allowing a silent interval in which the returning echo is given full play, since it is evident that the louder the clap the more intense also will be the echo and the shorter the clap the less the overlapping of echoes from slightly different points of the echoing object.

Radar works on a fundamentally similar principle. Very powerful and very brief electromagnetic pulses are sent out, *very powerful*, meaning several hundreds of kilowatt power content, and *very brief*, implying a few millionth of a second. The production of such pulses can be done by specially designed thermionic valves, capable of generating powerful ultra-high frequency oscillations. The sending out of these pulses must be done at exactly equal intervals, every 1/25 sec. say, if the measurement of long distances is involved, and every 1/5000 sec. for short distances. This is achieved by a suitable transmitter which works for about a micro-second and then remains inactive for the relatively long time of 1/25 sec. or 40,000 micro-seconds. The pulses are labelled on the basis of either frequency modulation or amplitude modulation. No active co-operation on the part of the object under investigation is required except that it must be able to reflect the pulses.

Fig. 4.18 : Prof. E. Appleton.

On account of the very great velocity of electromagnetic waves (the same as that of light), the tuning of, the radio pulse echo has to done in millionth of a-second, unlike in the case of sound echoes. The measuring of such extremely short intervals is readily carried out with a cathode ray oscillograph, incorporated in a receiver designed to register the pulse echoes. If a time/base of the desired number of microseconds is applied to the horizontal deflecting plates, while the pulse echoes, after passing through the radio-receiver, are fed to the vertical deflecting plates of a cathode ray tube, special curves with V-shaped "pips" corresponding to echoes received are traced on the fluorescent wall of the tube, from which the position of the object that sent out the echoes can be accurately determined as regards distance, direction, etc. Thus the principle of the radar consists essentially in the measurement of the echo delay time of the radio pulses. It was foreshadowed by Hertz's demonstration in 1886 that electromagnetic waves could be reflected by solid and liquid objects, but it might not have been achieved without Appleton's classical range measurement on the ionosphere with frequency modulated radio waves, successfully conducted in 1924, nor could it have reached its present practical perfection and universal application but for the discovery in 1925 of the amplitude modulated "radio-pulse" system by two Carnegie Institution scientists, Breife and Tuve.

Probably no other scientific or industrial development in the history of the world has undergone such progress in all phases simultaneously and on such a large scale as the radar. Research work, technique, actual production, training of operators—all these had to go on at the same time. The reason for this is found in the fact that the radar was first developed, on a large and intensive scale during the Second World War (1939-45) as a powerful weapon, both on the defensive and offensive phases, on account of its ability to direct warships and air-planes, aim and fire guns, pinpoint targets for bombing, help to destroy enemy planes, ships and fortifications, etc. Such a many-sided use of radar under the feverish impetus given by the war for homeland safety and final victory over the enemy largely contributed to the production of different radar systems which mark also the progressive evolution in the perfecting of the technique.

The C.H. or *"Chain Home" System* was the first radar to be introduced in 1935 for the detection and location of approaching aircraft bombers. The wavelength used in this system was about 10 metres. The transmitting array consisted of fixed horizontal aerials and reflectors suspended from 350 ft. masts which let the pulse spread out nearly uniformly over a wide sector in front of the station, but very little behind, in order to avoid con fusion with inland aircraft. The receiving aerial system, mounted on a wooden tower about 250 ft. high and a hundred yards away from the transmitting aerial, was made up of two pairs of aerials fitted with radio-goniometers, one for direction finding and the other for elevation measurements. This aerial fed the radio receiver-connected to the cathode ray tube "with the echo pulses coming back from an aircraft. The angle of elevation and the bearing angle of the approaching aircraft were measured by the use of the two pairs of aerials, while the distance was given by the echo delay time on the cathode ray tube. In this way the position of the aircraft could be completely fixed even when it was flying at a height o? 10,000 ft. and 100 miles away.

The C.H.L. or "Chain Home Low" System : The Chain Home system functioned very satisfactorily for the location of large formations of aircraft but was not so successful in dealing with small formations, particularly at low altitudes of a few hundred feet, on account of the extremely weak signals returned by them. The ocho strength was found to be directly proportional to the square of the product of the aerial height and of the flying height and inversely proportional to the square of the wavelength. To get long ranges on low-flying aircraft, if was therefore necessary to reduce the wave- length : this reduction in wavelength allowed the use of smaller aerials. These considerations led to the design in 1939 of the CHL system, in which the transmitters, capable of ranges exceeding 100 miles, operated on ultra-short waves of 1-6 metres. The dimensions of the aerial were of the order of a few yards; they were mounted on a turn-table and rotated about a vertical axis at a slow rate of six revolutions per minute, which sent out, in consequence, a limited beam of about 12° to 15° width. With such a rotating transmitting aerial, it was necessary to use a beam receiving aerial which should always be in line with the former, in order to secure maximum sensitivity. This was achieved by using the same aerial for both reception and transmission with the help of a duplexer, given the fact that transmission lasted for only short periods at a time, while reception could be made only during the intervals between the transmitted pulses. For the same reason, a radial time-base, which rotated at the same speed as and was in step with the beam, was used for the cathode ray tube. Small radar sets working on this principle were fitted to aircrafts, which patrolled to detect ships and aircrafts, ch:efly at night.

G C. I. or Ground Control Interception : This system was obtained by adding to the CHL set one of the most revolutionary and versatile devices in radar technique known as the PPI (Plan position indicator). The need for this improvement arose due to the insufficiency of the CHL system to assure success for night fighters. The CHL system became effective only within the range of about 3 miles from the target, and before that range was reached some device was needed to bring the night fighter in the right direction where the target was GCI provided that device by the use of PPI which consisted in using the received pulses to brighten the spot on the cathode ray tube screen, instead of letting them produce "pips" on the range scale. The special type of cathode ray tube used for this purpose was called the A-scope, in which the electron beam swept from the centre of the screen along' a radius to the periphery and back to the centre; the electron beam could also be rotated about the centre and the rotation was synchronised with the rotation of the aerial. With such an arrangement the target, when detected, produced a bright spot at a distance and in a direction from the centre of the screen which corresponded with the distance and direction of the target from the station—hence the name plan position indicator. As the radar equipment was rotated, the reflected pulses from different points gave rise to corresponding bright spots; this continued until a whole series of bright spots corresponding to the outline of all reflecting surfaces in the neighbourhood had appeared on the screen thus giving a complete "plan picture".

Centimetric Radar : All the systems described above suffered in a greater or less degree from a major defect due to the width of the radio beam, which caused on the screen not a real spot corresponding to any particular echo, but an extended patch, which rendered difficult accurate location of isolated targets, adequate separation of closely adjacent targets and elimination of background smudging of spurious echoes. Hence there was a long-felt need for a system in which a really fine pencil of radiation scanned systematically, point by point, the area under investigation, so that only the objects in a very small section sent back echoes at any one instant. Repeating this process sufficiently rapidly, a sharp and detailed picture could be obtained on the cathode ray tube screen showing with a clarity of display the plan position of the reflecting object. To produce such a fine pencil, aerials and reflector systems having an aperture of many wavelengths would have to be used, which meant large dimensions and heavy loads to be carried by aircraft or ships, unless the *wavelength of the radio pulses used was brought down to less than ten centimetres*, in which case aerials and reflectors could be correspondingly reduced in size and weight to practical limits. The use of centimetric wavelengths demanded, in its turn, high pulse power and high receiver sensitivity. The problem was successfully solved in 1939 by a team of English scientists under the leadership of Prof. Oliphant. Dr. J.T. Bandall produced the new *radar magnetron valve* which generated power far in excess of anything attained till then in centimetric wavelengths, by applying the resonant cavity technique to the relatively ineffective magnetron,

A magnetron is a diode consisting of a vertical cathode filament surrounded by a concentric cylindrical anode, to which is applied a homogeneous magnetic field in a direction parallel to the cylinder axis, hence perpendicular to the electric field existing between the cathode and the anode. The old type was first devised by A.W. Hull in 1921, but was improved to the *split anode* form by Okabe in 1924. Under the combined influence of the electric and magnetic fields, the electrons travelled in circular paths between the anode and the cathode with great rapidity and thereby produced high frequency oscillations of the order of 3,000 Me/sec. (10 cms.). In the actual design the anode was divided into two halves which were connected in push-pull fashion. The external circuit was in the form of LC resonant line, the length of which was ad justed to half a wavelength. When a P.D. was applied to the split anodes, the valve behaved as though it had a negative resistance and high energy oscillations, called dynatron oscillations, were produced. But the output power of the valve was still rather lows.

4.14 CATHODE RAY OSCILLOGRAPH

The simple form of the cathode ray oscillograph was devised by Braun in 1897 and was called the Braun tube for some time. It was an ordinary discharge tube type used by J.J. Thomson.

Fig. 4.19 : C.F. Braun.

Electric Field could be established in the path of the beam in a direction perpendicular to the axis of the tube by raising to a high P.D. two metal plates suitably arranged inside the tube. If the field applied is oscillatory the luminous spot will move up and down tracing a straight line. To observe directly the way in which the field varies as a function of time it is sufficient to spread out these oscillations in the perpendicular direction by some device which will give a uniform transverse motion, for instance, looking at the image of the spot by reflection in a plane mirror rotating about an axis parallel to the direction of deflection of the spot. The chief drawback of the Braun tube was that it required very high voltage for operation which depended on the emission of electrons from a *cold* cathode.

Subsequent important improvements made on the Braun tube evolved the modern relatively low voltage but more efficient cathode ray oscillograph.

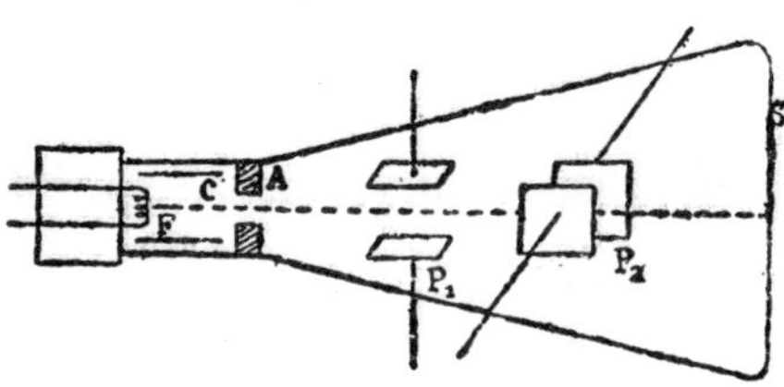

Fig. 4.20 : Cathode Ray Tube.

In this modern type, cathode rays are obtained from heating a tungsten filament F coated with an oxide of alkaline earth, which produces a good supply of thermionic electrons. The filament serves also as the cathode. The electrons are accelerated by means of a high P.D. of about 500 to 2,000 volts maintained between the filament and the anode A which is in the form of a disc with a hole in the centre. The electrons emerging from this central hole in the anode or the "anode gun", as it is called, form a narrow pencil of cathode rays which travels in a straight line and strikes the fluorescent screen S, producing a bright spot of light. The ray on its way to the screen can be deflected vertically or horizontally by means of electrostatic fields between two pairs of plates P_1 and P_2 disposed in mutually perpendicular planes or by magnetic fields produced by coils outside the tube. The cathode ray tube is a very sensitive arrangement by means of which it is possible to detect and record a transitory phenomenon even of the shortest duration; the basis of the possibility is the fact that the extremely minute mass of the electron enables it to acquire a relatively large velocity component in the extremely short interval of time during which a disturbing field can act upon it. The electron stream therefore follows quite faithfully and without inertia the variation of electric and magnetic fields, no matter how rapid these variations might be. The cathode ray oscillograph responds to periodic disturbances of frequency of the order of 200 millions/sec.

Focusing : This refers to the device of concentrating a great number of electrons in the thin pencil that striken the screen. Such a concentration is of great importance for obtaining a bright and well-defined spot on the screen. A certain amount of focusing in obtained by the use of a *Wehnelt cylinder* C, also called the *shield*, which is simply a metal cylinder surrounding the filament and maintained at a negative potential. The effect of this arrangement is to repel the electrons which leave the cathode in a diverging direction and thereby concentrate the stream along the central axial line, so that an intense beam of electrons is shot out of the anode gun. But as the shield by itself does not produce a really sharp spot of light on the screen, the mutual repulsion of the individual electrons in the beam leaving A causing the beam to diverge, some additional focusing device must be used. There are two principal methods of doing this: *viz., gas-focusing* and *electrostatic focusing*, according to which the instrument is classified as gas-focus or soft tube and electrostatic-focus or hard tube. In the gas-focus system, after highly exhausting the tube to remove the air and occluded gas, a small amount of an inert gas, such as argon or helium, is introduced. This residual gas is usually ionised by collision with the electrons; the resulting negative electrons add themselves to the stream while the heavy, slow-moving positive ions form a core for the beam. This positive core continually attracting the electrons in order to regain the electrons lost by it, produces an automatic concentration effect, the electrons clustering round it in a very compact pencil. The gas-focus tube has several *advantages* such as satisfactory operation at *low voltage* of about 300 to 500 volts, high sensitivity, in the sense that the beam can be deflected with ease even with a relatively weak deflecting agency and well-defined and brilliant spot of about 5 to 1 mm. diameter. But it suffers also; from the following defects—*dependence of focus on the intensify of the beam, i.e.,* on the number of electrons in the beam—if the beam current is altered by changing the shield potential, the focus is lost, which makes this type quite impracticable for television; failure of focus at high deflecting speeds—the positive ionic core remains within the beam only when the beam moves relatively slowly when deflected, so that with a high frequency voltage across the deflector plates the spot will be no longer sharp; limited life, due to the bombardment of the cathode by the heavy positive ions, which results in appreciable damage to the sensitive electron emitting surface of the cathode and thus limits the life of the tube to a few hundreds of hours.

In the alternative electrostatic focusing system instead of a simple disc for the anode, a series of anodes at increasing potentials is used, the potential of each being so adjusted that the resultant electrostatic field causes the electrons to be overgo. The effect of these anodes, which are usually two or three, is aniogous to the focusing of light rays by lenses and the arrangement is known as the *"electron lens"* or *"electron gun"*.

The focusing is controlled by the relative voltages of the anodes, being chiefly affected by that of the second A_2, while the first A_1 is kept at a fixed potential about one-fourth the value of the second. With this kind of focusing the tube can be evacuated completely so that no damage is done to the cathode by the bombardment of positive ions, which means that the life of the tube is considerably increased to several thousands of hours. Other *advantages* of this type of tube are:

(i) *Good focus up to quite high deflective fields* owing to the absence of heavy ions of the residual gas, and

(ii) *Possibility of controlling the brilliance of the spot* independently of the focus, since the voltage on the first anode alone affects the beam current and hence the brilliance of the spot, while the focus depends on the voltage of the second anode. The chief *disadvantage* of this hard tube lies in the necessity of using a *much higher anode voltage* than in the gas-focus soft tube as the focusing action cannot be obtained satisfactorily at low voltages; the anode voltage ordinarily ranges from 700 to 2,000 volts. This high voltage brings with it *reduction* in *sensitivity*, since the electrons accelerated by higher voltages move faster and in consequence are deflected to a similar extent from their path under the influence of the applied deflecting agency. It is also found that the *spot does not remain sharply focused over the whole screen*. But these defects have been eliminated to a great extent in the modern cathode ray tube which is of the electrostatic focusing or hard type

It is to be noted that the electron beam may also be focused by a magnetic field produced by a coil fitted outside the tube. This method is known as *magnetic focusing*, which has some advantages over the electrostatic focusing, such as, a very sharp focus and of no loss of focus at the extremities of the screen, provided the intensity and position of the focusing magnetic field are carefully adjusted. On account of these advantages, magnetic focusing is used in television, where a sharp focus over the whole screen is highly desirable.

The Deflecting System : The two pairs of plates employed to deflect the electron beam horizontally and vertically are usually called in terms of the deflections they produce. The pair producing the horizontal deflection of the spot is called the *horizontal* or X-*plates* although they are actually disposed in a vertical plane. Likewise, the second pair actually horizontal but causing a vertical displacement of the spot is called the vertical or Y-*plates*. It should be noted that the actual direction of movement of the spot depends on the position of the tube; hence the tube is usually mounted in such a manner that it can be rotated slightly in order that the deflection may be properly aligned. The deflection caused by the two pairs will be accurately at right angles to each other, although not necessarily horizontal and vertical but they can be rendered to become so by a slight rotation of the tube. The position of the deflector plates affects the sensitivity of the tube. For, the closer the plates are disposed to the beam the greater will be the deflection of the beam.

This is the reason why they are arranged inside the tube, even though it would be more convenient to have them outside. The plates should not be placed too close, however, for, then, the deflected beam may strike the plate and never reach the screen to produce the luminous spot.

Electromagnetic Deflectors : The method just described is called the electrostatic method of deflection of the spot. Another system is that in which the same effect is produced electromagnetically and this involves the use of coils outside the neck of the tube; such an arrangement makes the internal structure of the tube simpler since no deflecting plates are to be fixed inside the-tube. Two pairs of coils must be used to produce the horizontal and vertical deflections. It is important that the magnetic field is uniform, for the deflection produced is dependent upon the actual strength of the field. If, for a given current through the coil, the field produced is different at the two sides of the tube the sensitivities will differ. Hence it is essential that the coils are symmetrically disposed towards the tube.

Dimensions of the Tube : The cathode ray tube ordinarily consists of a long glass tube conical in shape, the screen being at the wider end. The length of the tube as well as the diameter of the screen are carefully chosen as a compromise between different factors such as space, sensitivity and brilliance of the spot. With a given arrangement of the deflecting system the sensitivity can be increased by increasing the length of the tube, since the displacement of the spot on the screen depends also on the distance between the deflector and the screen. In practice, however, a compromise has to be made between sensitivity and space. The longer the tube the larger will be the diameter of the screen on account of the conical shape. But the wider the screen the higher must be the operating anode voltage to maintain the brilliance of the trace on the screen. For the trace being the result of an illusion of the eye due to visual persistence following the repeated rapid movement of a single spot of light, the longer the trace made by the spot the less will be the apparent brilliance of the trace, unless the intrinsic brilliance of the trace is increased. This can be done by increasing the velocity and energy of the electrons in the beam, as the fluorescent screen merely converts the electron energy into light. Now to increase the energy of electrons the anode voltage which accelerates them must be increased. But as we have already seen, this increase of operating voltage reduces the sensitivity.

Hence, once again we have to compromise between sensitivity and brilliance. It is to be noted that the brilliance of the spot depends also on the colour of the spot, as certain colours are more sensitive to the eye than others. Cathode ray tubes are ordinarily made in various sizes, from one inch to six inches screen diameter, the length of the tube being two or three times the screen diameter for oscillograph work, while for television purposes still larger screens and longer tubes are used. A tube about 10 inches long with a screen of 3 to 4 inches diameter and operating at about 1,000 volts gives satisfactory results. Zinc silicate (willemite) is commonly used as the fluorescent material for coating the inside of the screen and gives a bright green spot easily visible since it corresponds roughly with the maximum colour sensitivity of the human eye.

In many applications of the cathode ray oscillograph, the movement of the spot in one direction, usually the horizontal, is required to be proportional to time. The arrangement used to achieve this is called a *linear time-base*; by employing this device, the base line of any pattern on the screen can be regarded as a time-axis and the pattern itself will be a faithful record of the variation of the quantity under study with *time*.

The *simplest time-base circuit*, shown in Fig. 4.21, is made by placing a condenser C across the terminals of a neon lamp N, charging the condenser through a resistance R from the D.C. mains and discharging it through the neon lamp. When the D.C. supply is switched on, the condenser charges up through the resistance until the voltage across it reaches the striking value of the neon lamp. This is usually of the order of 150 volts so that the D.C. mains voltage must be greater than that value. When the neon lamp glows, it short-circuits the condenser which thus discharges through the lamp. The latter is then put out. The process is repeated, the neon lamp flashing at regular intervals, depending upon the product of the values of resistance and capacity. If they are both large, the flashing rate will be low. It is customary to vary the resistance R and keep C constant. When used as a time-base it is the voltage across the condenser which is applied to the horizontal X-plates of the cathode ray tube. As the condenser charges up, the spot on the screen moves slowly in the horizontal direction until the condenser gets charged and then due to the discharge through the neon lamp the spot is brought back instantaneously to the starting point. This sweep of the spot is repeated at the flashing rate of the neon lamp. The frequency of the time-base can be calculated as follows: Assuming that the charging is linear, if E is the voltage of the D.C. mains supply, the charging current i = E/R. The charge on the condenser Q = it, where *i* is the time of charging. The voltage on the condenser, V, which is the same as the striking voltage for the neon lamp, is given by V = Q/C. ∴ t = Q/i = VCR/E and frequency f = 1/t = E/VCR. The frequency of the time-base device can be varied by varying the values of C and B within certain limits. In the study of oscillations, chiefly those met with in radio technique, whose frequency is very high, the time-base frequency must be *synchronised* with the frequency of oscillations under test by making the time of horizontal sweep of the spot an exact multiple of the period of oscillations under examination. When this is done the various traces coincide and we get the impression of a stationary wave on the screen, although actually it is the combined effect of a large number of waves superposed.

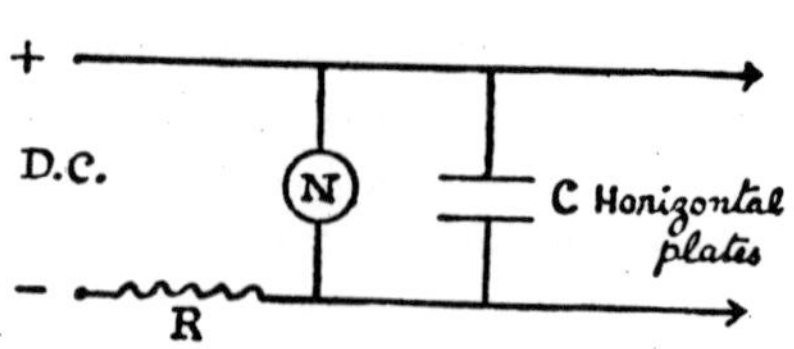

Fig. 4.21. Simple Time-base Circuit.

4.15 PHOTO ELECTRIC CELLS AND THEIR USES

Many types of photo electric cells have been developed based on photo electric effect. These cells can be used for the measurement of light intensity as also in various control devices. There are different kinds of photoelectric cells, according to their construction type.

The emission cell is also called *alkali-metal cell* on account of an alkali metal being used in it as emitter of photoelectrons. It consists essentially of a glass or quartz bulb, the inner surface of which is coated by *electrolysis* with one of the alkali metals, say potassium, rubidium or caesium. Further, the metal is sensitised in different ways. For instance, in the potassium or Clarendon cell, very much used in the study of stellar spectra, the metal is treated with hydrogen gas so that the sensitive surface is really a layer of potassium hydride. For the detection of white light a composite surface consisting of a caesium film deposited on a silver base which is covered with caesium oxide is employed and is found extremely good for sound reproduction in talkies. If a particular spectral region is to be studied increased sensitivity is obtained by using that alkali metal which has a maximum selective effect in that region. A window is left in the bulb through which light can enter the bulb and act on the photosensitive surface and cause emission of photoelectrons. The photoelectric current between the sensitive surface C and another electrode A introduced in the bulb and acting as anode is measured by connecting the sensitive surface to the negative terminal and the anode to the positive of a high-tension battery of about 100 volts. E is a sensitive electrometer to measure the small current.

The bulb may be highly evacuated (the vacuum cell) or may contain a gas such as helium, argon or neon, at a low pressure, same tenths of a mm. of mercury (the gas-filled cell). Vacuum cells keep a strict proportionality between the current and the intensity of light and have very little inertia, the emitted electrons moving freely, unhampered by the molecules of any residual gas. Further, the sensitivity remains unaltered for a very long time, provided the cathode is properly chosen. Hence, though not as sensitive as the gas-filled cells, they function with great regularity and constancy and are used for precise intensity measurements and for television purposes. Gas-filled cells produce a much more intense photoelectric current than the vacuum cells, due to ionisation by collision in the gas. But the proportionality between the current and the intensity is generally lost; also there is an increase of inertia which prevents these cells from following faithfully rapid variations in the light intensity; hence they are not quite suitable for television. The commercial gas-filled caesium cells, having a high sensitivity for red light, a low work function and the threshold far in the red region, are, however, very much used in practice for many other purposes.

The photoelectric current produced by the emission cell is very small. For instance, with an ordinary vacuum cell illumined with a lamp of 60 C.P. placed at a distance of 1 metre, the current strength is of the order of 1/100 of a microampere. In the gas-filled cells the current will be evidently greater, but still small, so that it cannot be used directly for practical purposes. Hence it is usually amplified by means of thermionic valves. The circuit details for the amplification of the photoelectric current are as shown in Fig. 4.22.

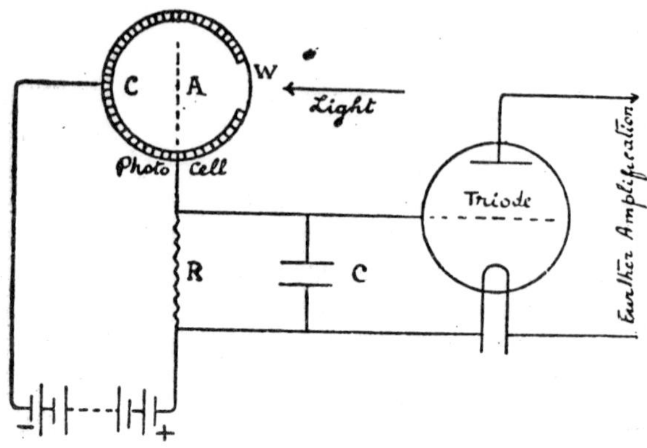

Fig. 4.22 : A Circuit for Amplifying Photoelectric Current.

When the intensity of the incident light varies, there will be a variation in the photoelectric current strength. This variation when applied to a resistance R in the battery circuit causes a variation in P.D. of the order of 1/100 to 1/10 volt between the ends of R. This voltage variation is applied through a condenser C to the grid of a triode valve, which will modify to an appreciable extent the plate current of the valve, preserving, however, the relative proportions of variation. Thus the small photoelectric current is amplified and passed on to farther stages of amplification. In order to obtain amplification without distortion, R must be much smaller than the internal resistance of the cell. As there is a condenser in the resistance circuit the variation of tension 'e' at the ends of B caused by a given illumination will attain its final value E only after a certain time (given by the expression $e = E(1 - e^{-t/CR})$. For distortionless amplification it is desirable to have a very small time constant CR, for which purpose C and R must, be made as small as possible. Under these conditions the inertia of the system becomes negligibly small, a point in favour of amplification without distortion. It is to be noted that the strength of the photoelectric current depends upon a large number of factors, such as the nature of the cathode, the nature and pressure of the gas present, the potential difference between the electrodes, the size of the window; the intensity of light, the wavelength, if monochromatic, and the spectral distribution in the case of composite light, etc. Hence for a quantitative use of a given cell several characteristic curves are obtained such as current vs. wavelength for a constant rate of energy reception, current vs. applied P. D. for a constant intensity of light, etc.

Rectifier cell works on the principle of the inner photoelectric effect. It is a true cell in the sense that it generates an *e.m.f.* without the application of any external potential, by merely making light fall on it. The rectifier cell consists essentially of a semi-conducting layer formed on the surface of a metal plate by heat treatment or cathode sputtering. Over the semi-conductor is a thin semi-transparent film of metal, which maintains electrical contact with the semi-conductor and at the same time allows light to illumine it. When light is incident on the semiconductor, electrons are emitted and travel in a direction opposite to that of the light rays. If a circuit be formed between the metal base and the surface film with a low resistance galvanometer, the current can be detected and treasured. The current is not proportional to the intensity of illumination in a simple linear manner but depends on the resistance in the circuit. For small values of the resistance the current, is found to be nearly proportional to the light intensity.

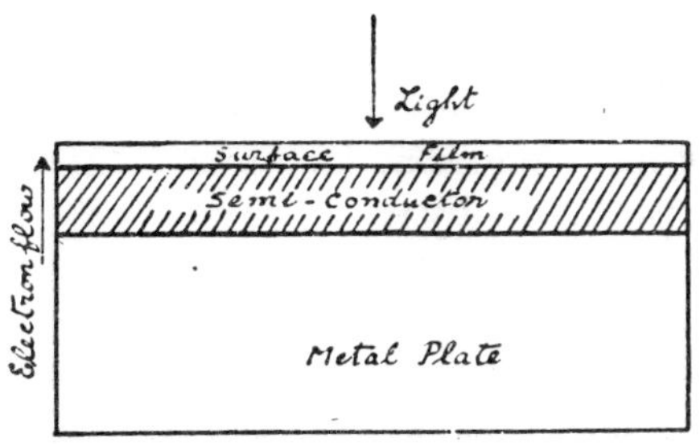

Fig. 4.23 : Rectifier Cell.

The action of the cell is not readily understood. Probably due to the presence of a "transistor" a *photo-active boundary* is formed in the semi-conductor, and when light, is absorbed at the surface of separation between the film and .the semi-conductor photoelectric emission takes place. The transistor was discovered in 1948 by three American physicists of the Bell Telephone Laboratories, J. Bardeen, W. Shookley and W. Brattain who have been jointly awarded in 1956 the Nobel prize for such an invention that represents a very significant land-mark in electronics. Intense researches are being actually carried on, since the device promises to play a very important part in radio-technology as an able competitor to thermionic valves. A transistor is essentially a semi-conductor which normally contains an equal number of electrons and "holes" in its crystal lattice. By the introduction-of a small-quantity of "impurity", an inequality is brought about between the numbers of holes and of electrons, which results in a unidirectional flow of electricity under suitable conditions.

The cell is called a *rectifier* cell, because Lange, in 1930, found for the first time that when a *metal plate rectifier* comprising a copper plate coated with cuprous oxide (Cu_2O) used for converting A.C. into D.C. was exposed to light, a relatively large current resulted. The first rectifier photo-cells were constructed with a copper disc in contact with a layer of the semi-conducting cuprous oxide. Since Lange's discovery much work has been done in developing this type of cells and a widely used modern form is the *selenium cell* (Weston photronic cell). By a special process an iron disc ia coated with a film of selenium on the surface of which a film of platinum is formed sufficiently thin to be translucent. When light rays strike the platinum film, they penetrate to the surface of the selenium layer and release electrons which flow in the direction opposite to that of the incident light. No external voltage is needed for this type of cells. They are robust and .cheap and are as sensitive as the gas-filled emission cells. But the current produced cannot be amplified by thermionic valves. With strong light sources, however, the current can directly operate galvanometers.

Conductivity cell also works on the principle of the inner photoelectric effect but due to a different immediate cause, namely, the change in resistance of certain substances when illuminated. The current obtained from this cell is probably not the primary photoelectric current, since the material used, such as selenium, thallous sulphide, being, semi-conductors, cause the electrons to accumulate at a boundary and set up a back *e.m.f.* which can ultimately prevent the flow of the small primary current. But in these substances a large secondary current flows in addition to the primary current due probably to the accumulation of primary electrons at the boundary which somehow diminishes the resistance of the substance and renders it more conducting. For the working of these cells an external potential of about 100 volts must be applied, which fact differentiates them from the rectifier cells. The light sensitivity of these cells is not linear and hence they cannot be used for accurate measurements of light intensity. They have also a considerable time-lag both in reaching the minimum resistance on illumination and in recovering again to the original value when the light source is cut off. They are therefore not very much employed except for operating relays to control illumination.

Uses of the Photoelectric Cell

The photoelectric cell has been termed the "magic-eye" and rightly so, for, it acts as a sort of substitute to the human eye, but with a sensitivity far superior, and finds extensive application in every walk of life. Study of astrophysical phenomena, such as the temperatures of stars and stellar spectra, correct control of the temperatures of furnaces and of chemical reactions, calorimetry and photometry use the photoelectric cell with great advantage to obtain the best results. In practical life protection against danger in working electrically driven tools, protection against thieves, fire, etc., automatic control of street lighting, of signals at level-crossings, of gates and doors, of the speed of cars and trains, automatic counting of objects made by a machine or persons visiting a place, etc., are rendered easy and effective by the use of the photo-cell. We shall not describe in detail how all this is done. But we should deal briefly with three modern applications of the photoelectric cell which combine in themselves the three important phases of human life, scientific, artistic and utilitarian. They are:

(i) sound films or talkies,

(ii) photo-telegraphy or telinogram, and

(iii) television.

4.16 ELECTRON MICROSCOPE

An electron microscope is a device to magnify minute objects, where a beam of electrons is employed.

In an electron microscope electrons can be focused by suitable electric and magnetic fields.

To appreciate fully the principle of construction and action of the electron microscope, it is necessary to note the various stages of evolution in the technique of magnifying instruments. Given the fact that direct observation of minute particles is of great importance for many purposes, while the human eye by itself can see only objects whose dimensions are of the order of about .02cm., microscopes came into use long ago as aids to overcome the weakness of human perception of extremely minute particles. First, the simple magnifying glass, then the compound microscope devised already three centuries ago and more recently the ultra-microscope gradually extended the range of detailed perception of smaller and smaller particles until a limit was reached at about .00002cm. The absolute theoretical limit of magnification of the optical ultra-microscope was about 3,000, while in practice the magnifying power of the order of 1,000 alone was obtained. This limit to magnification was soon recognised to be due not to imperfections in the design or construction of the instrument, but to the light used to illumine the object. There would be actually no difficulty in producing a set of lenses which would magnify much more than 3,000 times, but nothing at all would be gained by it. For the image would be larger, but it would be correspondingly blurred and no more detail would be revealed. It is the light, in fact, that sets the limit to magnification. The wavelength of visible light used for illumination in optical microscopes is of the order of 6×10^{-5} cm. Objects smaller in dimension than this wavelength could not be seen since light would pass between them without being scattered by them. Attempts were next made to improve this limiting factor by the use of short ultra-violet light and photographic plates, but with only a very small measure of success.

Then came the idea of substituting a beam of electrons for a beam of light. But before such an idea could suggest itself to scientists, certain other properties of the electrons had to be discovered first and recognised. These were:

(1) The wave nature of electrons, proposed on theoretical grounds, by Louis de Broglie in 1923 and confirmed a little later by the experiments of Davission and Germer, G.P. Thomson and others on the diffraction of electrons, and

(2) The 4 possibility of focusing electrons by electric and magnetic lenses. In 1926-27, Busehen Germany, worked out the theory of the electron lens represented by axially symmetric magnetic or electric fields basing himself on the experiments of Mac Gregor-Morris and others demonstrating the focusing properties of magnetic coils with respect to cathode rays. With these properties fully tested roundabout 1930, it was but a short step further to the practical design of the electron microscope. In the year 1932, with much patience and skill, the first instrument of its kind was constructed. With continuously perfected technique, the electron microscope at present gives an actual magnification of about 20,000 with an upper limit of about 100,000. Thus, we

have come to possess an instrument which is capable of bringing into visibiiity a whole range of detail of extremely minute objects, invisible even under the most powerful optical microscope.

Let us now see how the wave nature of electrons and the focusing of electrons by electric or magnetic lenses lead to such a high degree of magnification. For useful magnification two conditions must be satisfied : high resolution into detail and perceptibility of the resolved details. Hence the higher the resolving power of the instrument with a simultaneous greater perceptibility, the greater will be its magnifying power. If the perceptibility decreases with increase in resolution, the image, as its size increases, would get correspondingly blurred and hence no useful purpose is served. This is what sets the low upper limit to the magnifying power of an optical microscope.

In the electron microscope the resolving power can be greatly increased without detriment to perceptibility of detail. For, the resolving power of a microscope depends on two factors, *viz.*, the wavelength of the light used and the numerical aperture. The smaller the wavelength of light used the greater will be the resolving power, but the smaller the numerical aperture the smaller will be the resolving power. Now when electrons are scattered or diffracted, they behave, according to de Broglie, as light of wavelength given by $\lambda = h/mv$, where h is Planck's constant, and m and v are the mass and velocity of the electrons. For electrons accelerated by a P.D. of about 60,000 volts, v can be calculated from the relation $1/2mv^2 = eV$, and $m = 9 \times 10^{-28}$gm. The wavelength A turns out to be about 5×10^{-10}cm. This is 10s times smaller than that of visible light. Hence the resolving power of the electron microscope should be 100,000 times greater than that of the light microscope. But the numerical aperture in the electron microscope is usually smaller than that of the optical microscope on account of limitation set by lens aberrations. This factor will therefore reduce the resolving power. But even so, the resolving power of the electron microscope is far greater than that of the ordinary microscope.

5

THE QUANTUM THEORY

5.1 INTRODUCTION

In the year 1900 Max Planck of Germany introduced the quantum theory. We known well that when a body is heated it emits electromagnet further heating makes the metals (some) to emit radiation. The nature of radiation, depends on the temperature of the emitter.

As a result of scientific investigations, all material objects have been found not only to be built up of minute particles but also contain intangible constituent of energy, which may be set free from all association with matter, when it travels through space in the form of radiant energy or radiation. It has also been recognised that radiation manifests itself in different types such as the electromagnetic Hertzian waves, infra-red, visible, ultra-violent. X-rays, γ-rays, etc., but all of them having essentially the same nature and travelling with the same speed in vacuum. According to the "quantum theory", the radiant energy of a system or the exchange of radiant energy between different systems occurs not in a *continuous fashion* permitting all possible values, as demanded by the wave theory, but *in a discrete quantified form, as integral multiples of an elementary quantum of energy*. This means that radiant, energy should be considered as atomic in structure, like matter. Einstein was forced to admit such a conception in the case of light, as we have seen in connection with his theory of relativity, where ether, the essential support of light waves, had to be set aside as futile in the interpretation of physical phenomena. The successful application of the same idea to a context experimental fact such as the photoelectric effect, which he was able to make, lends therefore an indirect support to his more universal idea of relativity.

In the present chapter we shall first briefly state the different classical theories of light that were proposed before the advent of the quantum theory, in order to have before us a historical background that will make us appreciate the need that was felt for the introduction of this new theory. We shall then study the origin of the quantum theory, *i.e.*, the immediate cause that led to its formulation and the further development, made by Einstein and others, that ultimately crystallised into the more comprehensive theory of "light-quanta" or "*photons*". Finally we shall apply the new conception of radiation to the two important phenomena which gave the first experimental evidence of its validity, *viz.*, the photoelectric effect and the Compton scattering of X-rays.

5.2 THEORIES ABOUT VISIBLE LIGHT

In 1680 a Dutch Physicist argued that light might be some wave disturbance transmitted by some iridium. Newton had proposed corpuscular theory. But Huyghens challenged him by certain ways.

The wave theory of light, according to which light was considered a wave disturbance in a medium called ether that filled all space uniformly, was first proposed by Huyghens and then perfected by Fresnel. The latter was able not only to dispose of the objections arising from Newton's corpuscular theory but also adequately explain the phenomena of diffraction, interference and polarisation on the basis of the wave nature of light. According to him, light is propagated from place to place when the particles of the medium, ether, are set in simple harmonic vibration in a direction at right angles to the direction of propagation and the resulting disturbance or energy travels out as a *transverse wave motion*.

Fresnel

Fig. 5.1

With such a wave propagation of light, *diffraction* which is fundamentally a bending of light round very small obstacles and *interference* which arises from the agreement of phase of the waves or the contrary are readily explained. Like-wise polarisation can be accounted for by the Transverse nature of the waves; reflection and refraction of light also receive a satisfactory interpretation with the additional assumptions of the principle of conservation of energy and continuity of displacement. Even rectilinear propagation can be understood by the very small size of the waves involved. Other experimentally observed facts, such as the independence of the velocity of light from the speed of the source and the decrease in the velocity when light passes through a medium denser than air, are completely in favour of the wave theory of light, while directly contrary to the demands of the corpuscular theory. Towards the middle of the 19th century, therefore, the wave theory was accepted as a final and complete explanation of the nature of light.

But soon a serious difficulty arose as regards the nature of the very medium, ether, waves in which constituted light. The velocity of propagation of waves, such as light was conceived to be, is given by the relation $v = \sqrt{E/d}$, where v is the velocity of propagation, E the elastic modulus and d the density of the medium. Now since v for light is very high, *i.e.,* 3×10^{10}cms./sec., it follows that the elasticity of ether must be very great, many timed greater than that of steel while its density very low *even lower than the lightest gas*, which is further confirmed by the free and unhampered motion of celestial bodies through space. On the other hand, to account for the transverse nature of light waves, which is convincingly demon strated by the phenomenon of polarisation, the same medium should be considered as a most *rigid elastic solid.* Thus, ether seems to possess contradictory properties. Hence, it could easily be surmised that something was wrong with Fresnel's ether wave theory of light. It was at this stage that Maxwell came forward with his famous electromagnetic theory which not only solved the above-mentioned difficulties to a great extent but even achieved a marvellous synthesis of the different types of radiation establishing an essential identity between light and electricity.

The Classical Electromagnetic Wave Theory

Maxwell, in 1864, starting with a novel and original idea of a "*displacement current*" in the dielectric medium in which electrical phenomena take place, worked out the electromagnetic field equations on the assumption that this dielectric current obeyed the same law as the ordinary conduction current within certain limits. The final differential equations representing the electromagnetic field were found to be of the same type as those for any progressive wave motion. A more detailed analysis showed that in a progressive plane polarised electromagnetic wave the electric and magnetic intensities were at right angles to each other and both were confined to a plane which was perpendicular to the direction of propagation. Further, the electromagnetic wave was propagated with a velocity equal to $1/\sqrt{k\mu}$ where k and μ are the specific inductive capacity and permeability of the medium respectively. This quantity $1/\sqrt{k\mu}$ had already been shown, in the theory of units, to be equal to the ratio of the electromagnetic and electrostatic systems, having the dimensions of velocity (LT^{-1}). By measuring the same electrical quantity in the two systems of units, their ratio could be experimentally determined, as was done by a great number of investigators, and $1/\sqrt{k\mu}$ was found to be 3×10^{10}cms./sec., which is the same as the velocity of light. This naturally led Maxwell in 1865 to conclude that light also was essentially an electromagnetic phenomenon and thus the electromagnetic theory of light was born.

James Clark Maxwell

Fig. 5.2

According to this new conception, light is considered as the result of rapidly alternating displacement currents in the medium which give rise to magnetic effects, similar to those associated with conduction currents. The two fields, the electric and the magnetic, inseparably associated, the one varying proportionately with the other and the variation of one giving rise to the other, urge each other forward with a finite velocity, *viz.,* that of light. The fact that the mutually perpendicular electric and magnetic fields are always confined to a plane perpendicular to the direction of propagation accounts very well for the transverse nature of the light waves. The replacing of the vibrations of a 'mechanical ether' of the older wave theory by the vibrations of the inseparably connected electric and magnetic vectors at once removed many of the difficulties of the "elastic solid" theory of Fresnel, while retaining practically all its advantages; ether came to be considered whence forward as merely a medium whose displacements constituted the electromagnetic field.

All experiments designed to detect the ether directly having failed, many scientists at the present day have given up the idea of postulating ether even as a mere carrier of electromagnetic waves, and have preferred to interpret electromagnetic waves in terms of what is known as "waves of probability", as we shall see later. Whatever be the ultimate fate of ether, Maxwell's great discovery of the essential identity between electrical and optical phenomena still remains intact. But when it was first proposed, few were inclined to accept it without further research and confirmation.

5.3 QUANTUM THEORY OF RADIATION

In the heat radiation it is possible to obtain, thermal equilibrium. This can be done by keeping the radiation, at constant value. If the body has no reflecting power, then it absorbs all incident radiation, and appears black. The radiation emitted the body is known as black body radiation.

During the years 1893 to 1897 Lummer and Pringshein carried out number of experiments, on the black body radiation. The experiment was repeated for temperature between 1000°C to 1650°C.

As early as 1859 Kirchon" discovered that:

(i) A black body not only completely absorbs all the radiation falling on it, but also, conversely, behaves as a perfect radiator when heated, and

(ii) The radiation from such a body depends only on the temperature to which it is raised and not to all on the nature of the body. In other words, the interchange of energy between the body and the surrounding space in which the body is situated will lead to the establishment and maintenance of a state of equilibrium completely donned by the temperature. In this state of equilibrium the enclosed space will be filled with radiation whose spectral distribution and total intensity are functions exclusively of the temperature.

In the study of black body radiation a special interest was attached to the spectral distribution of energy in the emitted radiation. *How was the energy distributed among the various wavelengths in the spectrum and at what wavelength was most of the energy emitted* ? Naturally, the electromagnetic theory, then prevalent, was applied to the problem but it led to wrong and inconsistent results not agreeing with experimental data.

In 1884 Stefan and Boltzmann, using the idea of the pressure exerted by radiation according to the electromagnetic theory and thermodynamical principles, showed that the total energy density, *i.e.,* the energy of radiation in unit volume of space due to all the different wavelengths in the spectrum was proportional to the fourth power of the absolute temperature of the black body. This is what is known as *Stefan's Fourth-Power Law* of black body radiation, which evidently does not throw any light on the energy distribution among the different wavelengths in the thermal spectrum but refers only to a globular effect.

Kirchoff **Boltzmann.**

Fig. 5.3

In 1893 Wien, in order to find the actual distribution of energy in the thermal spectrum, considered the case of an enclosure full of black body radiation expanding *adiabatically* with a velocity small compared with that of light and proved by thermodynamical reasoning that after such an expansion the radiation still remained full but was characteristic of the new temperature. From a knowledge of the work done by the

Prof. Wien

Fig. 5.3(a)

radiation pressure during the adiabatic expansion the new distribution of energy could be obtained. In this way he was able to establish two laws known as *Wien's displacement laws*.

In 1900 Bayleigh and Jeans tackled the problem of the energy distribution in a different manner. According to the electromagnetic theory, a black body radiator emits radiations of *continuously variable wavelengths* from zero to infinity. This radiation is imagined as broken up into monochromatic wave trains. Next the number of such wave trains or independent modes of vibration lying within a certain wavelength range, say λ to $\lambda + d\lambda$ is determined by applying the laws of statistical mechanics. Finally, using the theorem of equipartition of energy, the energy carried by each mode of vibration and hence the energy distribution can be obtained. The expression for the energy distribution thus derived is

$$dE = B\lambda^{-4}\, Te^{-a/\lambda T}\, d\lambda \text{ (Bayleigh-Jeans formula)},$$

where B and a are constants.

These *different formulae* of energy distribution derived on the basis of *classical theory led* to wrong and even absurd conclusions. Thus, for instance, in Wien's formula, even when $T = \infty$, E is still finite, which is in open contradiction with the experimentally verified Stefan's fourth power law. Likewise, in Rayleigh. Jeans formula, the energy radiated in a given wavelength range $d\lambda$ increases rapidly as λ decreases and

Sir James Jeans **Lord Kayleigh**

Fig. 5.4

approaches infinity for very short wavelengths, which, however, cannot be true. Further, if the expression be integrated over the whole range of wavelengths from zero to infinity on the basis of the classical continuous emission, the total energy thus obtained turns out to be infinite for all temperatures except absolute zero, which is clearly an absurd conclusion. Their results are represented graphically, where the amount of energy emitted in a narrow wavelength range is plotted against the wavelength of each range. The different curves are obtained for the different temperatures at which the black body is maintained.

The total energy of radiation at any one temperature is given by the area between the curve corresponding to that temperature and the horizontal axis. The area is found to increase according to the fourth power of the absolute temperature. Thus, Stefan's Law is verified. The wavelength corresponding to the maximum energy represented by the peak of the curve shifts to shorter wavelength side as the temperature increases, as a to be expected from Wein's displacement law : λT = a constant.

Paschen showed that Wien's energy distribution formula agreed with the experimental curves for short wavelengths but not for long wavelengths. On the other hand, Rayleigh Jeans formula could be made to fit the curves, as Rubens and Karlbaum showed, in the region of long wavelength only. None of the theoretical formulae could therefore account for the shape of the radiation curve over its entire wavelength range. However, since the theoretical derivations were free from error, an anomalous situation had, to be accepted in the disagreement of theory with experiment, unless one assumed that the fundamental assumptions of the classical theory were at fault. This was what exactly occurred to Planck, who, in consequence, proposed in 1901 a new revolutionary hypothesis known as the *theory of quanta*, by means of which he was able to derive the correct law of thermal radiation.

5.4 PHOTONS

Einstein was the first man to recognise about the importance of radiation. Einstein was able to apply the quantum conception of radiation; In 1913 Bohr used the theory effectively. In 1922 Compton applied quantum theory of radiation to the phenomenon of scattering of X-rays.

The picture of radiation appears very much like a return to Newton's corpuscular theory but it is not quite so. For, while Newton thought that radiation was solely made up of discrete minute material particles, the photons are considered as discrete quanta of energy given by hv, which therefore involve the frequency v of the radiation. This means that the photons, unlike the light corpuscles of Newton, include in their very concept the wave nature also of radiation.

As a matter of fact, photon is a synthesis of the dual nature of radiation, wave and corpuscular, which aspects are now regarded as complementary rather than antithetical. The reason for this is that the wave nature of radiation cannot be completely given up since that alone and not the other quantum idea can account for the phenomena of diffraction and interference, to explain which precisely the wave theory was postulated. On the other hand, only the quantum conception of radiation can explain adequately the phenomenon of black body radiation, photoelectric effect, etc. Hence, if we wish to interpret satisfactorily all the radiation phenomena, we are forced to attribute a double aspect to radiation. We shall now consider some of the characteristic properties of radiation under this most comprehensive idea of photons.

The evidence which experimental facts provide for the real existence of photons is of the same general nature as that which we have from experiments proving the existence of electrons. In each case, experiment suggests an indivisible entity having definite quantities associated with it—'e' and 'm' for the electron and 'h' and 'c' for the photon. Fractions of a photon are as unknown as fractions of an electron. Photons may be considered as retaining their *individual identities* through all changes except that of being completely absorbed into or emitted out of an atom or molecule. They may change their energies, but then they adjust their frequencies to their new energy contents so that each photon remains a complete unit. The radiation with which we usually deal in Atomic Physics is produced by disturbances in single atoms and it is found to be a general law that every such disturbance produces one and only one complete photon.

The energy content of a photon is always found to be in terms of the quantum represented by hv. This quantum unit, however, is not the same for all kinds of radiation, since it depends on the frequency v which varies with different radiations, as we have already seen in connection with the electromagnetic spectrum. In the original quantum theory the energy of the photon was always considered to be an integral multiple of hv as hv, $2hv$, $3hv$......nhv. But in the new quantum or wave mechanics it is further refined to a value $(n + ½)hv$, n being an integer, which means that the limiting value of the energy of a photon is not zero but $½hv$. *The energy of the photon is independent of its intensity, depending only on its frequency,* a concept contrary to classical ideas where radiation is considered purely as waves and energy estimated by the intensity of the wave disturbance, dependent on the physical properties of the medium. Intensity, according to the quantum theory, simply gives the total number of photons in a beam and has no relation whatever with the energy of the individual photons in the beam. The energy depends only on the intrinsic properties of the photons.

Since the photons are all propagated with the same velocity of light, the theory of relativity can be legitimately applied to them. One of the very important consequences of this theory is the mass energy equivalence according to which, therefore, the photon must possess mass of its own given by the relation $m = h\nu/c^2$, since $h\nu$ is the energy of the photon. Further, as the photon is always in motion, *it must also have a momentum* Just as in the case of material particles in motion with this essential difference, however, that all photons move with the same velocity while c material bodies move with variable and different velocities which can never reach c but can only approach it asymptotically.

The experimentally observed Compton effect in the scattering of X-rays offers direct evidence of the existence of the mass and momentum for the photon, since with these assumptions alone and using the quantities $h\nu/c^2$ and $h\nu/c$ to represent the mass and momentum of X-ray photons, the different characteristics of the phenomenon are completely interpreted.

Photon Statistics

The different properties of radiation which is made up of photons can be readily understood by applying the laws of statistics to a large assembly of photons, as is done in the kinetic theory of gasses. Radiation is pictured as a sort of gas made up of photons that are endowed with random haphazard motions like the molecules of a gas. There is an important difference, however, between an ordinary gas and the "photon" gas. Whereas the molecules of a gas rush about with different velocities and different energies, the photons all travel with the same velocity c, though with different energies $h\nu$. Just as the pressure of a gas can be imagined to be due to the impacts of the molecules, *the pressure of radiation* can be considered as resulting from the impacts of its photons. Just as the density of a gas at a point is a statistical concept and is proportional to the chance of finding a molecule at that point, so also intensity of radiation may be statistically defined as proportional to the chance of finding a photon at the point considered. *The temperature of radiation* like the temperature of a gas is also a statistical idea. One cannot speak of the temperature of a single photon any more that of a single molecule. *The energies of the separate photons* conform to the statistical law known as Planck's law, just as the energies of the molecules of a gas follow Maxwell's law.

The "*black body*" radiation dependent solely on temperature can be imagined as a crowd of photons moving equally and indiscriminately in all directions, just as the molecules of a gas in temperature equilibrium. With this picture, Stefan's fourth power law can be readily deduced. While the black body radiation arises from a random crowd of photons, a beam of radiation can be thought of a regular shower of photons all moving in parallel paths and its property of polarisation can be accounted for by the angular momentum in the photons which constitute the radiation. Planck's constant '*h*' has the same physical dimensions as angular momentum and supposing that all the photons in the beam spin with the same angular momentum $h/2\pi$, which may be clockwise or anti-clockwise, the different types of polarisation, circular, elliptic and plane as well as non-polarisation can be readily explained.

The photons that constitute radiation are electrically neutral and are not therefore affected by electric or magnetic fields; hence also they do not ionise directly by themselves. The ordinary methods used in the case of charged particles cannot be used directly to detect photons or make measurements of their energy or intensity. As a matter of fact, all the instruments used in their analysis are sensitive not to them as such, but to the secondary effects produced by their interaction with matter.

5.5 THE MAGNETIC SPECTROGRAPH METHOD

The magnetic spectrograph method, involves the measurement of energies of the emitted photoelectrons, by making them, traverse a circular path. Maurice de Broglie in 1921 undertook the verification of Einsteins equation in the case of high frequency radiations.

A narrow beam of X-rays of frequency v enters through a small window, a highly evacuated brass box B, and falls on the target T of the material under consideration, which is in the form of a thin narrow strip arranged on the side of a heavy metal block D. Photoelectrons are expelled from T in all directions and with different velocities. The whole apparatus is placed in a uniform magnetic" field acting at right angles to the plane of the figure. Some of the photoelectrons pass through the narrow slit S and are deflected by the magnetic field to describe a semicircular path and eventually strike and photographic plate PP placed horizontally on the metal block D. If the electrons leaving T have velocities v_1, v_2......they will move in circles of radii ρ_1, ρ_2... given by the general relation $Hev = mv^2/\rho$, and will strike the plate at K, L, etc. By suitably adjusting the position of the plate and the width of the slit S it is possible to get a certain focusing effect, electrons of the same velocity but slightly different initial directions being made to strike at the same point such as K, L, etc. and the diameters of the respective circular paths being the distance between S and K, S and L, etc.

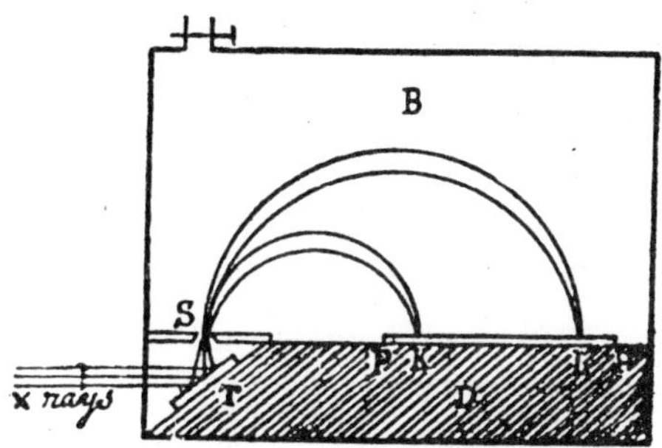

Fig. 5.5 : Magnetic Spectrograph.

Thus, using a radiation of a given frequency ν, which can be determined by measuring the wavelength of the X-ray line employed, it is found that several linear traces appear on the plate as seen in the adjacent photo obtained by Maurice de Broglie, where the traces marked 4, 5, 6, 7, have been identified with the K and L photoelectron of the silver atoms. From the geometry of the apparatus the radius p of the circular path followed by the photoelectrons which made each one of the traces can be readily found. The intensity H of the uniform magnetic field used can be measured. Using these values of ρ and H in the relation $Hev = mv_2/\rho$ or $v = H\rho\,(e/m)$ and assuming-the value of e/m, the velocity v of the photoelectrons and hence their kinetic energy $= \frac{1}{2}\, mv^2/\rho$ corresponding to each trace can be evaluated. As we are, however, dealing with radiation of high frequency the photoelectrons can move with velocities comparable to that of light so that relativity correction has to be applied. This means that instead of $1/mv^2$ we must use the relativistic formula to evaluate the kinetic energy of the photoelectrons. Having thus determined the kinetic energies of the photoelectrons ejected, from the different energy levels in the atom and knowing $h\nu$, the relations

Fig. 5.6 : Magnetic Spectrum of Photoelectrons Produced by X-rays from a Tungsten Anticathode Falling on a Silver Target.

$$\tfrac{1}{2}mvk^2 = h\nu - w_k,$$
$$\tfrac{1}{2}mv_L^2 = h\nu - w_L,$$

etc. can be verified. The values for w_k, w_L. etc. thus determined agree well with those obtained by spectroscopic methods. This establishes therefore the validity of Einstein's equation for very high frequency radiation also.

5.6 THE COMPTON EFFECT

Compton in 1925 proposed an adequate explanation, on the basis of the quantum theory of radiation. He applied, the laws of conservation of energy.

According to the classical theory of X-ray scattering we have seen that:

(i) The scattered X-ray should have the same wavelength as die incident one,

(ii) The scattering coefficient a should be independent of the wavelength of the incident radiation having a constant value 0-2, and

(iii) The scattered radiation should be symmetrical as regards the distribution of intensity. But experimental research on scattering of high frequency radiation, as hard X-rays and γ-rays, brought out the following points of discrepancy from classical expectation:

(a) The scattered radiation was found to possess a greater wavelength than that of the incident radiation;

(b) The scattering coefficient a varied with the wavelength, of the incident radiation diminishing as the wavelength decreased;

(iv) The distribution of the scattered radiation was not symmetrical, scattering taking place practically in the forward direction, *i.e.*, in the same direction as that of the incident radiation.

These were the experimentally observed facts which were unintelligible on the principle of the classical theory.

If a photon of energy $h\nu$ strikes an electron it will communicate kinetic energy to the electron and hence will itself lose energy. The scattered photon will therefore have a smaller energy $h\nu$ and, in consequence, a lower frequency or greater wavelength than that of the incident photon. The observed change in frequency or wavelength of the scattered radiation is known as Compton effect which, according to Compton, offers a crucial test for the validity of the quantum theory of radiation, not merely as regards absorption of radiation as in the photoelectric effect, but even when radiation is freely propagated in space.

Compton, applying the laws of conservation of energy and momentum to the scattering phenomenon of high frequency radiation, considered as a collision between the photon and the electron, calculated the change in frequency that would take place in the process as follows:

For the sake of simplicity, it is assumed for the present that the electron is free and at rest before the collision. This evidently does not represent the actual state of affairs, since it can be shown that the electron is not free but is bound to the atom, nor at rest, but subjected to a high speed revolving motion round the centre of the atom. After establishing the main points of the theory, we shall see what modifications these facts will introduce.

In framing a general theory tint will hold good even for every high frequency radiations, capable of imparting to the electrons, involved in the scattering, velocities comparable to that of light, relativistic dynamics will have to be used.

6

THE NUCLEUS STRUCTURE AND PROPERTIES

6.1 THE STRUCTURE OF THE NUCLEUS

The nucleus of an atom has compact structure. It is capable of ejecting different particles like α-particles. Practically nucleus of an atom has no mass. But it plays the important role of a discrete carrier of energy.

According to Proton-neutron theory nuclei: are composed of protons and neutrons. In general representing nuclei by the two important numbers that characterise them *viz.,* the mass number A which is the integer, nearest to the mass of the nucleus and atomic nucleus.

Several theories have been, proposed which may be called, according to the elementary particles selected for the nuclear constitution, the *proton-electron, proton-neutron, neutron-positron* and *negative proton-neutron theories. Of these, the proton-neutron* concept alone has found general acceptance, as possessing a number of distinct advantages over the others.

The Proton-Electron Theory

The proton-electron constitution of the Jouleus was the first to be proposed and was in vogue for a time, until the neutron was discovered.

The discovery of the '*whole number rule*' by mass spectrum analysis *justified the age-long idea that the different nuclei are built up from the same simple element, hydrogen* (*proton*). The slight discrepancy of the actual mass of a free proton (1.00813) from the whole number was accounted for as either due to the arbitrarily chosen standard (1/16 of the mass of an atom of oxygen) or due more probably to the *'packing effect*', *i.e.,* according to the laws of nuclear reactions, a small amount of mass disappears when several protons are packed together to form a stable nucleus, as will be explained when dealing with mass defect.

The electrical neutrality of the atom as a whole argued, in its turn, that electrons also should enter into the constitution of the nucleus. For, except hydrogen, in all other atoms, it was found that the number of extra-nuclear negative charges fell short, by very much, of the supposed number of positive charges in the nucleus. For instance, considering the helium atom, since its mass is nearly four times as great) as that of hydrogen, it may be legitimately supposed that its nucleus is composed of four protons. The fact that 4 free protons are slightly heavier than a helium nucleus can be explained by the *mass defect* arising from the packing together of the protons, as stated above. Under these conditions, the helium nucleus carries four positive charges. On the other hand, it is experimentally proved with certainty that there are only two electrons outside the nucleus. Then how to account for the electrical neutrality of the. atom as a whole? The discrepancy between the positive and negative charges becomes all the more marked as one proceeds to the heavier elements of the periodic table. Thus, carbon atom should have 12 protons and hence 12 positive charges, while it has only 6 peripheral electrons and hence only 6 negative charges; oxygen atom 16 protons but only 8 peripheral electrons. In many elements the number of peripheral electrons amounts only to about half the assumed number of positive charges. A natural explanation of this state of affairs would be to introduce the negative charges required by the conditions of neutrality of the atom into the nucleus itself, which would balance the excess positive charges found there. As the mass of the electron is relatively very small, electrons introduced thus into the nucleus would not greatly affect the total mass of the atom while they would effectively render the atom neutral. Thus, in the case of helium, two electrons would be found inside the nucleus to neutralise the two excess positive charges; carbon would have six electrons inside its nucleus, oxygen eight and 30 on.

The β-ray emission by natural radioactive elements seemed to confirm the existence of electrons in the nucleus, since the emission of electrons in such cases is essentially a nuclear process.

The β-ray emission or electron disintegration of radioactive substances, which appeared to give a good experimental support to the previous theory, is accounted for in this theory, in a very ingenious manner, as follows : *The electron does not pre-exist in the nucleus but is formed just at the instant of emission, caused by the transformation of a neutron into a proton* : $n \rightarrow p + e^-$. The positron emission is likewise due to the converse process, *i.e.,* when a proton transforms itself into a neutron : $p \rightarrow n + e^+$. As protons and neutrons can be converted into each other in the nucleus, they are regarded as *two alternative states of a single heavy nuclear particle*, to which the name of *nucleon* has been given.

This theory has the merit of being based on two actually existing fundamental particles, the proton and the neutron, while it dexterously removes the difficulties encountered in introducing formally electrons in the nucleus as in the previous theory. It has the additional advantage of reducing nuclear constitution to a single particle, the nucleon, though of complicated characteristics. This conclusion is very close to Prout's old hypothesis.

The proton-neutron theory leads also to a more refined classification of elements than classical physics or chemistry can achieve. In chemistry, a rough classification is made by the easily changeable accidental physical and chemical properties, depending upon the peripheral electronic configuration, while here a highly perfected differentiation based on the more intrinsic nuclear properties, less prone to changes, is obtained. Atoms are classified into three different categories as follows:

Those that have different mass numbers A, but the same atomic numbers, are called isotopes which having the same physical and chemical properties cannot be distinguished w separated by the ordinary physical or-chemical methods. The isotopes represent elements in the strictest sense, while the chemical elements that occur in nature are mixtures of several isotopes. Since the essential individuality of an element is determined ultimately by the nucleus, we may say that the number of simple elements is very much greater than the 92 elements of the periodic table, and should include about 300 different isotopes that are found in nature. The factor that distinguishes the different isotopes is the relative number of neutrons in them, since Z which gives the number of protons is constant, while A, the total number of particles, varies and hence the number of neutrons, given by (A – Z), changes from isotope to isotope.

Atoms that have the same mass number A, but different atomic numbers Z, are called isobars which naturally exhibit different chemical and physical properties. Forty-four pairs of isobars are known, excluding some ten cases for which one of the two members is doubtful. Most of the isobars have even mass number and even atomic number, the last differing by two units, while those of odd mass numbers and odd atomic numbers differing by ono unit are rare. It is evident that in isobars both the number of protons and the number of neutrons are different. Isobars with even number of protons and odd number of neutrons occur rather frequently.

Atoms which, though having the same mass number A and the same atomic number Z, are distinguished by certain differences in the internal structure of the nucleus, manifested by different decay periods in the case of radioactive nuclei, are called isomers. A number of such atoms has been brought to light in the course of a careful analysis of artificial radioactivity. We have already given an account of the experimental study and theoretical explanation of nuclear isomerism.

It has been already pointed out that the theory of proton-neutron constitution of the nucleus fits in well the theoretical explanation of a great number of nuclear phenomena.

6.2 NUCLEAR MASS

The mass or weight of the nucleus is often called, the atomic mass. It means the weight of the whole atom.

The mass of an atom of atomic number Z includes beside the weight of the nucleus that of the Z orbital electrons.

On account of the isotopic constitution of elements, the *atomic mass considered here is not the same as the atomic weight ordinarily used in chemistry*. The former is the individual weight of each of the isotopes contained in one and the same element, while the latter is the mean combining weight of an element as found in nature, to which every one of the isotopes in the element makes contribution in proportion to its relative abundance.

To measure accurately nuclear masses one should, first of all, separate the different isotopes contained in one and the same element and then determine their masses individually. This cannot be done by the ordinary chemical methods based on physical and chemical properties. But the wonderful high-precision instruments known as the mass spectrographs have achieved the task to perfection. The weight of the individual isotopes thus measured is ordinarily termed the atomic mass. To get at the exact nuclear mass from the above quantity, one must pay attention to the atomic number of the element and to the degree of ionisation. Because, positive rays containing atoms singly or multiply ionised are used in mass spectrographs. If Z be the atomic number of the element and singly ionised beam is used, the mass of the (Z – 1) electrons must be deducted from the experimentally measured isotopic mass to obtain the actual nuclear mass. If doubly ionised beam is used, then the weight of (Z – 2) electrons must be subtracted from the atomic mass to get the nuclear mass and so on.

It is to be noted, however, that in ordinary nuclear reactions no error, is introduced if atomic masses are used instead of nuclear masses, since the nuclear charge must always balance up in every such reaction which means the number of electrons contained in the atoms does not change in any given reaction. The only exception is positron radioactivity, where to obtain the correct energy balance of the nuclear reaction, twice the mass of an electron must be deducted from the difference of the atomic masses of the original and, product atoms, due to the special nature of the process.

Accurate measurement of isotopic masses with mass spectrographs which are capable of reaching a very high degree of precision, correct up to even the fifth decimal place, has led to the recognition of a very important quantity, closely related to the mass of the nucleus and ordinarily known as the *mass defect.*

The mass of every nucleus is approximately an integer, which gives support to the hypothesis that any nucleus is constituted of fundamental particles of unit mass; this when combined with the additional fact of *exactly integral nuclear charge,* which is always smaller than the *nearly integral nuclear mass*, argues to the proton-neutron constitution of the nucleus, since among the constituent.

The difference between the measured mass and the mass number A is known as the mass defect: $\Delta = M - A$. This quantity which is negative for the majority of nuclei, except the very light and very heavy ones, is far outside the experimental error, being for the lightest nuclei about 100 times and for the heaviest ones about 10 times the probable error. Furthermore, the departure from the whole number is too systematic to be disregarded. On the other hand, the magnitude of the mass defect is much too small and depends much too regularly on the mass number A to justify the giving up of the whole number rule.

An adequate solution of the difficulty has been proposed on the basis of Einstein's mass-energy relation. *The grouping together of the elementary particles (protons and neutrons) into a stable nucleus involves a certain interchange of energy*. This is known as the "*binding energy*," *i.e.*, the energy liberated or absorbed when the constituent particles are *bound* together into a nucleus. In a real sense, we may think of the phenomenon of nuclear formation as analogous to chemical reactions in which energy is released or absorbed. This justifies expressing nuclear reactions by equations very much like those used in chemical n actions. Now, according to Einstein's mass-energy relation ($W = mc^2$ or

$m = W/c^2$), any energy is equivalent to a certain mass. Hence the binding energy involved in the actual formation of a nucleus can be legitimately assumed to be derived from the mass available, which results in the mass defect. On account of the factor c^2 in Einstein's relation a considerable amount; of energy corresponds to even a very small mass defect. In any nuclear formation, a change in the sum total of the masses involved implies a proportionate change in the sum total of the rest-energies. If rest-mass disappears, a corresponding amount of energy must appear in other forms, as the rest energy of particles newly created, as the kinetic energy of such particles, or even as electromagnetic radiation (γ-rays). If, on the other hand, the rest-mass increases there must be a corresponding conversion of energy into rest-mass.

6.3 NUCLEAR MODELS AND NUCLEAR METHODS

The important types of nuclear models are as follows:

1. Alpha-particle model
2. Liquid drop model
3. Shell model.

The contents of the nucleus is very much complicated to study. Now let us understand the following models.

(1) *Alpha-particle model,* based on the experimental facts that α-particles are ejected by nuclei in disintegrations, both natural and artificial. Wheeler, Weizsacker and Fano have proposed the alpha- particle model, according to which the nucleus contains *α-particles*, at least as *substructures*, although they cannot maintain their identity for a very long time inside condensed nuclear matter, but will dissolve into more elementary particles.

(2) *Liquid drop model*, first proposed by Bohr and Kalckar, then accepted by Heisenberg, Majorana, Wheeler and others, compares the nucleus to a liquid drop, the nucleons corresponding to the molecules of the liquid, due to several points of similarity, such as large interaction between constituent particles, nearly constant density, surface tension effect, etc. This model has been utilised with a certain amount of success in the interpretation of intra-nuclear forces and of nuclear transformations and, in particular, nuclear fission.

(3) *Shell model*, proposed by Haxel, Mayor, Feenberg, Nordheim and others, assimilates the nucleus rather to a *gas, made of independent particles* with small interaction, caused by their

movement in a common field of force and influencing one another only by means of the requirements of the Pauli principle which excludes identical particles from occupying the same quantum state and thus endows the nucleus with a shell-structure, as in the case of extra-nuclear electron family. This model has been successfully used to explain certain nuclear phenomena such as stability, spin, magnetic moment, etc.

These various models prove insufficient, if an attempt is made to put them on some solid theoretical foundation or to subject them to a really detailed comparison with experimental data. They should be considered as not exclusive of one another, but complementary in scope in the understanding of the extremely complicated behaviour of nuclei.

Intimately connected with the nuclear structure is the *electrostatic field* of the positively charged nucleus, ordinarily known as the *potential barrier*. The contents of the nucleus are, so to say, walled in and well defended against interference from outside by this potential barrier which therefore plays an important role in natural and artificial disintegrations.

The special characteristics of the potential barrier have been understood as follows : The experiments on α-particle scattering showed that up to a very short distance from the centre of the nucleus the Coulombian law of inverse squares holds, but that there is a critical distance within which it becomes ineffective. *In the region of the inverse square law*, all positively charged particles, which are directed towards the positive nucleus, will be repelled by the latter, the greater being the force of repulsion the closer the approach. On the other hand, the negatively charged electrons, surrounding the nucleus in different peripheral orbits in an atom, will be kept in their proper positions, without flying off at the slightest provocation, by the same Coulombian law of forces, thereby endowing the atom as a whole with a stable structure. The electrons that are closer to the nucleus will experience a greater force, of attraction, which explains why more work has to be done to remove an electron close to the nucleus than another in the extremities.

Within the critical distance, the simple inverse square law be comes less important, giving place to a *new and more complicated law* which is essentially characterised by *attractive forces*. The existence of such a non-Coulombian attractive force is readily inferred from natural radioactivity and artificial transmutation. Given that both the nucleus and the α-particle are positively charged, one cannot understand how the a-particles could remain inside the nucleus for any appreciable time unless attractive forces exist between the α-particles and the rest of the nucleus. Artificial disintegration experiments, in their turn, show that even positively charged particles can be assimilated by the positive nucleus to form a new stable nucleus provided the external particles approach close enough to the nucleus, *i.e.,* within the critical distance. The same attractive force is also implied in the generally accepted proton-neutron constitution of the nucleus. For, otherwise, one cannot explain how the *uncharged neutrons and charged protons* can unite together intimately to form a stable nucleus.

The important nuclear methods can be mentioned as follows:

1. α-disintegration method.
2. Interaction method.
3. α-particle scattering.

As the mean life and the energy of the α-particles can be experimentally measured in the case of any given radioactive α-emitter, the radius of the nucleus of that element can be readily evaluated. The nuclear radii thus obtained for elements belonging to the same radioactive series are comprised between 7×10^{-13} and 10×10^{-13} cm., in good agreement with the order of magnitude of the values given by other methods. It is also found that the nuclear radius is proportional to the cube root of the atomic weight. The method can be used only for *heavy radioactive elements* whose mass numbers A are greater than 208 and which are α-emitters. Further, it is not clear which particular radius, the average or of the outermost limit of density distribution, is measured in this method.

The nuclear radii have been determined from observations made on the interaction of light nuclear particles such as α-particles, protons and neutrons with the nuclei of the elements under study. Of the nuclear particles mentioned above, neutrons alone are uncharged. In the interaction between a charged nuclear particle and a nucleus, when the two are at a considerable distance apart, there is an electrostatic repulsion between them, given by Coulomb's Law. But, when the particle gets very close to the nucleus, say due to its high speed, it is found that deviations from Coulomb's law set in and finally an effective attractive force comes into play, as indicated by the particle being absorbed by the nucleus, in spite of their like charges. In the case of neutron, the interaction is evidently nil at large distances, but, at small distances of the order of the nuclear radius, it is 'specifically nuclear'. We shall now consider briefly the researches made on the size of the nucleus, using each one of the above-stated nuclear particles.

Rutherford s simple classical scattering formula was delivered on the assumption that the nucleus of the scatter was so heavy that its motion during the interaction (considered as a collision) might be neglected. It led to a rough estimate of the radius of the nucleus, which was found to be of the order of 10^{-13} cm.

The Scattering of Protons

Very significant results have been obtained by investigating the scattering of protons by proton, have in obtained by investigating the scattering of protons by protons, as regards not only the nuclear radius, but also the nature of the intranuclear force between proton and proton. Mott and Massey, in 1933, worked out the wave mechanical theory of the scattering of protons by protons on the assumption of the Coulombian repulsive forces between the two identical particles.

Experiments conducted, in 1936 by White and by Tuve, Hafstad and Heydenburg by projecting protons in hydrogen gas and measuring the intensity of scattering at various angles, as the energy of the incident particle was given increasingly large values, led to results in disagreement with theoretical predictions. White showed that the theory was valid only for energies less than 600 KeV. At higher energies and for scattering angles greater than 20°, many more particles were scattered than predicted. Tuve and his associates found that for 900 KeV protons the number scattered at 45° was four times greater than that expected from theory. Breit, Condon and Present were able to show that the experimental results could be reconciled with theory only under the assumption that there existed an attractive short range nuclear force between the protons at very dose contact, besides the Coulomb repulsive force. The experimental data also showed that the interacting particles must have approached to a distance less than 5×10^{-13} cm. which would therefore give the sum of the radii of two protons.

Very recent experiments of D.M. Chase and F. Rohriich of Princeton University on the scattering of 18-6 MeV protons gave $R_0 = 1.42$. It is to be noted that experiments using charged particles as probes are difficult to interpret, because of the electric intention which is added to the nuclear one.

Blackett and many other Scientists employed the cloud chamber to study the phenomenon of collision, of α-particles. They used different light nuclei.

Elements of lower atomic weights show anomalous scattering. Since the angular distribution of the scattered α-particles in their case is found to depart.

Chadwick has examined carefully the case of scattering of α-particles, by helium Nuclei, where as mott in 1928 worked out the wave mechanical theory of the, scattering of α-particles.

6.4 MAGNETIC RESONANCE

The magnetic resonance method, provides direct information about nuclear magnetic moments, nuclear electric moments, molecular, rotational magnetic moments etc.

In 1938 Rabi, Zacharias Milliman and Kusch, developed magnetic resonance method, which made possible accurate spectroscopy in the radio frequency range.

The principle on which the method is based applies not merely to nuclear magnetic moments, but rather to any system which possesses angular momentum and a magnetic moment. If a particle having an angular momentum J and a magnetic moment μ is placed in a magnetic field H, it will execute a processional motion about the field with the Larmor frequency $\nu_L = \mu H/J$. The resonance method consists essentially in adjusting the strength of the magnetic field until the precession is in resonance with an impressed oscillating magnetic field H whose frequency is in the radio-frequency range. Under this condition, *viz.,* when $\nu_L = f$, the particle will suffer reorientations, which can be detected by a suitable device. The smaller the ratio H'/H, the sharper the effect will be on its dependence on the exact agreement between ν_L and f. The magnetic moment μ can be evaluated from the known frequency f of the auxiliary oscillating magnetic field and the strength H_r of the primary-magnetic field required to produce resonance in the processional frequency, provided J is also known.

In the case of more complicated systems containing two or more coupled angular momenta which interact with each other as well as with the external field, it is simpler to view the resonance reorientations as taking place when the frequency of the oscillating field is in resonance with the frequency given by the Bohr relation $h\nu_m = W_n - W_m$, where W_n, and W_m represent the energies of the two states of the whole molecular system in the magnetic field the selection rule which governs these transitions in the case studied $\Delta m = \pm 1$, where m is the magnetic quantum number of the system. It may be pointed out here that the method detects not only transitions from state *m* to *n*, but also the reverse transition n to m. One of these corresponds to absorption of radiation and the other to stimulated emission. As Einstein has pointed out, the two processes are equally probable.

Molecular states in which all possible and more powerful compelete to the nuclear moments are removed, as in molecules in the $^1\Sigma$ state where the energy of the nucleus in the external field depends only on its own orientation,, independent of all other molecular interactions, This corresponds to a simple system of angular momentum and associated magnetic moment, in which reorientations will occur if the Larmor frequency of precession ν_L (= μH/I) and the frequency *f* of the oscillating field are in resonance. Rabi, Guttinger and Majorana have considered the transition probability in such a case on a quantum mechanical basis and have shown that a maximum number of orientations takes place when $f = \nu_L$. Under these circumstances, the nuclear magnetic moment can be determined, as indicated above, if the value of nuclear spin I is known.

Molecular states in which the various interactions between the molecular constituents are effective, i.e., when the energy of the nucleus in the external magnetic field depends not only on its own orientation in the external magnetic field, but also on the orientation of other nuclei and on the orientation of the rotational angular momentum of the molecule ("spin-orbit" interaction). Analysis of strong field resonance reorientations in such cases has led to the determination of the value and sign of *nuclear electric quadrupole moments and of molecular rotational magnetic moments.*

On account of the existence of nuclear spin, the ground states of many atoms, such as those of the alkali metals, consist of a set of closely spaced energy levels. Each level of this hyperfine structure corresponds to a value of the total angular momentum of the atom. The spacings are caused chiefly by the feeble interactions of the magnetic and electric fields of the electrons with the nuclear magnetic moment and the electric quadrupole moment respectively. Left to themselves, the atoms would radiate this energy in the form of electromagnetic radiation given by the Bohr formula and settle down to their lowest energy state. The region of frequency in which these radiations are emitted lies approximately between 1.5×108 and 1.2×10^{10} cycles per sec. Because of these low frequencies the life-time of a hyperfine structure level is very long and the intensity of spontaneous emission very feeble. In consequence, direct observation of this radiation would be very difficult. But it is possible to irradiate the atom with electromagnetic radiation of the correct frequency and of such intensities (provided by the oscillating field) as to cause it to absorb, or, by the Einstein process of stimulated emission, to emit a quantum of this frequency in a reasonably short time of about 10^{-4} sec. If such a process is detected, it offers a *direct method of measuring hyperfine structure*, which has many advantages over the optical methods. Firstly, the results are simple to interpret since only one atomic energy level is involved; secondly, the accuracy is very high, as only the measurement of a radio-frequency is to be made; thirdly it enables extremely small energy separations to be measured.

The experimental arrangement deigned by Rabi and his associates is diagrammatically shown in the Fig 6.1. The molecules of the substance under test are produced in the oven O of small dimensions maintained at a temperature such that the vapour pressure of its contents is a few tenths of a mm. of Hg. Of the molecules that emerge in all directions from such a source, a very small fraction forming a very fine beam passes through a cool spacing slit S and reaches a detector D. In the absence of any in homogeneous magnetic deflecting fields these molecules traverse straight line paths OSD and form the so called "direct" beam. The main portion of the apparatus must be highly exhausted (10^{-5} mm. Hg.) in order to render the beam collision-free and two slits S_1 and S_2, several times wider than S, are used near the source to isolate the oven chamber were relatively high pressures are required to secure sufficient intensity of the beam.

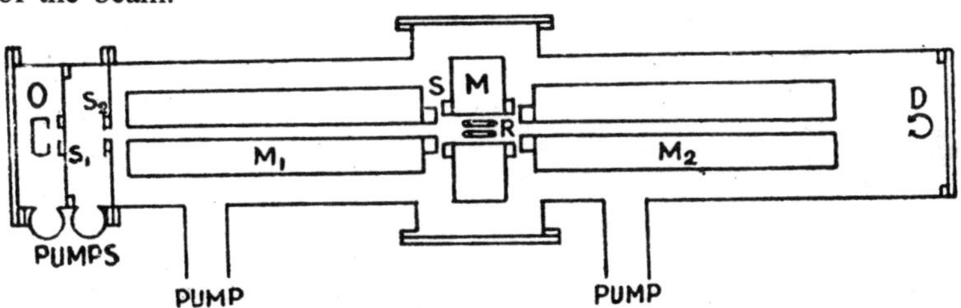

Fig. 6.1 : Apparatus Used in the Magnetic Resonance Method.

Two strong in homogeneous magnetic fields, whose gradients are in opposite directions are established using two electromagnets M_1 and M_2 with especially shaped pole-pieces, similar to those used in the Stern and Garlach apparatus. The first field due to M_1 (which is 25 cms. long and begins at 10 cms. from the source slit S_2) splits the beam into a number of polarised beams according to different moments and velocities of the molecules. The second field of opposite gradient due to M_1 (which is 30 cms. long and begins at 52 cms. from the source) will refocus the divergent beams from the first field at a certain point (92 cms. from the source) where the detector is placed. This refocusing takes place only for those molecules which have remained in the same quantum state through their passage in both the fields. It is found experimentally that when the two fields are properly adjusted the number of molecules reaching the detector is almost the same whether the fields are on or off. Now a short homogeneous field of variable strength H obtained with an electromagnet M and a superimposed oscillating field, perpendicular to H, produced by a high frequency current flowing through two parallel copper tubes bent in the form of a 'hair pin' and inserted in the gap between the pole-pieces of M are arranged between the two inhomogeneous fields. If H reaches a value for which the Larmor frequency ν_L of the molecules is equal to the frequency f of the oscillating field, resonance sets in, causing reorientations of the molecules which "flop over" into other quantum states. These flopped molecules are no more refocused by the field due to M_2. The resonance, therefore, produces a decrease in intensity at the detector, which thereby offers a means of knowing when the reorientation effect occurs. Each observed minimum indicates and certain Larmor frequency and allows the calculation of the corresponding magnetic moment.

The detector must necessarily be a very sensitive device capable of giving a linear response to the intensity of a slender beam molecules falling on it, as well as of measuring very small changes of intensity. There are in actual use, three different types of detector, known as the *Pirani gauge, the ionisation gauge and the surface ionisation gauge.* This last one is most sensitive and hence commonly used, whenever possible., It consists of a heated thin tungsten filament placed at the refocusing point of the molecular beam. The impinging neutral molecules (or atoms) re-evaporate from the filament as positive, ions, each having given up an electron to the surface of the filament. If the tungsten surface is kept sufficiently hot and if the difference between its work function and the ionisation potential of the molecule or atom is appreciably greater than kT practically all the molecules falling on the filament re-emerge as positive ions. These ions are collected by a plate which surrounds the filament and which is kept at a potential of about −10 volts with respect to the filament. The ion current thus obtained is amplified by means of a suitable vacuum tube. With the normal plate current of the tube balanced out, the change in this current is a direct measure of the number of molecules or atoms impinging on the tungsten surface. The surface ionisation device is able to detect an ion current of about

10^{-15} ampere, or about 6,000 molecules or atoms per sec. Measurements are made with it with ease and speed, a complete resonance curve being obtained in ten minutes or less. This, in turn, makes the data more reliable, since changes in experimental conditions are less likely to occur in short-time intervals than in long period.

6.5 PARITY AND STABILITY OF NUCLEI

Parity is a nuclear property, which involves, and requires quantum mechanical considerations.

For a particle of spin one the wave function is a vector. We can notice two types of vectors:

1. Polar Vector
2. Pseudo Vector.

Polar vector is always shown by a straight line with an arrow on it. Pseudo vector is shown by a simple line.

Nuclear states are characterised by a definite parity, which may be different for different states of the same nucleus. The conservation of parity has an important bearing on nuclear transformations. Thus, according to the parity law, two particles that are mirror images of each other must obey the same physical rules.

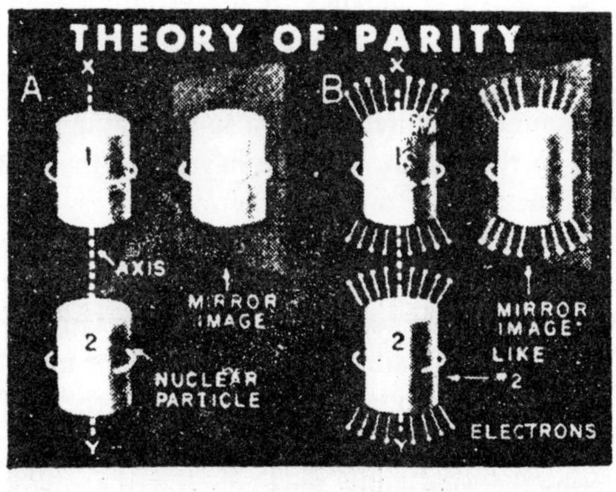

(a) (b)

Fig. 6.2 : Illustration of the Law of Parity.

This law of parity has been very much utilised in erecting, the structure of modern nuclear physics, even to the extent of summarily rejecting any theory that seemed to violate it. Very recently, however, such a fundamental .law has been seriously called into question, as we shall see in a later section.

Search for the causes of stability of nuclei have been made in the following four directions:

(i) Relative abundance of elements as well as of isotopes that occur in nature.

(ii) Mass defect curves.

(iii) Decay constants of natural radioactive substances.

(iv) Magic numbers.

(i) *Relative abundance of elements and isotopes,* the basic supposition made in this study ia that elements and isotopes which occur in nature *quite, abundantly are very stable.*

The relative plenty or scarcity of the different elements in terrestrial and other cosmic matter has been for many years a problem of the greatest interest among chemists, geologists and astrophysicists. It has now become a subdivision of a wider topic, the relative abundance of the different isotopes. Better said, there are now two branches of research, *viz.,*:

(i) The relative abundance of elements with respect to one another, and

(ii) The relative abundance of the different isotopes within each element.

As regards the first and older problem, *viz.,* the relative abundance of elements found in nature, it is in a somewhat unsatisfactory state and seems likely to remain so. For, we have only the earth's crest, a few meteorites, some nebulae and the outermost layers of stars available for study; the nebulae and the stars only by spectroscopic methods, of which the results are not always easy to interpret. The interior of the earth and of the stars remain impenetrable to us. The relative abundance of the elements in the few accessible regions is by no means the same and gives us no sure basis for guessing what it might be in the inaccessible regions.

Nevertheless, since the problem is one of the few possible sources for formulating theories about, the evolution of elements, it has received serious attention. An enormous number of chemical analysis of igneous rocks and other geological materials has been made. More recently, the more powerful and convenient method of X-ray analysis has been applied by Hevesy, Noddack and others to the problem with noteworthy success. As a result of the vast data thus acquired certain conclusions have been arrived at for the elements in the earth's crest, which are so strong that one is very much tempted to extend them to the whole of Nature. They are:

(a) *Heavier elements are much more rare than lighter ones*, which means that the light elements are much more stable than heavy ones.

(b) There is a *great predominance of elements of even atomic number (Z) over elements of odd,* first pointed out by Harkin, hence called Harkin's rule which has been beautifully confirmed by X-ray analysis, even in the case of the rare earths. This finding means that *atoms of odd, Z are much less stable than those of even Z.*

6.6 FUNDAMENTAL PARTICLES

The proton and the neutron are the bricks with which nuclei are built. The most fruitful source of fundamental particles has been the cosmic rays. These rays consist of energetic protons and heavier nuclei. Artificially produces mesons are now available in various laboratories. All the fundamental particles are unstable except the electron and the proton.

The neutron, is the longest lived of the unstable particles.

Fundamental particles, being 'simple' forms of matter, cannot be described by shape, colour, etc., which have further no significance in a realm where Heisenberg's uncertainty principle holds sway. They are usually designated by six important constants, *viz.,* mass charge, mean life, spin, parity and magnetic moment. If these constants of a particular particle are known, they provide sufficient data for a prediction of the particle's behaviour under most circumstances. Hence, great deal of experimental researches has been devoted to estimating the values of these constants for the different particles which exist.

It is almost certain that the table of fundamental particles given here will soon be out of date, at least in some respects, as further evidence is collected. It is to be hoped that a clearer picture of the elementary particles will lead to a better understanding of the already known facts and an accurate prediction of new phenomena. We now turn our attention to the relation of the fundamental particles to electromagnetic and meson fields. To get at this relation we need first to know the modem conception about fields.

Modern Theory of Fields

It is usual to distinguish two kinds of fields, *viz., classical and quantum fields.*

The classical field is a kind of tension or stress which can exist in empty space in the absence of matter. It reveals itself by producing forces which act on any material object that happens to lie in the space the field occupies: *e.g.,* electromagnetic, gravitational. The chief characteristic of a classical field is that its strength at a given point varies *smoothly* as the point moves around in space. With the inclusion of the principle of relativity, the classical field theory gives a satisfactory explanation of all large-scale phenomena, *i.e.,* which do not take into account that the universe is built with elementary particles; but it fails completely in interpreting atomic phenomena, *i.e.,* in describing the behaviour of individual atoms and elementary particles.

The quantum field is based on the existence of a specified list of elementary particles with definite masses, spins, charges and interactions with one another. The law of this field is the uncertainty principle and the consequent universal principle of statistical fluctuation. Hence the strength of a quantum field at a point can never be accurately measured. The quantum field energy can exist only in discrete units—quanta, which have precisely the properties of the elementary particles of which the universe is built. Some twenty qualitatively different quantum fields exist. Each fills the whole of apace and has its own particular properties. There is nothing else except these fields. Between various pairs of these fields there are different kinds of interactions. Each field magnified itself as a type of elementary particle. The particles of a given type are always identical and indistinguishable. The number of particles of a given type is not fixed, for particles are constantly being created or annihilated or transmuted into one another. The properties of the interactions determine the roles for creation and transmutation of particles. In this picture, all fields in Nature are quantum fields. The so-called classical electromagnetic and gravitational fields are simply special large-scale manifestations of a quantum field. The particle that corresponds to the electromagnetic field is the *photon* which appears to be different from other elementary particles only because its laws of interaction make it especially easy to create and annihilate. The particle corresponding to the gravitational field is the graviton which has not been observed as individual particles. The electromagnetic and gravitational fields are *long range*, while the fields corresponding to other particles are *short range*, less than the size of atom. In consequence, the photon and graviton have no rest-mass and always trawl at a constant velocity *c*, while the other particles, such as the electrons, protons, mesons, etc., have a rest-mass and can travel with any velocity not exceeding *c*.

The quantum field may or may not carry an electric charge; *e.g.*, the electron field and the electromagnetic field. Any charged field must be represented by two-types of particles which are alike in all respects, except that one has a positive charge and the other negative, and which under suitable conditions can be created or annihilated together in a single event, *e.g.*, electron and positron, proton and anti-proton.

The linking of certain particles to a field must not be understood as if the field ia composed of such particles. Actually field and particles are, so to say, merely different aspects of one and the same concept. The elementary particles may be considered as transient fluctuations and concentrations in their respective fields, or according to a recent suggestion of Heisenberg, as discrete stationary states of non-linear field equations which describe matter in a very general way.

Of the quantum fields described above, the two which are of immediate interest to us are the electromagnetic and meson fields. In building a cogent theory for nuclear forces, the central problem to be solved is the *pion-nucleon interaction*, since as we have seen, the pion is chiefly responsible for nuclear forces. To achieve this, the photon-electron interaction of quantum electrodynamics has shown us the way.

Quantum electrodynamics, which deals with the interaction of a pair of quantum fields, *viz.*, the electron and the electromagnetic field, in a manner consistent with the special theory of relativity, was framed immediately after the birth of quantum mechanics. But its mathematical treatment always gave rise to certain infinite quantities which could not be readily interpreted. In 1947, the theorestical physicists, F.J. Dyson, B.F. Feynman, J. Schwinger and S.I. Tomonoga developed the so-called "theory of renormalisation" which effectively solved the difficulty of these infinities by showing that the quantities which used to turn out infinite should be interpreted as a change of the mass and charge of the particle : the renormalisation procedure gave also a particle of selecting between possible and impossible expressions for fields and their interactions. Thanks to this principle, we can now predict to almost any degree of accuracy the results of experiments which are entirely electromagnetic in nature. One of the most accurate tests at present is certain extremely small spottiness of hydrogen spectral lines called the *Lamb-Rutherford shift*. The complete success of modern quantum electrodynamics has greatly encouraged the attempt to construct the theory of pion-nucleon interaction which is called the "pseudoscalar meson field theory."

Appearing in the photon-electron interaction of quantum electrodynamics is the wave function of the photon. This function is a polar vector, since the photon has spin one and positive or even parity. The photon amplitude is multiplied by a known quantity which makes the expression for the total energy of the photon-electron system a scalar to the case of the meson field, we use the fact that the pion is a pseudoscalar. It turns out that the pion wave amplitude, appearing in the total energy expression, can be multiplied by two possible quantities. Both of these quantities contain the nucleon wave amplitude, which mean that pion is 'coupled' to the nucleon. The term which arises from this coupling is called the pion-nucleon interaction. Now the two possible methods of coupling are pseudoscalar and pseudovector, P.T. Mathews and A. Salam, extending the theory of renormalisation to the meson field, have found that the pseudoscalar coupling can be renormalised, but not pseudovector coupling. Hence, it is assumed that the pion-nucleon coupling is pseudoscalar.

In 1963, a young French physicist, Maurice Levy, made a very important discovery, *viz., a direct consequence of the pseudoscalar interaction is that the forces between two nucleons are strongly repulsive at close distances*. Previous theories, which assumed that the forces between two nucleons are always attractive, could not really account for the stability of nuclei, since the strong force of attraction between the two nucleons should make them fall into each other. Levy's discovery saved the situation, because he showed that there was at small distances a very strong repulsion which prevented the two nucleons from falling into each other. By a consistent use of this kind of pseudoscalar interaction. Levy was also able to account for several phenomena observed in nuclear forces; for instance, the insignificant contribution of mesons heavier than the pion to nuclear forces under normal conditions, the saturation characteristic of nuclear forces, the dependence on spin the existence of a quadrupole moment in deuteron, etc.

Although, the pseudoscalar meson field theory has led to results in good qualitative agreement with experimental data, yet it still remains a quantitatively unsolved problem. The chief difficulty arises in determining the appropriate coupling constant. In electrodynamics, the constant multiplying the photon-electron interaction term is e, the charge of the electron. In the theory it is called the photon-electron coupling constant. It is found useful to introduce a dimensionless quantity α, known as the fine structure constant $\alpha = 2\pi e^2/ch = 1/137$, since the equations of electrodynamics can be solved only as a power series in a. Each term in the series will be about 137 times smaller than a previous one, so that after a first few terms, the contribution of a becomes negligible. In the mason field theory, the quantity analogous to *e* is the pion-nucleon coupling constant g and the dimensionless quantity analogous to the fine structure constant a is $2\pi g^2/ch$. The numerical value of this constant, instead of being a small fraction, is about 15. Physically, this implies that the nuclear forces are about 100 times stronger than atomic forces. Mathematically, however, it implies that any attempt to solve the equations of meson field as a power series in this constant is quite hopeless, since successive terms get larger rather than smaller as the series proceeds. We may sum up the situation in the meson field theory today by saying that we have a set of complicated equations which have to be solved by some means that does not involve a series expansion in $2\pi g^2/ch$. Even when these equations are solved—which is going to be a feat of mathematical skill—a great deal more has to be done in the shape of fitting the pion-nucleon interaction into the experimentally studied complicated phenomena, such as the scattering of mesons by nucleons and of nucleons by nucleons, the production of mesons by the interaction of photons with nucleons and of nucleons with nucleons etc., before we can arrive at an adequate understanding of the nature of the intranuclear forces.

7

METALS AND CONDUCTORS

7.1 CONDUCTIVITY

In metals the atoms are in ionized state, and they vibrate about their equilibrium positions. Some electrons move randomly inside the metals.

H.A. Lorentz, first time proposed the free electron theory. This theory explained thoroughly the electrical conductively of the metals.

When a metal is heated, the heat energy supplied to it can be distributed in three different ways. A part goes to increase the kinetic energy of the vibrating atoms while another part goes to increase the vibrational potential energy of the atoms due to the increase of their amplitudes of vibration. Finally, a part of the supplied heat energy is shared by the free conduction electrons whose random velocities are thereby increased. According to the law of equipartition of energy, the A mean kinetic and potential energies of the metallic atoms at the lattice sites at the absolute temperature T are $3kT/2$ each where k is Boltzmann constant. In addition, the mean kinetic energy of the free electrons is also $3kT/2$. Assuming the number density of the free electrons in the metal to be equal to the number density of the atoms at the lattice sites, the total internal energy of the metal per mole at the temperature T will be

$$U = N_o\left(\frac{3}{2}kT + \frac{3}{2}kT\right) = \frac{9}{2}RT$$

where N_0 is the Avogadro number and R is the universal gas constant. Hence the molar specific heat of the metal should be

$$C_v = \left(\frac{\partial U}{\partial T}\right) = \frac{9}{2}R = 9\text{cal/mole.}$$

temperatures re-found to be 6 cal/mole in agreement with Dulong-Petit's law. In other words, the metallic specific heat has the same value as would be obtained by neglecting the contribution from the motion of the free electrons altogether and is equal to the lattice specific heat $C_l = 3N_o k$.

To explain the above anomaly it was pointed out by W. Pauli and A. Sommerfeld that we cannot apply the classical statistics of Maxwell and (Boltzmann to the free electron gas in metals on the basis of which the equipartition law mentioned above is deduced. According to them, the electron gas in metals is degenerate and is governed by the Fermi-Dirac quantum statistics which gives the number of electrons in the energy interval d e at e per unit volume at the temperature T.

$$n(\varepsilon)d\varepsilon = \frac{\sqrt{2m^3}}{\pi^2\hbar^3} \cdot \frac{\sqrt{\varepsilon}\,d\varepsilon}{\exp\,(\varepsilon - \varepsilon_f)/kT + 1}$$

Here m is the electron mass, $\hbar = h/2$rt where h is Planck's constant ε_f is a constant known as the *Fermi energy*.

Since the electrons obey Pauli's exclusion principle, each energy level can be occupied by only two electrons with spins aligned oppositely. Thus the electrons in the metal fill up all the energy levels starting from when the metal is heated from 0 K to a few hundred kelvins, the energy gained by an electron is on the average kT which is of the order of one tenth of an electron volt or less for $T \sim 1000$K. Since all the lower energy levels upto u are filled up, the electrons in these levels have no available empty levels to which they can go by absorbing such a small amount of energy. Only the electrons near the top of the filled levels, *i.e.*, near the upper end E_0 of the energy distribution curve can absorb the amount of energy kT to be raised to an upper level with energy greater than ε_f. Thus only a small fraction of the supplied heat energy can be taken up by the free electrons, the bulk being taken up by the atoms at the lattice sites. The number of the electrons which are thus raised to the higher energy levels can be calculated in the following manner.

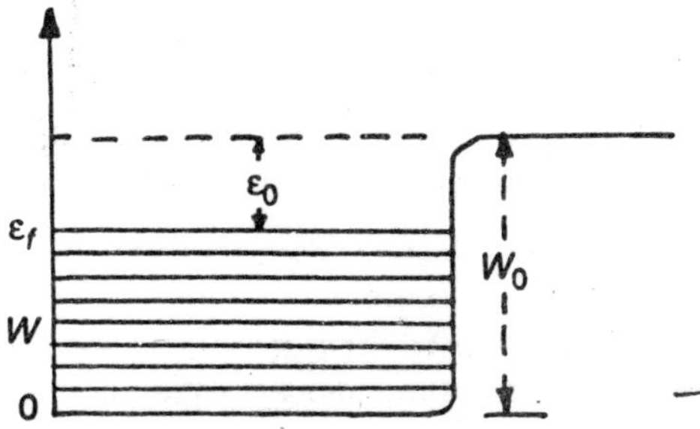

Fig. 7.1 : Electronic Levels in a Metal.

If there are N electrons filling up the levels upto ε_f, the number of levels in the interval 0 to ε_f will be N/2 since each level is occupied by two electrons. Assuming the levels to be equipaced, the energy gap between the successive level will be

$$\Delta\varepsilon = \frac{\varepsilon_f}{N/2} = \frac{2\varepsilon_f}{N}$$

7.2 FREE ELECTRON THEORY

As we know metals are good conductors of heat and electricity. In 1990 Paul Drue, a German first time tried to explain, about free electrons, in metals.

Latter, H.A. Lorentz studied the velocity distribution of the electrons.

According to Drude-Lorentz model, the free electrons move through the lattice structure of the metal and suffer repeated random collisions with the lattice vibrations, defects and impurities, much like the collisions between the gas molecules in a given volume of a gas. For this reason, the free electrons in the metals are termed *electron gas*. The mutual repulsion between the electrons is neglected in this model.

Because of the randomness of the motion of the electrons in a metal, there is no net electric current flowing in the metal unless an electric field is applied. When an electric field is applied across a metal, the free electrons acquire a *drift motion* in the direction of the applied field in addition to their random thermal motion. This gives rise to the flow of an electric current through the metal.

If a potential difference V is applied between the two ends of a metal rod of length L and cross-sectional area A, then the force acting on each electron carrying a charge e is

$$F = -\frac{Ve}{L} = -Xe,$$

where, $X = V/L$ is the electric field. The resulting acceleration of the electron in the field direction is

$$f = -\frac{Xe}{m}$$

The negative sign is due to the electron charge being negative.

If the electron motion under the action of the applied electric field were unopposed, the drift velocity of the electron due to the above acceleration would go on increasing indefinitely. Actually this does not happen, since according to Drude-Lorentz theory the electrons suffer collisions with the lattice vibrations, defects and impurities. The drift velocity gained by an electron due to acceleration by the applied electric field which is superimposed upon its random thermal velocity is lost completely as a result of the collision and only the random velocity remains. After this, the electron again begins to get accelerated by the applied field and loses the drift velocity thus gained at the next collision. The process goes on repeating so that we can talk of a constant average drift velocity υ_d by the action of the field when a large number of collisions is considered. We can look upon the whole process by supposing that a resistive frictional force F_r acts upon the electron, opposing its gain of drift velocity due to the accelerating field.

The probability of the electron suffering a collision in time *dt* may be written as dt/τ where *t* is taken to be a constant in the first approximation having the dimension of time. The rate of change of the drift velocity due to the action of the field may be written as

$$\left(\frac{d\upsilon_d}{dt}\right)_F = f = -\frac{Xe}{m}$$

The rate of change of υ_d due to collisions at the lattice sites will be

$$\left(\frac{d\upsilon_d}{dt}\right)_C = -\frac{\upsilon_d}{\tau}$$

since $1/\tau$ is the probability of collision per second. Hence in the steady,

$$\left(\frac{d\upsilon_d}{dt}\right)_F + \left(\frac{d\upsilon_d}{dt}\right)_C = -\frac{Xe}{m} - \frac{\upsilon_d}{\tau} = 0$$

$$\upsilon_d = -\frac{Xe\tau}{m} = -\frac{Ve}{mL}.\tau = -\frac{Ve}{L}.\frac{\tau}{m}$$

If the electric field X is suddenly withdrawn at some instant t = 0 when the electrons have acquired a drift velocity υ_d, then due to repeated collisions at the lattice sites, the drift velocity gradually falls to zero and the electron velocity becomes completely random again. How soon this will Happen is determined by the time T. In this case, since $(d\upsilon_d/dt)_F = 0$, we can write

$$\frac{d\upsilon_d}{dt} + \left(\frac{d\upsilon_d}{dt}\right)_C = -\frac{\upsilon_d}{\tau}$$

which gives on integration

$$\upsilon_d = \upsilon_{d0}\exp(-t/\tau)$$

where υ_d is the average drift velocity at $t = 0$. Because of the exponential fall of υ_d with time, τ has been given the name *relaxation time*. For good conductors, *e.g.*, copper, silver etc., $\tau \sim 10^{-14} s$.

It can be shown that T is equal to the mean time interval between successive collisions of the electrons at the lattice sites.

Let P (t) be the probability that an electron does not suffer any Collision upto the time t after suffering a collision at $t = 0$. Since the probability of suffering a collision in an interval of time dt is dt/τ (see above), the probability that the electron does not suffer any further collision between t and $t + dt$ is $(1 - dt/t)$. So the probability of the electron not suffering any collision upto the time $t + dt$ is given by

$$P(t + dt) = P(t)\ (l - dt/t)$$

Expanding the left hand side by Taylor Series we get, after neglecting the higher order terms

$$P(t) + \frac{dP}{dt} = P(t) - \frac{P(t)}{\tau} dt$$

$$\therefore \qquad \frac{dP}{dt} = -\frac{P(t)}{\tau}$$

Integrating we get $\qquad P(t) = \exp(-t/\tau)$.

This gives P (0) = 1 as it should be. The mean free time between collisions will then be

$$< t > = \int_0^\infty t\ P(t)\, dt = \int_0^\infty t \exp(-t/\tau)\ dt = \tau$$

The above result is true if it is assumed that there is no persistence of velocities after the collisions.

If the mean free path of the electrons between successive collisions is λ and the mean velocity of the electrons due to thermal motion is $< c >$, then assuming $\upsilon_d << (< c >)$ we get

$$\tau = \lambda/<c>$$

which gives $$\upsilon_d = -\frac{V_e}{mL}\tau = -\frac{Ve}{mL}\frac{\lambda}{<c>}$$

Since the electric field $X = -V/L$, the mobility of the electron, defined as the drift velocity gained in a unit electric field, will be

$$u = \frac{\upsilon_d}{X} = \frac{e\tau}{m} = \frac{e\lambda}{m<c>}$$

If there are n free electrons per unit volume of the metal, the electric current flowing through the rod due to the applied electric field will be

$$I = ne\,A\upsilon_d = \frac{ne^2\,A\lambda}{mL<c>}\,.\,V$$

So the current density is

$$j = \frac{I}{A} = \frac{ne^2\lambda\,V}{mL<c>}$$

From Ohm's Law, we have $I = V/R$ where R is the resistance of the metal rod which is thus given by

$$R = \frac{m<c>}{ne^2\lambda A} = \rho\,\frac{L}{A}$$

where ρ is the resistivity of the metal. So we get

$$\rho = \frac{m<c>}{ne^2\lambda} = \frac{m}{ne^2\tau}$$

The conductivity of the metal is

$$\sigma = \frac{1}{\rho} = \frac{ne^2\lambda}{m<c>} = \frac{ne^2\tau}{m}.$$

In the above treatment τ has been assumed constant. Actually τ is a function of-the velocity of the electrons. A more rigorous treatment using Boltzmann transport equation gives in place of the following expressions for the mean drift velocity υ_d

$$\upsilon_d = -\frac{X_e}{m}\,\bar{\tau}$$

where the mean relaxation time T is defined as

$$\bar{\tau} = \frac{<\upsilon^2\tau>}{<\upsilon^2>}$$

Here it is assumed that the electron gas obeys Maxwell-Boltzmann statistics.

The electrical conductivity then comes out to be

$$\sigma = \frac{ne^2}{m}\,\bar{\tau}$$

If we assume to a first approximation that τ on the r.h.s. is independent of the velocity, then we get by writing

$\tau = \lambda/\upsilon$ where λ is also assumed to be independent of υ

$$\bar{\tau} = \frac{<\upsilon^2\tau>}{<\upsilon^2>} = \frac{<\upsilon^2\lambda/\upsilon>}{<\upsilon^2>} = \frac{\lambda<\upsilon>}{<\upsilon^2>}$$

But $$<\upsilon> = \sqrt{8kT/\pi m}$$

according to Maxwell-Boltzmann statistics. Also since m $< \upsilon^2 >/2$ $=3kT/2$ we get,

$$\bar{\tau} = \frac{\lambda < u >}{(3kT/m)}$$

$$= \frac{\lambda m}{3kT}\sqrt{\frac{8kT}{\lambda m}}$$

$$\sigma = \frac{ne^2\lambda}{3}\sqrt{\frac{8}{\lambda mkT}}$$

7.3 SCHOTTKY EFFECT

Increase of the anode potential causes, the increase of the saturation thermionic current. (i_s)

In 1914, this phenomenon was observed by W. Schottky, which was latter named as, schottky effect.

We have seen that for the emission from a metal surface, the electron potential energy has to change from $U_0 = -(\varepsilon_o + \varepsilon_f)$ inside the metal to $U = 0$ outside in vacuum. Here $\varepsilon_0 = e\phi$ is the thermionic work function. This causes an abrupt change of the potential energy. However in 1 the actual case, the change in potential energy is more gradual. This is due to the fact that at the time of emission, an attractive image force acts on the electron tending to hinder the emission. According to the electrostatic image theory, if a point charge *e* is at a distance x from a plane metal surface, the total effect of the opposite charges induced by it on the metal surface is equivalent to that of an equal and opposite point charge $-e$ situated at an equal distance x behind the metal surface such that the line joining the two charges is perpendicular to the surface. The force acting on the charge e due to this image charge is known as the image force. The potential energy of the electron due to the image force is

$$V_e(x) = -\frac{e^2}{16\pi\varepsilon_o x}$$

This gives rise to a more gradual change in the potential energy of the electron in coming out from the metal. Quantum mechanical theory shows that reflection effect at the wall of the potential well which introduces the factor is considerably less in this case than in the case of an abrupt change of potential at the metal surface if the image force is not considered.

Normally, the attractive image force, which begins to become prominent at a distance of a few angstroms beyond the surface, goes to zero at infinity. However if an anode at a positive potential is placed facing the heated emitter surface, the electron experiences an outward force due to the anode potential which acts opposite to the image force. The two forces cancel one another at some distance from the surface of the emitter. So if an electron is able to come out from the emitter upto this point, it will be emitted. Since this point is nearer to the emitting surface than in the absence of the anode potential, an electron with lesser initial energy can be emitted from the heated surface in this case. This means that the work function is effectively reduced due to the action of the positive anode potential. Higher the positive potential on the anode, greater is the reduction in the effective work function.

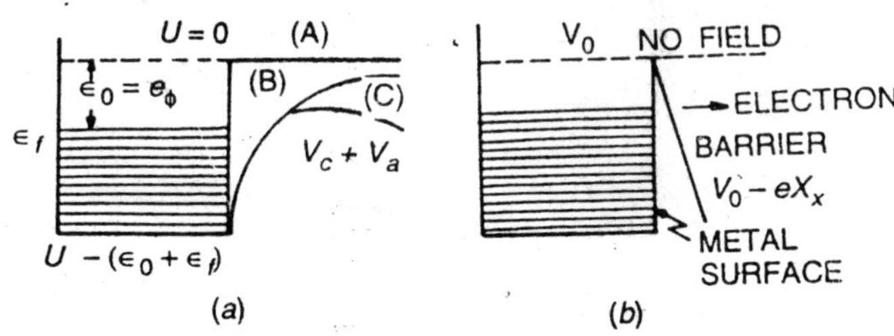

Fig. 7.2 : (a) Effect of Image Force on the Potential Energy (A) of an Electron in a Metal (Curve B) and with an Anode Potential (C). (b) Effect of Applied Held (10^8 V/m) on the Shape of the Barrier Wall.

7.4 BAND THEORY

While studying behaviour of the different types of solids, it is necessary to consider the nature of potential field inside the lattice. Band theory consider, large number of interacting particles.

Consider an isolated atom of a monovalent element, *e.g.*, sodium (Z = 11). The variation of the potential acting upon the electrons in this atom due to the Coulomb field of the positively charged nucleus. The various energy levels of the atom are also shown in the figure which form a predicted by quantum mechanics. The occupation the atom are also shown in the figure which form a discrete set as predicted by quantum mechanics.

Now consider two 'sodium atoms placed side by side at a large distance from one another. The potential energy diagram of the two atoms. As long as the two atoms are separated by a large distance, the resultant potential energy has the appearance of that due to two completely isolated atoms, there being a potential barrier separating the two which prevents the electrons of one of the atoms to interact with those of the other.

If now, the two atoms are brought closer together, the potential energy curves overlap and as a result there is a minimum in the resultant potential energy curve midway between the two atoms. This reduces the height of the potential barrier between them.

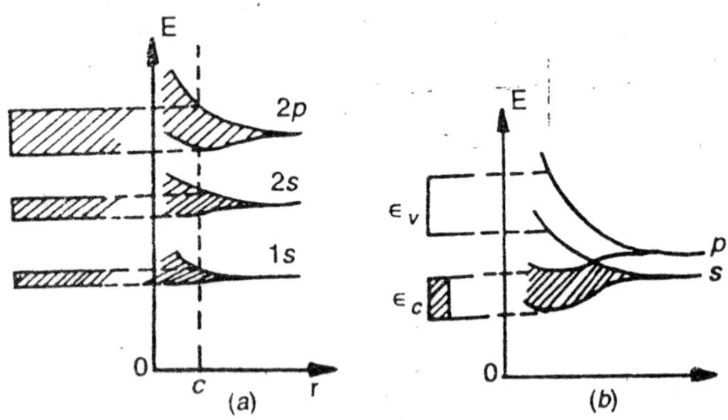

Fig. 7.3 : (a) Variation of Bandwidth with Distance Between Lattice Atoms. (b) Overlapping Between Higher Energy Neighbouring Bands for a Diamond Crystal.

The potential energy diagram for an array of sodium atoms which is found to change periodically as we go from one lattice ion to the next. Actually the potential energy is infinite at the lattice sites. Midway between two lattice sites, the potential energy attains a minimum value since the attractive forces due to the two neighbouring atoms are equal at these points.

Now, if there are n atoms in an array there will be $2n$ electrons in the $1s$ level for the whole system, since each atom has two $1s$ electrons with opposite spins. Since according to Pauli's exclusion principle, there n cannot be so many electrons in a particular level, we have to assume that the $2s$ level for the whole system splits up into n close lying levels which can accommodate 2n electrons with pairs having oppositely aligned spins in each of the split levels without violation of the exclusion principle. These levels thus constitute an energy band known as a permitted band. The other levels of the individual atoms, *e.g.*, $2s$, $2p$, $3s$ etc. are also similarly split up and give rise to energy bands.

The widths of the energy bands depend upon the spacings between the lattice sites (lattice constant) and on the energy spreads of the individual atomic energy levels. They are independent of the total number of atoms in the lattice. The bands associated with the inner orbits of the individual atoms (*e.g.*, $1s$, $2s$) have usually smaller band-widths because of the very weak influence of the neighbouring atoms on the motion of electrons in the corresponding orbit of the individual atoms. On the other hand the width, of the higher energy bands are much larger since the influence of the neighbouring atoms on the electrons in the upper energy levels (*e.g.*, $2p$, $3s$) of the individual atoms is much stronger.

In an actual crystal, the spacings between the lattice atoms is a constant and cannot be changed. However if we imagine that it is possible to change these spacings, then we can study the variation of the band-widths with the change of the distance between the lattice atoms; When this distance is large, the bands have the appearance of the individual atomic levels, having very little spread in energy. As the distance between the lattice atoms is decreased, the splitting of the levels begins to increase which results in an increase of the bandwidth. In the inset of the figure at left are shown the nature of the bands when the distance between the lattice atoms is equal to the actual lattice constant.

The gaps between the lower energy bands are usually relatively large. These gaps are known as *forbidden zones.* There can be no electrons in the solid with energies lying in these zones. They can only have energies in the permitted band regions.

The gaps between the higher energy bands are usually much smaller. In some cases, the widths of the bands may be so large that there may be overlapping between neighbouring bands. Such overlapping bands are shown in the case of a diamond crystal. In this case the bands arising out of the $2s$ and $2p$ levels overlap in such a way that they give rise to two other bands as shown. In between these two there is a forbidden zone. The lower band is known as the *valence band* and the upper band as the *conduction band.*

Since the number of electrons in a particular energy level within a band can at most be 2, the maximum number of electrons in the band can be twice the number of levels constituting the band. If all these levels are filled with electrons, then we have a *filled energy band.* Usually the lower energy bands are of this type. The higher energy bands may be either completely filled or partially filled. If there is no electron in a band then we have an empty band. Partially filled bands may arise either from the atomic levels which are partially filled or due to the overlapping of a completely filled band and an empty band just above it

We have stated earlier that there cannot be any electron in the forbidden zone. However if some impurity atoms are present in a crystal, then there may be permitted energy levels within the forbidden zones. The action of *impurity semi-conductors* depends on the presence ef such impurity levels.

7.5 CLASSIFICATION OF SOLIDS

According to band theory crystalline solids can be classified into three classes.

1. Conductors.
2. Insulators.
3. Intrinsic Semiconductors.

1. Conductors

If the uppermost energy band in a solid is partially filled with electrons with the lower energy bands completely filled, then the application of even a very small electric field in the solid causes some of the electrons to make transition to the empty higher energy states within the band. As a result they are able to move about freely within the metal. As explained before such partially filled bands are formed due to partially filled energy levels of the atoms as in the case of alkali metals. They may also be formed due to the overlapping of filled and empty (or partially filled) bands.

Partially filled bands are characteristics of metals which are good conductors of electricity and heat.

2. Insulators

We know that most nonmetals are bad conductors of heat and electricity so that they constitute the insulators. In an insulator, the uppermost energy band (valence band) is completely filled with electrons. So due to the exclusion principle an electron in any level within this band cannot make transition to another level in the same band. They can only go to a level in the upper empty band (conduction band). However in insulators, the gap between the valence band and the empty conduction band is of the order of several electron-volts. So even when an electric field is applied to the insulator, the electrons in the uppermost filled band do not acquire sufficient energy to conductivity of the semi-conductors increases with increase of temperature. This behaviour is just the opposite of what is observed in metals. The conductivity of the latter decreases with increase of temperature. The conductivity of semi-conductors is also found to increase when they are exposed to light.

The semi-conductors of the type discussed above are known as *trinsic semi-conductors*. It may be noted that in these semi-conductors, as the electrons in the filled valence band are raised to the empty conduct ion band (at T > 0K) some vacant sites are created in the of accepting electrons (solid points) from amongst those remaining in the valence band. When an electric field is applied, not only are the electrons raised to the conduction band, but also some of the electrons in the valence bands will be able to move about freely under the action of the applied field, thereby giving rise to an electric current.

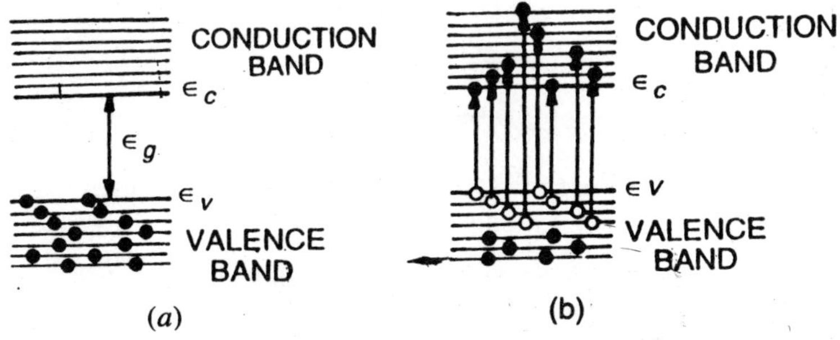

Fig. 7.4 : Conduction in an Intrinsic Semi-Conductor, (a) All Electrons in Valence Band. (b) Some Electrons Raised to Conduction Band.

With very narrow forbidden zone, the number of electrons raised to the conduction band from the valence, band as the temperature is raised may be substantial, which will increase the electrical conductivity of the crystal quite appreciably. Thus for germanium with $E_g = 0.66eV$, the electron density in the conduction band at room temperature is $n_i \sim 10^{19}$ per m^3 and the specific resistance $\rho_i \sim 0.48$ ohm-m. On the other hand for diamond with E, = 5.2 eV, n_i, is only about 10^4 per m^3 and correspondingly $\rho_i \sim 10^8$ ohm-m at room temperature. As the temperature is raised to about 600 K, the electron density in diamond however rises by many orders of cross the relatively broad forbidden zone and hence cannot make transition to levels in the upper empty conduction band. So the application of an electric field does not induce any electric current to flow in an insulator. Again if an insulator is heated, the electrons in the valence band do not gain sufficient thermal energy to make transitions to the empty conduction band. So they are bad conductors of heat as well.

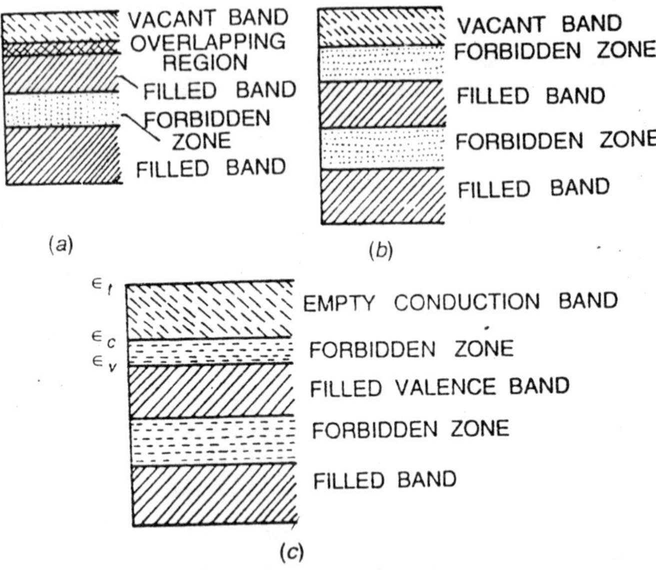

Fig. 7.5 : Classification of Solids According to Band Structure; (a) Conductors; (b) Insulators; (c) Semi-Conductors.

3. Intrinsic Semi-Conductors

In some substances, *e.g.,* silicon, germanium or grey tin in group-IV of the periodic table, the gap between the uppermost completely filled valence band and the next higher empty conduction band is quite small. At the absolute zero of temperature, they behave like insulators since they do not posses sufficient energy to make transitions to the upper conduction band. However, at ordinary room temperature, some of the electrons in the valence band have sufficient thermal energy to make transitions to the conduction band. When an electric field is applied to such a substance, these electrons can move about freely within them. So they exhibit limited conductivity. They are called semi-conductors. When they are heated, more electrons are transferred to the conduction band so that the magnitude and the specific resistance becomes comparable to that of germanium at room temperature.

Hole Conduction

We have seen above that in an intrinsic semi-conductor, a fraction of the electrons from the completely filled valence band at 0 K is raised to the empty conduction band when the temperature is raised. As a result some vacant sites, also known as "holes", lie near the top of the filled band. When there is an electric field, the current associated with all the electrons in a completely filled band is zero so that we can write

$$I = -e\sum_i v_i = -\sum_i{}' v_{i-ev_j=0}$$

where Σ' refers to the sum over all electrons except $i = -j$; $-e$ is the electronic change and v, is the velocity of the *i*th electron. If there is one hole in the valence band due to the absence of they *j*th electron, the current becomes

$$I' = -e\sum_i v_i = +ev_j$$

Thus, the current in the valence band due to one "hole" is equivalent to that due to the motion of a single positively charged particle of the same charge as the electron. Such a hole has a positive effective mass equal numerically to the negative effective mass of an electron which initially occupied a state close to the top of the valence band. Only then the current due to the "hole" will be equivalent both in magnitude and direction to that due to the electron of the almost completely filled band.

If an external electric field X is applied, then the velocity v, will be changed so that the current I' will also change. The rate of charge of with time is given by

$$\frac{dI'}{dt} = e\frac{dv_j}{dt} = \frac{e^2}{m^*}X$$

It is possible to produce radical change in the conductivity of these materials by the addition of a small quantity of some *impurity* in them. The resulting material is known as extrinsic or *impurity semi-conductor*.

Consider the two intrinsic semi-conductors germanium (Z = 32) and silicon (Z = 14), both belonging to group IV(a) of the periodic table. Their atoms contain two electrons in the outermost $4p$ and $3p$ orbits respectively and are tetravalent.

Now an intrinsic semi-conductor made of Si or Ge always contains some impurity atoms, however pure they may be. These impurity atoms have their own energy levels (*impurity levels*) which may be either in the allowed or in the forbidden bands or in both at different distances above the top of the valence band or below the bottom of the conduction band.

Similarly, when impurities are intentionally mixed with an intrinsic semi-conductor, changes in the band structure occur.

As we studied the impurity atoms donate electrons to the material, and it is known as a donor impurity and the levels that the additional electrons occupy are called donor levels.

When the electron is raised from the donor level to the conduction band, a hole is created in the immobile arsenic atom, so it does not take part in the electrical conductivity.

The impurity, atoms which accept electrons from the valence band, are known as acceptor atoms and the corresponding levels are known as acceptor levels.

7.6 THEORY OF SUPER CONDUCTIVITY

S. Bardeen, L.N. Cooper and J.R. Schrieffer, have developed theoretical understanding of superconductivity. They gave more concentration about, microscopic theory.

E. Fröhlich in 1950 proposed a model for an electron moving through a metal lattice continuously emitting and reabsorbing *virtual phonons*. According to him, the electrons perturb the neighbouring atoms causing them to oscillate. These lattice perturbations (phonons) then react on the electron. He pointed out that such *electron-phonon interaction* might produce a ground state of energy lower than that of the completely filled Fermi sea of noninteracting electrons. There will be an energy gap between such superconducting ground state and the normal conducting states of the metal.

Cooper in 1956 suggested that the *electron-phonon-electron interaction* whould make the Coulomb repulsion between two electrons smaller (or even absent), in the superconducting state than in the normal state. When an electron deforms the lattice in its vicinity a second electron sees the deformation and adjusts itselfs to take advantage of it to lower its energy. Thus the second electron interacts with the first via the lattice deformation. We may say that due to the deformation of the lattice by the first electron, phonons are created which are propagated through the crystal. The second electron is affected by absorbing this phonon. Since the phonon exchange is *virtual*, the energy is conserved for the crystal but the energy of the interacting electron pair need not be conserved before and after the exchange (c.f. pion exchange between nucleons). Thus Cooper showed that for a pair of electrons just above the Fermi surface a bound state could result if the phonon exchange gave rise to an attractive interaction.

Later Bardeen, Cooper and Schrieffer showed that under certain circumstances, there might be condensation of electrons of opposite spins into bound pairs having opposing values of the wave vectors for zero current density. Such pairing involves electrons in states within the energy kT_D (k = Boltzmann constant) of the Fermi energy. At $T = 0$ with zero current density, the ground state of a superconductor is a highly correlated state. All the states near the Fermi surface are thus filled with pairs of electrons of opposite wave vectors (**k**) and spins to the fullest possible extent. These are known as Cooper pairs. The ground state of all such cooper pairs is represented by a single wave function. A Cooper pair ($\mathbf{k}_1\uparrow$, $-\mathbf{k}_1\downarrow$) by an exchange of a virtual phonon can go over to another unoccupied pair position ($\mathbf{k}_2\uparrow$, $-\mathbf{k}_2\downarrow$). There are strong experimental evidences in support of the existence of such pairs.

There is a significant gap between the energy of a Cooper pair and the energy of two single unpaired electrons which is of the order of millivolts. This is many times larger than the energy required to break up the pair.

Each Cooper pair is symmetric about $k = 0$ for a superconductor with zero current density. If there is a nonvanishing persistent current, the Cooper pairs have non-zero net momentum. As we know that, in a normal conductor, the conductivity is prevented from being infinite due to the scattering of the electrons by phonons and by localized defects. This does not happen in the superconducting state because the above energy gap prevents the Cooper pairs against any change of net momentum. Thus the infinite D.C. conductivity of a superconductor emerges as a natural consequence of the BCS theory.

The existence of the energy gap between the superconducting ground state and the states of normal conduction in metals as proposed by the BCS theory has been confirmed experimentally.

The energy gap (ε_g) decreases towards zero as the temperature approaches T_c since the number of paired electrons decreases with rising temperature. The ratio ($\varepsilon_{gT}/\varepsilon_{g0}$) of the energy gaps at the temperatures *TK* and *OK* is a function of (T/T_c). The functional relation

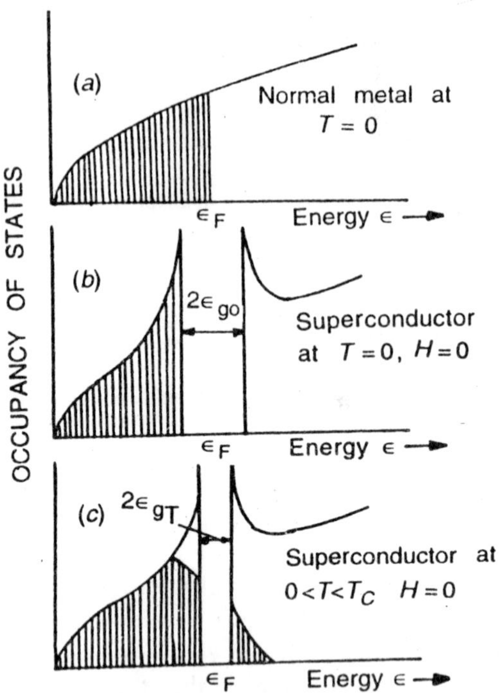

Fig. 7.6 : Variation of Energy Gap Between Superconducting Ground State and Normal States in a Metal; (a) Nonconducting State; (b) and (c) Superconducting States at $T< T_c$ and T = O K. Between the Two Expected on Theoretical Ground has been Confirmed Experimentally for In and Ta.

Super conductivity has many important application. For the construction of super conductivity magnets it is used. Super conductivity is utilized to design audio-frequency modulators, detectors of high frequency commutators without contact switches, cyclotrons etc.

7.7 JOSEPHSON EFFECT

The freely moving copper pairs can make transition from one superconductor to the other. A finite supercurrent may flow in the circuit containing the Josephson junction even with zero voltage, across the junction. This phenomenon is known as, D.C. Josephson effect.

If ψ_1 and ψ_2 are common wave functions, then Schrodinger equation V = 0.

If a D.C. voltage V is applied between the two sides of the insulator then this causes a radio-frequency current to flow across the Josephson junction. This is known as the A.C. Josephson effect. An external r.f. voltage applied with the D.C. voltage can then cause a D.C. current to flow across the junction.

An electron-pair experiences a potential energy difference qV in passing across the junction where $q = -2e$ is the charge of each Copper pair. Assuming that a pair on one side has the potential energy $-eV$ while it on the other side has the p.e. + eV we can write

$$i\hbar\frac{\partial\Psi_1}{\partial t} = K\psi_2 - eV\Psi_1$$

$$i\hbar\frac{\partial\Psi_2}{\partial t} = K\psi_1 - eV\Psi_2$$

As before, substitution for ψ_1 and ψ_2 gives

$$\frac{\partial n_1}{\partial t} = -\frac{\partial n_2}{\partial t} = \frac{2K}{\hbar}\sqrt{n_1 n_2}\sin\delta$$

$$\frac{\partial\theta_1}{\partial t} = \frac{eV}{\hbar} - \frac{K}{\hbar}\sqrt{\frac{n_2}{n_1}}\cos\delta$$

$$\frac{\partial\theta_2}{\partial t} = \frac{eV}{\hbar} - \frac{K}{\hbar}\sqrt{\frac{n_1}{n_2}}\cos\delta$$

For $n_1 \approx n_2$, $$\frac{\partial\delta}{\partial t} = \frac{\partial}{\partial t}(\theta_2 - \theta_1) - \frac{2eV}{\hbar}$$

Integration gives

$$\delta(t) = \delta(0) - \frac{2eV}{\hbar}t$$

and $$J = J_0 \sin\{\delta(0) - \omega_0 t\}$$

where $$\omega_0 = \frac{2eV}{\hbar}$$

Since H is a small number (compared to ordinary voltages and times), the oscillations are very rapid and the net current is very small.

With V= 1 μV, we get the frequency $\nu_0 = \omega_0/2\pi = 483.6$ MHz. When an electron-pair crosses the barrier, a photon of energy $\hbar\omega_0$=2eV is emitted or absorbed.

We have seen that with V = 0, there is D.C. current across the junction. But if a D.C. voltage V is applied then this goes to zero.

The frequency ν_0 is called the *Josephson frequency*. If e.m. radiation from an outside source having frequency ν falls on the Josephson junction, beats will be produced between the oscillating supercurrent mentioned above and the incident radiation whenever ν is an integral multiple of ν_0 : $\nu = s\nu_0$ where s is an integer. Under this condition V will be an integral multiple of $\hbar\omega_0$=2e = hν/2e so that $V = sh\nu_0/2e$. This shows that the values of V will appear in steps in the current-voltage curve. Such steps were observed by S. Shapiro using a niobium oxide-lead junction cooled to 4.2 K. These steps also exhibit the quantum nature of *super-conductivity*.

Josephson effect was used by Parker, Langenberg, Denenstein and Taylor to determine the ratio *e/h* very accurately, by using junction of tin-tin oxide-lead at a temperature ~ 1.2K. The junction was exposed to radiation of frequency 104 MHz (λ ~ 3 cm). The following values of 2*e/h* were obtained from emission and absorption respectively.

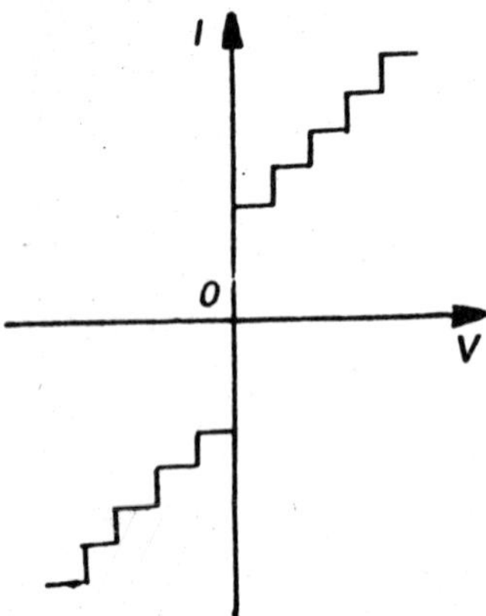

Fig. 7.7 : Current-Voltage Curve for Niobium Oxide-lead Josephson Junction in Microwave Field.

The superconductivity quantum interference effect between two Josephson junctions has been utilized in the construction of extremely sensitive magnotometers. The sensitivity can be further improved by using a larger number of junctions.

The method by which the brain analysis is carried is known as magneto-encephalography.

8

PHYSICS OF MOLECULES

8.1 INTRODUCTION

A molecule may be defined as the smallest particle of a substance, which is capable of existing permanently in a free state.

In 1811 Avogadro put forward his ideas about molecules. According to him, a molecule is an aggregate of a definite, number of atoms capable of existing in a free state.

A study of molecular spectra finds its proper place at this stage. For, in the first place, the postulates and principles involved in the elucidation of atomic spectra can be used also, *mutatis mutandis*, in the interpretation of the different characteristics of molecular spectra. Secondly, analysis of this type of spectra leads to the realisation of certain important structural properties of the nucleus, such as the isotopic constitution, spin, etc., with which we have to deal in the following chapters.

For a clear understanding of molecular spectra, a knowledge of the structure and properties of molecules is very helpful, although many points concerning molecular structure have been actually gathered from a scrutiny of molecular spectra.

8.2 CLASSIFICATION OF MOLECULES

Depending on the constituent atoms molecules can be classified as follows:

1. Monatomic molecules (one atom).
2. Diatomic molecules (two atoms).
3. Triatomic or polyatomic molecules (three or more atoms).

Monatomic molecules can have only translatory motion, while diatomic and polyatomic molecules can have also rotational and vibrational motions.

Molecules have been classified from considerations of their rotational motions.

The moments of inertia about the internuclear axis is zero and the other two about axes perpendicular to the internuclear axis are equal, we have the *diatomic* or a *linear polyatomic molecule*, *e.g.*, H_2, D_2, N_2, O_2, HCl, HBr, CO_2, N_2O, HCN, etc. If A = B = C, we have *spherical molecules*, which have no preferred axis, *e.g.*, CH_4 (methane), SiH_4 (silane), where the four hydrogen atoms are at the corners of a regular tetrahedron. If two of the moments of inertia are equal, the molecule is called *symmetric rotator* or *top*. The motion of this type can be represented as a pure rotation about the axis of either the greatest or least inertia, with the axis making a precession round that of total angular momentum. If the two smaller moments of inertia are equal, *i.e.*, A = B, we have the *oblate symmetric top* (a disc in the extreme case); if the two larger are equal, *i.e.*, B = C, we have the *prolate symmetric top* (spindle-shape). A largo and important group of molecules, such as NH_3, methyl halides, ethane, etc., are *symmetric rotators*. Finally if A ≠ B ≠ C, we have asymmetric rotator or top which can be described as executing pure rotation about the axis of either the greatest or least inertia, while this axis makes a precession and a nutation round the axis of total angular momentum. It is a spinning top whose angle to the vertical is oscillating. The majority of molecules found in nature belong to this class. Simple triatomic molecules, such as. H_2O and SO_2 are *asymmetric tops*. The details of such a classification of molecules have been studied with the help of the observed molecular spectra and appropriate theoretical considerations. The available experimental data in this study have greatly increased in volume and importance with the advent of the microwave technique of analysis of pure rotational spectra of molecules.

Molecules have been classified into different categories with respect to the *vibrational motions* of the constituent atoms. Thus we have:

(i) The *triatomic linear molecules* with two sub-groups *viz.*, those with central symmetry, like CO_2 and CS_2, and those without central symmetry, like N_2O, COS, HCN, CICN, etc.,

(ii) *The triatomic triangular molecules* (YX_2) a special and interesting case of which is the type of molecule having an axis of symmetry that bisects the angle XYX, *e.g.*, H_2O, H_2S, O_3, NO_2, SO_2, etc.,

(iii) *The four-atom linear molecules*, ($Y_2 X_2$) a well-known example of which is acetylene (HCCH) that has also a centre of symmetry,

(iv) *The four-atom pyramidal molecules*, (YX_3) such as ammonia (NH,) in which the N atom is at the apex of the pyramid and the H atoms at the corners of an equilateral triangle form the base,

(v) *The five-atom molecules* (ZYX_3) *e.g.,* methyl halides (CH_3F, CH_3Cl, CH_3Br. CH_3I), and methane (CH_4) as *n* special case, where the H atoms are at the corners of a tetrahedron and the C atom at its centre, with a consequent high degree of 'spherical' symmetry, and so on. More complex molecules with a greater number of atoms may be regarded as made up of simpler molecules coupled together.

This method of differentiation of molecules is based on the analysis of the vibrational frequencies of molecular spectra. The general principle used may be stated as follows : If a molecule has n atoms and three degrees of freedom are assigned to each atom, the total number for the whole system is $3n$. Of these, three are accounted for by the translational motion of the molecule as a whole and three more are referred to its free rotation, The remaining $3n - 6$ degrees of freedom must therefore represent modes of vibration of the system. In the case of linear molecules, since only two degrees of freedom are involved in rotation, $(3n - 5)$ will be the degrees of freedom for their vibrations. According to this postulate, linear triatomic molecules ($n = 3$) should have four fundamental frequencies, triangular triatomic molecules three fundamental frequencies, four-atom linear molecules ($n = 4$) seven, pyramidal molecules six, five-atom molecules ($n = 5$) nine and so on. These internal modes of vibrations are substantiated with data obtained from the observed molecular spectra, due attention being paid to degenerate cases arising from symmetry conditions.

When atoms combine to form molecules, they bring with them their peripheral electrons and the governing factor in the stable formation of molecules is the valencies of the constituent atoms, determined by the number and nature of the electrons in their outermost incomplete shells. Hence a molecule also has a configuration of electrons which form one coherent system, as in the case of an atom, with closed shells, each containing definite number, as permitted by Pauli's principle; the electrons outside the closed shells act as the 'optical' and 'valency' electrons, responsible for the observed molecular spectra and the formation of more complex molecules, elements and compounds. But it appears that commonly the constituent atoms retain the electrons in their K-shells and that the electronic structure external to these is but regarded as a large number of shells, each containing two electrons only and each having a distinct ionisation potential or term value.

A very important consideration in molecular structure is to see how the individual separate atoms are bound together into a molecule. This is ordinarily known as *chemical binding*. Quite early in the study of the properties of chemical compounds, it was found necessary to distinguish two types of molecules. The first type, well represented by common gait (NaCl), forms electrically conducting aqueous solutions, in which the molecule splits up into ions. Even in the solid state, the charges are distributed unequally between the constituents, as seen from X-ray analysis of crystals like NaCl. Substances of this type usually have relatively high boiling points and typically saltlike properties. The second type, which is represented by HOI, BeO, CO, CN and organic compounds like methane (CH_4) and benzene (C_6H_6), is characterised by low electrical conductivity, relative volatility and complete lack of salt-like properties. Abegg, the chemist called the first type heteropolete and the second *homopdar*. It is to be noted that this classification fails to cover 'elementary' molecules as H_2, N_2, Cl_1, O_2, Cl_2, I_2 etc., which are non-polar. Nowadays, following Franck, according to whom the decisive criterion is whether a molecule in dissociating tends more readily to split up into ions or into atoms, the first type is called *ionic molecule* and the second *atomic molecule*. The above-mentioned elementary molecules can now be put under the category of atomic molecules. Consideration of the interchange of electrons by the constituent atoms in the formation of the molecule has enabled us further to designate the ionic type as *electrovalent* and the atomic type as *covalent*.

There are compounds which have intermediate types of binding; others cannot be assigned to either of the above two classes. Of these, the following two are now sufficiently well recognised:

(i) *Cohesion binding*, *i.e.*, loose binding without saturation of valency, due to the Van der Waals forces, and

(ii) *Metallic binding* which is effective in the lattice formation of metals. We shall first consider, somewhat in detail, the two prominent types, *viz.*, the ionic and atomic molecules and then state briefly the chief characteristics of the other two, as they are also of interest and importance from the physical stand point.

8.3 COVALENT LINKAGE

In 1923 Lews put forward his idea that, the formation of stable covalent molecule is frequently associated with the existence of an electron pair, which is such that each of the pair is shared by or belongs to both of the two constituent atoms.

On account of their opposite spins, the two electrons neutralise each other. Those electrons, which possess no partner in the above sense, may obviously become neutralised by pairing off with a new electron. It is these unpaired electrons which are responsible for chemical combination. Covalent binding usually occurs when the constituent atoms can exist in a state of high valency. An important example is a structure with a valence of four, as in carbon, silicon, etc., which occur in a crystalline form where each atom is surrounded by four nearest neighbours located at the cottiers of a regular tetrahedron. In this structure each of the four valence electrons of an atom enters into an electron-pair or covalent bond with electrons from nearest neighbouring atoms and all valances are saturated. As a result, such substances have high electrical resistance and are very hard and strong.

Assuming therefore that the covalent linkage arises from the sharing of an electron pair by the constituent atoms, we have next to inquire into the inner mechanism of such a sharing in order to understand the nature of the linkage involved. It cannot be electrostatic in nature, as in the case of ionic molecules, since, according to the classical theory, the pair of electrons should always repel each other and hence cannot form a link between the two atoms. Nor can the linkage be affected by the magnetic forces due to the spin of the electrons, which calculation ^shows to be too weak.

It may be recalled here that we met with a similar difficulty in connection with the large energy difference between the term-system of the helium atom and that Heisenberg solved it by invoking an *exchange energy of wave-mechanical resonance* that gives rise to a strong interaction between the two electrons. Heitler and London, in 1927, employing the same wave-mechanical resonance principle, were able to account satisfactorily for the covalent linkage in the H_2 molecule.

Heitler and London considered the Hg molecule, which is built up of two equal nuclei and two electrons, as the simplest case of covalent linkage that can be analysed with relative-ease. The structure of Hi molecule is assumed to involve the two nuclei, still separate, but surrounded by common electron atmosphere corresponding to the two electrons. The application of wave mechanical resonance to such a system is easily understood by the following analogy : If two electrical oscillating circuits, having the same frequency ν_0 are brought close to each other, the coupling throws them to some extent out of tune, ν_0 being split up into two different frequencies, one a little greater than ν_0 and the other a little lower, and beats are produced. The conditions in the hydrogen molecule are similar : the electrons revolving round the nuclei of the two separate hydrogen atoms correspond to the two identical oscillating circuits. When the two atoms approach close to each other and are coupled, the coupling puts them out of the tune a little, giving rise to two frequencies, one slightly higher than ν_0 and the other slightly lower. Since, to every frequency ν_0, corresponds an energy $h\nu_0$, the total undisturbed energy $2E_0$, of the two separate hydrogen atoms gives rise on combination to a somewhat lower and a somewhat higher energy of the coupled system.

8.4 DISSOCIATION OF A MOLECULE

In 1925 Franck attempted to explain the mechanism of photochemical dissociation. Later condon was able to develop Franck's ideas quantitatively.

In molecular spectra, three types can be distinguished: pure rotation spectra which lie in the far infra-red and are caused by quantum changes in the rotational energy, rotation-vibration spectra in the near infra-red, produced by quantum changes in both vibrational and rotational energies and electronic band spectra in the visible and ultra-violet, effected by quantum changes in the electronic, vibrational and rotational energies. Now, when molecular spectra result from *light absorption*, the continuous absorption region, which is an indicator of the dissociation of the molecule, is observed only in the third type, *i.e.,* electronic bands, but not in the other two, *viz.,* pure rotation or rotation-vibration bands. Theoretically speaking, any adequate supply of energy should give rise to dissociation even by the increased rotation of the molecule which is responsible for the pure rotation spectrum or by increased vibration of the constituent atoms, which causes the rotation-vibration bands. In practice, however, an electronic transition, which is responsible for the electronic bands, is required for the dissociation of the molecule by light absorption. Why cannot light-energy be absorbed as rotations or vibrational energy alone in sufficient quantity to dissociate the molecule?

The solution proposed is as follows : Molecules without all electric moment cannot be affected by the electromagnetic field of the incident light and hence their rotational energy cannot be increased-by absorption at all. Molecules with electric moment can be influenced by the incident light; but the selection rule governing the acquisition of rotational energy permits that only one rotational quantum can be absorbed, which is evidently not sufficient for the splitting up of the molecule. Concerning vibrational energy, molecules with no electric moment cannot increase their vibrational energy by absorption of light for the reason given above; but a molecule with electric moment which may be regarded as an unliarmonic oscillator, can, in theory, absorb more than one vibration quantum, unlike in the case of rotation; but changes in vibrational energy of increasing magnitude diminish rapidly in probability, so that only the first few vibrational quanta are absorbed, which cannot apparently lead to dissociation.

As regards rotational energy associated with electronic transition a niolecnie can also increase its rotational energy to the extent of dissociation. The final decomposition may, however, arise from two dissociation causes:

(a) *Mechanical Instability* : Oldenberg has given a strict criterion for the occurrence of mechanical instability, *viz.*, when the rotations are large and the energy of dissociation small, and has applied it to mercury hydride. The bands corresponding to transitions from higher electronic states to the fundamental state all break off at a definite rotational quantum (different for the different vibrations) of the lower state. These rotational levels all lie much higher than the energy of dissociation of the state, as can be shown from the potential curves.

(b) *Predissociation* : This term was first introduced by Henri, in 1924, to describe a phenomenon observed ia the absorption spectrum of sulphur (S_2). The rotational fine structure became suddenly diffuse at shorter wavelengths. This means that for values of vibrational energy equal to or greater than a certain critical value, quantisation of rotational energy ceases or becomes imperfect. Further researches by Henri and others showed that

   (i) The occurrence of predissociation in absorption spectra is common with both simple and polyatomic molecules,

   (ii) Predissociation may set in suddenly or gradually,

   (iii) In many oases more than one region of predissociation occurs.

8.5 MOLECULAR SPECTRA

In 1919 band spectra was observed by Imes. The theoretical foundations of the isotopic effect in band spectra were laid down in 1925 by Mullikan.

The Band Spectra of Elementary Diatomic Molecules

A particular significance is attached to the band spectra produced by such molecules and among them that of hydrogen occupies a special position. The spectrum of the hydrogen molecule does not resemble the usual band spectrum, but appears as a "*many-line spectrum*", stretching from the infra-red to the ultra-violet, the band-heads being absent altogether. The *Fulcher hands* have been known the longest of all, one in the red and one in the green, both having only a few lines: four bands discovered by Croze have about twelve lines each. But the sequence of the lines in the bands is so widely separated that the lines appear, at first sight, to have no relation among themselves. Merton, however, has tested these lines as regards their behaviour with varying pressure, temperature, etc. and shown that they belong to the same bands.

The reason for such an effect with the H, molecule is naturally due to the small moment of inertia which renders the intervals between adjacent lines in a band proportionately so great that they present the appearance of not grouping together into a band. This serves, as a limiting case in the general theory of band spectra. We have an instructive intermediate stage, between the many-line spectrum of hydrogen and the ordinary band spectrum, in the spectrum of helium, discovered by Goldstein and Curtis and measured for the first time by Fowler. Whereas in the hydrogen spectrum the band character seems to have disappeared entirely, in the helium spectrum it can still be recognised, but by no means so strikingly as in the case of the cyanogen bands.

Alternating Intensity in Band Spectra

Another peculiarity of the band spectra of diatomic molecules consisting of like atoms, *e.g.,* H^1, O^{16}, etc., is the alternating intensity of the rotational lines, *i.e.,* either every second line is missing or the lines are alternately strong or weak. Heisenberg correctly ascribed this phenomenon to a *wave mechanical resonance effect* of the two identical atoms and showed that if the nuclear spin vanishes, alternate lines drop out, whereas if there is a resultant spin, the lines will alternate in intensity.

Band spectra research has now developed into a science of considerable dimensions. For the interpretation of the extensive experimental data which have been collected, ingenious mathematical methods, such as group theory, consideration of symmetry, etc., have been called into service. The investigation and analysis of band spectra give valuable information regarding molecular structure, as we have already seen; they offer a very delicate and accurate means of studying the isotopic constitution of elements; they have formed also a convenient background for the study of the Raman effect.

Since the three principal components of the total internal energy of the molecule, *viz.,* the rotational and vibrational and electronic, depend upon the masses of the atoms that constitute the molecule, it follows that the isotopic nature of the atoms will affect the above mentioned component energies, modify the frequencies of the spectral lines and thus make its appearance in the band spectra by supplementary lines and bands separated and deformed from the normal lines and bands. Thus for instance, the doubling of each line in the rotation-vibration band spectrum of HCl can be shown to be due to the existence of two isotopes of Cl of masses 35 and 37. The isotopic effect will make its appearance in all the three types of band spectra, but not to the same extent or degree of importance since the dependence of the mechanism of production of each type on the masses of the atoms is not the same.

8.6 RAMAN EFFECT

When a beam of monochromatic light was passed through organic liquids such as benzene toluene etc., the scattered light contained other frequencies in addition to that of the incident light. This phenomenon was studied by Sir. C.V. Raman, while studying the scattering of light by liquids.

A round bottomed glass flask was filled with dust-free toluene or benzene and the liquid strongly illumined by the 4358 line from a mercury are, suitably filtered and concentrated by a lens. The scattered light was examined by means of a spectroscope placed transversely, *i.e.,* in a directional right angles to that of the incident radiation. In the spectrum of the scattered light, a number of new lines was observed on both sides of the main line. Those on the low- frequency side were more numerous and more intense than those on the high-frequency side. Most of these new lines were strongly polarised and their spacing was symmetrical about the main lino. They are now generally referred to as *Raman lines*, or more specifically, those on the low-frequency side as Stokes' lines and those on the high as *anti-Stokes' lines*.

Raman having thus observed a set of new discrete frequencies in the scattering process, contrary to the expectations of the classical theory, was able to establish that they constituted a new phenomenon, distinct from both the simple Rayleigh or coherent scattering and the more complex fluorescent scattering. He argued that in Rayleigh's scattering no frequency change was caused, but only some of the already existing frequencies were selected, whereas in the phenomenon under study new lines of different frequencies appeared even when a single frequency was scattered.

Nor could the observed effect be assimilated to fluorescence, although there was a certain amount of superficial resemblance between the two, in so far as, with a given exciting line, new lines made their appearance in both. For, in the *first* place, the frequencies in the fluorescent spectrum were always less than the incident frequency, while the Raman lines had frequencies both greater lines) and less than that of the incident line. *Secondly*, the frequencies of the fluorescent lines were really independent of that of the exciting line, provided the latter was able to produce fluorescence; but in the case of the Raman lines, their frequencies were directly related to that of the incident light, since if the latter was varied the former changed at the same rate. *Thirdly*, while the frequencies of the fluorescent lines were fixed by the nature of the scatterer, it was the frequency shifts of the Raman lines (known also as Raman frequencies) that were determined by the scatterer rather than the frequencies themselves. *Fourthly*, the lines observed by Raman were strongly polarised unlike the lines in the fluorescent spectrum. *Finally*, the most important fact that confirmed Raman's discovery as a new effect was that the Raman frequencies were either actual infra-red frequencies in the absorption spectrum of the scatterer or differences in such frequencies, which was not the case in fluorescent scattering. This fact proved further that the observed effect was a molecular phenomenon. Hence, the modified frequencies observed in the scattering process were a new type of secondary radiation, to which the name *Raman effect* was given and was initially considered as the *optical analogue* of the *Compton effect*, since both belong to the same category of *incoherent scattering* of radiation, which can be explained only on the basis of the quantum theory.

The Raman effect has been extensively studied by a great number of workers. The *general technique* used in these researches is to illumine the substance under investigation with an intense roonochrottiatic source of light and photograph the scattered radiation by means of a spectrograph arranged in a transverse direction. But the technical details vary according to the nature of the substance under test, *i.e.,* liquid, solid or gas, the chief purpose being to obtain best and quick results.

The original simple arrangement of Raman was not quite efficient and required very long exposures of about hundred hours and more to obtain good records of the Raman spectrum. Hence, improvements were made as regards the container of the substance, the source of radiation, filter, spectrograph, etc.

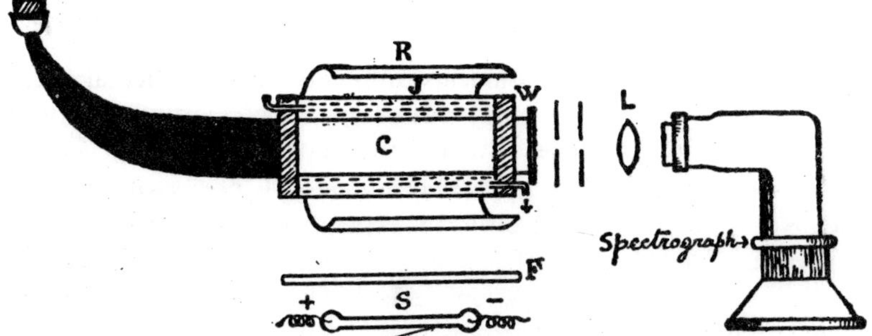

Fig. 8.1 : Apparatus for the Study of Raman Effect.

An ideal *source* S would be light from a helium discharge tube filtered by nickel oxide glass, giving a strictly monochromatic line of wavelength 3888 Å. But on account of the many technical difficulties involved in the construction and manipulation of this source, it is not widely used. The source ordinarily employed is the mercury arc, the next" beat available, from which it is possible to get single wavelengths by the use of suitable filters. Thus, for instance, to obtain the 4358 line, slightly acidulated quinine sulphate solution contained in a novial glass vessel is used as filter, which nuts off all the other lines except 4358Å. To get the 4046 line a solution of iodine in carbon tetrachlorido contained in a novial glass coil is "found to be a very satisfactory filter. The filter solution may be arranged either to surround the Raman tube or in front of the arc. The mercury arc is placed as, close to the Raman tube as possible, which results in large intensity of the incident light. A semi-cylindrical aluminium reflector R enhances the intensity of illumination still further.

The chief features of a *spectrograph*, suited for the study of the Raman spectra, are:

(1) large light-gathering power,

(2) special prisms of high resolving power, and

(3) a short-focus camera. A lens L in front of the piano window W directs the scattered radiation upon the slit of the spectograph which is carefully aligned along the axis of the Raman tube and screened from the direct rays of the arc. The intense Raman lines of a liquid such as CCl_4 can be photographed in about an hour with a small spectrograph, but the recording of the complete spectrum may require up to ten or fifteen hours, depending largely on the intensity of the incident light, the speed of the

spectrograph and the intrinsic brilliance of the Raman lines. It may be noted that instruments of high resolving power such as gratings are not used with advantage, on account of the poor luminosity which necessitates long exposures.

The intensity of the light scattered from gases is very weak but this difficulty has been overcome by intense illumination of the gas under high pressure and the use of spectrographs of great light-gathering power. Wood employed a very long tube of HCl gas and obtained its Raman spectrum at atmospheric pressure using a specially made mercury arc, which was placed in contact with the gas tube, hollow, cylindrical reflectors enclosing both of them. The illumination produced in this way was very intense as the light from the arc was returned back and forth between the walls of the reflectors. Rasetti was the first to develop the technique of exciting Raman effect in gases under high pressure, which shortened considerably the time of exposure. He used a thick-walled quartz tube, 20 cms. long and 2-2 cms. internal diameter, which could withstand pressures of 10 to 15 atmospheres. With the 2537 line of the mercury ate as the exciting radiation, he was able to obtain the Raman spectra of several gases under pressure.

Fig. 8.2 : Dr. S. Bhagavantam

It is seen that a number of new lines and bands, exhibiting a variety of characters of intensity, width, polarisation and fine structure are recorded on either side of the exciting radiation. There is also some unresolved continuous radiation, which generally appears as wings extending slightly unsymmetrically on either side of the parent line. This continuous spectrum shows great variations in intensity with different substances. Each line in the incident spectrum, if of sufficient intensity, gives rise to its own set of lines or bands and associated continuous spectrum. We shall now outline the main results obtained from researches made on the effect with such photographic records.

About a hundred liquids, so far examined, show the phenomenon in an unmistakable manner. The frequency shifts of the Raman lines produced with benzene correspond to an infra-red wavelength 3.27μ, in which region benzene exhibits a strong band in its absorption spectrum. A close examination, however, of the infra-red absorption spectrum and the Raman spectrum of benzene shows that none of the Raman lines are represented in infra-red absorption and *vice-versa*.

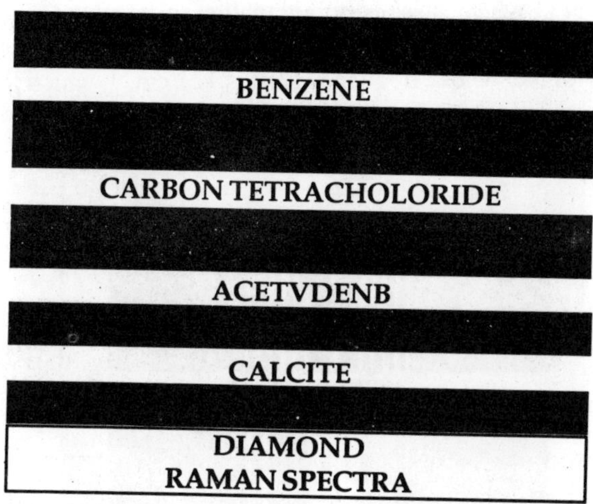

Fig. 8.3 : Raman Spectra of Different Substances (Bhagavantam).

Solutions of salts in water give the Raman up extra characteristic of the salts and the water. Goneaan and Venkateswaran found that the bands due to water in aqueous solutions of H_2SO_4, HCl and HNO_3. acids become sharper with increasing concentration. The similarities exhibited by solutions of carbonates of different metallic radicals and the similarity of the sulphates and nitrates appear to support the view that the characteristic frequencies are those of the ionised acid radical.

Oxygen, hydrogen and nitrogen give a pattern of equally-spaced lines. The intensities of the individual lines alternate in the case of nitrogen and hydrogen, while with oxygen the alternate lines are absent, as seen in the adjacent figure, whore the Raman spectra of oxygen and nitrogen, obtained by Rasetti, are reproduced. These peculiarities in intensity of alternate lines have led to very significant conclusions as regards molecular structure and nuclear spin. McLennan carried out a series of experiments on the Raman effect with liquid oxygen, nitrogen, hydrogen and nitrous oxide. His results with liquid oxygen suggested that the normal mode of vibration of the molecule is the one involved in the production of the four Raman lines observed, while those for liquid hydrogen supported the view that hydrogen at low temperatures must be regarded as a mixture of two distinct types of molecules known as the *para* and *ortho* hydrogen. The Raman spectrum of CO_2 which is similar to that of oxygen indicates a symmetric structure for its molecule, while that of N_2O where there is no alternation in intensity of the lines suggests an unsymmetric molecular structure, Rasetti has observed a Raman line with NO whose frequency shift is 121 cm.$^{-1}$, which has been classified as having an electronic origin.

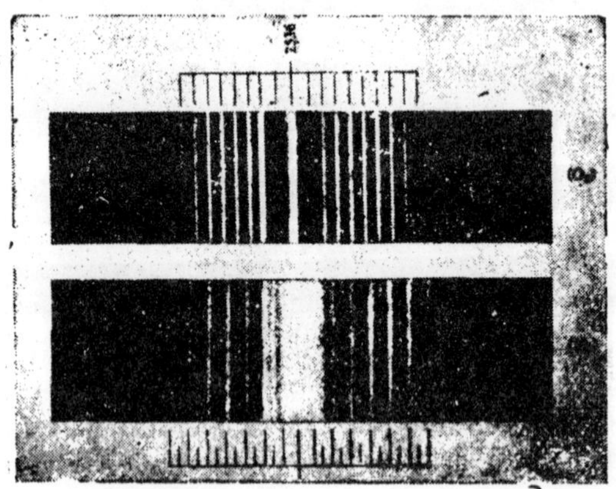

Fig. 8.4 : Raman Effect in Oxygen and Nitrogen (Rasetti).

On account of the numerous experimental difficulties, comparatively a few solids only have been studied so far. Mandelstam, Landsberg, Baer and Menzies were the pioneer workers in this- branch of Raman effect Other investigators were Schaefer, Matossi and Aderhold, Miss Osbome, Cabannes and Canals, Nedungadi and Bhagavantam. The substances analysed are gypsum, quartz, calcite, sodium nitrate, potassium nitrate, ammonium phosphate, ammonium chloride, diamond and a few others. The Raman lines obtained with crystals are sharp, becoming diffuse with rise of temperature. In calcite ($CaCO_3$) two lines which are nearest to the parent line have been definitely identified with the oscillations of the crystal lattice, while the others are due to the vibrations of the CO_2. groups. With gypsum ($CaSO_4$, $2H_2O$) which contains two molecules of water of crystallisation, Krishnan found, in addition to the wavelengths which could be attributed to the SO_4 radical, three sharp lines at $\lambda = 2.8\mu$, 2.9μ and 3.0μ; which are evidently due to the water of crystallisation and are practically in the same position as the components of the band observed with water. Diamond exhibits a strong and sharp line of a comparatively large frequency shift, which has to be ascribed to a lattice oscillation. In solid benzene, the intense continuous spectrum obtained with liquid benzene is replaced by bands.

Intensity of Raman Lines

The experimental determination of the intensity of Raman line is beset with many difficulties on account of their extreme weakness. The intensity of a Raman line, when expressed as a fraction of the parent line, is usually a few hundredths in liquids and a few thousandths in gases. No accurate data are available as regards the absolute intensities of Raman lines in liquids or gases. More reliable results are obtained in the determination of relative intensities of a given set of Raman lines excited by a given parent line. To obtain good results, photographic plates which have great and uniform sensitivity in the region of investigation should be used. The densities obtained for the lines under comparison are measured with a microphotometer. On each plate, a set of calibration spectra are recorded, with the help of which density curves for each wavelength can be drawn, and the intensities corresponding to any density can be obtained from such curves. The measured intensities should be corrected for various causes of error, such as absorption in the body of the substance in the Raman tube, the oblique refraction at the prism surfaces in the spectrograph, etc.

In spite of the numerous experimental difficulties, results of fundamental importance have been obtained. Very faint Raman lines as well as very intense ones are met with, *e.g.*, the band in NaCl recorded by Rasetti and the anti-Stakes' line in diamond obtained by Bhagavantam are very faint, while the principal Raman line of diamond at 1332 and the line in benzene having a frequency shift of 992 are very intense. These variations in intensity as we pass from line to line and from substance to substance are of great significance in the study of molecular structure and chemical constitution.

Since the intensity of scattering increases with the fourth power of the frequency, it is advantageous to use, where the absorption of the substance under investigation permits, light of as short a wavelength as possible. Hence the ultra-violet mercury resonance line 2537 Å is often used. The Stokes' lines are always more intense than the corresponding anti-Stokes' lines. The anti-Stokes' lines grow more intense and, in addition, all the Raman lines move inward towards the parent line as the temperature is increased.

Just as the Raman lines vary greatly in their intensities, so also their states of polarisation. The fact that different lines are differently polarised is probably connected with their relative intensity.

The experimental arrangement used for determining the polarisation of Raman lines is essentially, but with the following modifications. The light from the source is concentrated by means of a condenser into the substance contained in the Raman tube. A suitably oriented double image prism whose function is to separate the vertical and horizontal components in the scattered light is placed in front of the slit of the spectrograph, so that two images, one above the other, are formed on the slit, which are simultaneously photographed.

The state of polarisation of a Raman line is measured by a quantity known as the *depolarisation factor* which is simply the ratio of the intensities of the horizontal and vertical components when the incident light is vertically polarised. This ratio is readily obtained from the traces photographed as described above, by one of the usual methods employed for comparing the intensities of two beams of the same wavelength. In order to get fairly accurate values of the depolarisation factor, the following precautions should be taken :

(i) Crystalline quartz should not be used for condenser, spectrographs or windows, since its optical activity complicates the phenomenon of polarisation;

(ii) The window of the Raman tube through which the scattered light emerges should be strain-free and plane,

(iii) Errors arising from oblique refraction at the prism surfaces, want of transversality in the incident beam and slit width should be eliminated.

From the many and varied experimental data, it is clear that the Raman effect is a molecular phenomenon. In the case of free molecules scattering light, three different kinds of Raman effect can therefore be distinguished, viz; a rotational effect, a vibrational effect and an electronic effect. A mixed "rotation-vibration" effect can also take place under certain circumstances. Solids can exhibit yet another type of Raman effect in which the crystal lattice as a whole takes the place of the molecule.

Since the rotational energies involved are small relative to the vibrational energies, the pure rotational Raman lines lie correspondingly closer to the parent line and are often masked by the intense light of the parent line on the photographic plate. They are obtained separately only in the case of certain light gases, such as hydrogen, deuterium, oxygen, nitrogen, etc. In heavier gases, the lines are much closer and instruments of greater resolving power are required to separate them.

Most of the observed Raman lines and bands with moderate or large frequency shifts are due to the vibrational effect, corresponding to various normal modes of vibration of the molecule or the crystal lattice. The fainter lines may be due partly to overtones or combination tones. The lines arising from an oscillating crystal lattice are characteristic of only the solid state and are not present in liquids or gases.

Raman lines due to the electronic effect are rarely observed, as in the single case of NO obtained by Rasetti.

Analysis of the continuous spectrum under high dispersion reveals that it cannot be separated from the parent line and that its intensity is maximum near the parent line. Although it is difficult to decide upon the exact nature and origin of the continuous spectrum in all oases, very probably it is due to an unresolved rotational effect.

In the solid state, it is replaced by broad bands, while in the gaseous state by separate lines, which may be ascribed to the rotation of the molecule. In certain cases the broad bands which replace the continuous spectrum have been identified with the oscillations of the crystal lattice

The rotational and vibrational Raman spectra closely resemble the far infra-red and near infra-red absorption spectra of molecules respectively. Further, in many cases the Raman frequencies are equal either to the actual infra-red frequencies of the absorption spectrum of the scatterer or to the differences of such frequencies. But there seems to be little or no correlation between the relative intensities of the Raman lines and the intensities of the corresponding infra-red bands. In addition, not all Raman lines have their corresponding infra-red bands, nor all infra-red bands their corresponding Raman lines. Thus the most intense Raman lines in benzene and toluene, whose frequency shifts are $10^{-3}\mu$ and 10.2μ respectively, have no corresponding intense infra-red bands. On the other hand, the intense absorption bands of benzene and toluene at wave-lengths 9.75μ and 6.86μ respectively have no corresponding Raman lines. Raman lines have also been observed corresponding to infra-red transitions which are forbidden by the selection rules, as in the case of HCl gas, already cited. In such cases, Raman effect provides a very valuable complement to the study of infra-red absorption spectra in the detection of energy levels of molecules.

8.7 IMPORTANCE AND APPLICATIONS OF THE RAMAN EFFECT

The Raman effect is of immense practical importance on account of its many useful applications in physics and chemistry. The Raman effect is a very convenient and powerful tool of research in problems concerning the intimate structure of matter, chiefly as regards the constitution of molecules, their number arrangement and motion in gaseous liquid and solid states.

The Raman effect has been put to very great use in the study of the structure of molecules. Raman spectra are, in general, determined by those factors which affect most the nature of vibration, *viz.,* the number of atoms in the molecule, the masses of the atoms and the strength of the chemical bonds between the atoms.

Taking a diatomic molecule, its natural frequency of vibration is given by $\nu_0 = (1/2\pi)\sqrt{f/\mu}$, where F is the restoring force per unit displacement and μ the resultant mass. This means that:

(a) There is only one vibration frequency in diatomic molecules,

(b) A molecule containing light atoms should have a higher frequency than one containing heavier atoms, and

(c) A molecule in which the force binding the atoms is great should have higher characteristic frequency than one in which the force is weak. This force depends on the nature and strength of the inter-atomic bonds. Thus a diatomic molecule with a double bond should have a higher frequency than one containing only a single bond. The Raman lines are also expected to appear with great intensity when the bond is of the covalent type (homopolar or electrically non-polar molecules) while the reverse is the case when the bond is of the electrovalent type (polar or hoteropolar molecules). The reason for this is to be found in the fact that the Raman lines essentially depend upon the symmetry of the molecules and the extent to which the polarisability is affected by the oscillations. In covalent molecules the binding electrons remain common to the nuclei, so that the polarisability of the molecule is considerably modified by nuclear oscillations and this variation in polarisability gives rise to Raman lines. In electrovalent molecules the binding electrons definitely change over from one nucleus to the other in the formation of the molecule so that the polarisability of the molecule is little affected by nuclear oscillations, which means no Raman lines will occur.

Considering *polyatomic molecules,* which are constituted by more than two atoms, the Raman spectra will naturally be much more complex. First of all, more than one characteristic frequency is to lie expected : *e.g.,* triatomic molecules will, in general, have three such frequencies. Secondly, the arrangement of atoms, such as symmetrical or unsymmetrical, linear or non-linear symmetry, etc., is an additional factor by which the intensity of the Raman lines is essentially determined. Hence, from the number and intensity of the observed lines in the Raman effect, in conjunction with infra-red data, it is possible to draw important conclusions about molecular structure. Theory leads to the following important rule, known as *the rule of mutual exclusion.* For molecules with a centre of symmetry, transitions that are allowed ia the infra-red are forbidden in the Raman spectrum and *vice-versa.* This rule implies that for molecules without a centre of symmetry there are transitions that can occur both in the infra-red and the Raman effect.

The Raman effect offers us a new method of studying crystals, which, in a sense, is complementary to X-ray diffraction. For, it furnishes us with just the kind of information that X-rays do not Sir give concerning the strength and nature of the forces which hold the crystal together. Furthermore, this information is of a precise character, as the Raman lines are very sharp and can be measured very accurately. The frequencies deduced from the Raman spectra and the known positions and masses of the atoms in crystals enable us to determine the binding forces in the crystals.

Although, the study of Raman effect in crystals has been initiated only recently and has not much advanced till now, yet some important, results have already been obtained. The case of diamond studied, by Ramasamy and Bhagavantam, showed clearly that it is not necessary for a crystal to contain molecules in order that it should exhibit Raman effect, for diamond is a non-polarsubstance in which it is impossible to identify any particular group of atoms as forming an ion or molecule. The Raman spectra of a series of crystalline nitrates analysed by Krishnamoorthi and Gerlach established that the frequency characteristics of the NO_3 ions are notably influenced by the presence of the kations. This suggests that the assumption of complete ionisation in the solid crystal may be invalid.

The Raman effect in crystals has been studied also with reference to two other fundamental problems in crystal physics, *viz.,* the existence of a "*mosaic*" *structure* and *thermal agitation* in crystals. From the

Raman spectra of crystals, the existence of discrete mono-chromatic infra-red frequencies has been clearly demonstrated. In a quantitative study of the phenomenon, Raman has shown that a crystal has, in general, $(24p - 3)$ fundamental modes of atomic vibration with monochromatic frequencies, where p is the number of interpenetrating atomic lattices of the crystal. Of this number $(3p - 3)$ modes may be ascribed as oscillations of the p interpenetrating lattices, while the remaining $21p$ modes are oscillations with respect to each other of various important planes of atoms in the crystal. The number of monochromatic frequencies is shown to be considerably reduced when the crystal has a high degree of symmetry. But the number of frequencies and the geometry of the modes have been fully worked out from considerations of symmetry. For instance diamond has been shown to have eight fundamental frequencies, rocksalt nine, and fluorspar fourteen. The fine structure of the infra-red spectrum of crystals actually observed is readily and adequately accounted for on the basis of the above-mentioned analysis.

The Raman effect has been applied also in the study of certain aspects of nuclear physics, such as the spin, *statistics* as well as the *isotopic* constitution of the nucleus. The rotational Raman lines of homonuclear diatomic molecules like H_2, D_2, O_2, N_2, etc., characterised by their alternating intensities, have been most fruitful in giving precise results. Thus, for instance, in the case of *hydrogen*, the Raman lines arising from odd rotational levels are three times more intense than those coming from transitions between even levels, which can be shown to be due to the fact that the hydrogen nucleus has a spin of half a unit, obeying Fermi-Dirac statistics and that the statistical weight of the odd rotational levels is three times that of the even rotational levels. With *deuterium*, the stronger of the alternating Raman lines correspond to transitions between even levels and their intensities are twice that of the weaker ones arising from odd levels. These observed data lead to the results that the deuterium nucleus has a spin of one unit, conforming to Bose-Einstein statistics, and that the statistical weight of the even levels is twice that of the odd levels. In the case of oxygen the entire rotational spectrum consists of lines arising from transitions between odd levels only.

The study of the Raman effect in a large number of the halides by Krishnamoorthi has established beyond doubt that Raman lines appear with great intensity in covalent compounds while the reverse is the case in electrovalent compounds. Thus, for instance, NaCl, KCl, etc., which are electrovalent compounds, exhibit no Raman lines corresponding to their infra-red frequencies, whereas HCl, the chlorides of non-metals, etc., which are covalent compounds, give vibrational Raman lines.

It is of interest to note that the vibrational Raman line is observed in HCl, gas and liquid, but not in aqueous solution of HCl, which consists of hydrogen and chlorine ions. In aqueous solutions of salts, where ionisation is complete, one finds no Raman lines characteristic of the salt as a whole but only those corresponding to the molecules of the separated ions and water. These facts confirm what we have stated above about the dependence of the intensity of the Raman lines on the chemical nature of the linkage, *viz.,* Raman effect appears most strongly when the atomic structure under consideration is bound by covalent bonds and tends to weaken when the bonds change their character and pass into electrovalent linkages. The fact that most of the organic compounds, characterised by covalent linkages, give rise to strong Raman spectra may be considered as further evidence in support of the above conclusion.

The strength of chemical bonds is also duly reflected in the Raman effect, so that, calculating the force constants with the help of the Raman frequencies, the chemical bonds can be classified into triple, double and single.

Complex compounds, mixed molecules and water of crystallisation are some of the other important problems which have been tackled with the help of the Raman effect.

The Raman effect has been found of great value in organic chemistry also. It has been used in determining the presence or absence of specific linkages in a molecule, identifying impurities of certain types, estimating quantitatively the relative proportions in which the constituents of a mixture are present fixing up the structure of important molecules and studying the different types of isomerism. It is interesting to note that the chemist's classification of organic substances into aliphatic or open chain compounds and *aromatic* or closed chain compounds is clearly reflected in the character of the respective Raman spectra.

Very many interesting problems appertaining to this branch of chemistry, such as the transition from crystalline to amorphous state, electrolytic dissociation, hydrolysis, etc., have been investigated with the Raman effect. Some of the more important results obtained are:

Although, there is a general correspondence between the Raman spectra obtained in the crystalline and amorphous states of a substance, the latter state gives rise to broad and diffuse bands in the place of sharp lines produced by the former.

Raman effect affords an ideal method for investigating the phenomenon of *electrolytic dissociation*, as spectra can be obtained both for the pure substance and for the aqueous solution of it at different concentrations. The observed positions and the variation of intensity of the lines enable one to determine directly the nature and number of ions produced and thus decide whether the dissociation is complete or only partial.

The problem of *hydrolysis* is similar to that of electrolytic dissociation. If a salt, possessing a Raman spectrum different from that due to the base or the acid, is hydrolysed by water, the degree of hydrolysis can be readily estimated by measuring the relative intensities of a set of lines characteristic of either the base, the acid or the salt.

Thus the Raman effect has come to stay as a powerful weapon in the hands of scientists for the analysis of those intricate phenomena which reveal the intimate structure of matter.

9

POSITIVE RAYS

9.1 INTRODUCTION

A German physicist Gold stein was the first to investigate the nature of the positive ions. He studied that, if there was a small hole in the cathode of the low pressure discharge tube, then a luminous, ray was, observed behind the cathode. So he called the rays as 'Canal rays'.

These rays, consist positively charged particles. So, later, they were named as *positive rays.*

Since the mass of the electron is very small, the mass of the positive ion will be almost equal to the mass of the atom.

Additional interest is attached to positive ray analysis, as it has been able to settle decisively between a clash of opinions among scientists regarding the problem whether all the atoms of any one chemical element are identical in all respects including their mass. Dalton, the founder of the atomic theory, thought that the various laws of chemical combination could be readily explained by supposing that each element consisted of identical atoms. Prout, on the other hand, with the very enticing idea of building all elements from a single primordial element, hydrogen, and of expressing the atomic weights of different elements by integral multiples of that of hydrogen, taken as unity, suggested in 1815 that the atoms of a given element need not be identical in mass and that the chemical atomic weight of an element is only an average value of the different weights of the atoms" of that element. Careful determinations of atomic weights, however, showed that the atomic weights of many elements were not whole number multiples relative to that of hydrogen. Prout's hypothesis was therefore discarded for some time in favour of Dalton's. But with the discovery of natural radioactivity towards the end of the 19th century and the study of the radioactive elements produced in the process, sufficient experimental evidence was available to support Prout's idea that the atoms of an element need not be identical in mass.

As a matter of fact several elements were found in the radioactive series having identical properties but different atomic weights and the name of isotopes was given to them, since they would occupy the same place in the periodic classification according to chemical properties. Efforts were then renewed to find whether atomic masses could be represented by whole numbers and whether there were isotopes among the non-radioactive elements also. It was at this point the positive ray analysis entered the field with a definite solution of the problem in favour of the "whole number rule" and isotopic constitution of elements.

J.J. Thomson initiated this line of research in the year 1911 by his classical experiments on the electric and magnetic deflection of positive rays, ordinarily known as the *parabola method* of positive ray analysis. Aston, at the Cavendish Laboratory, Cambridge, took over the problem from Thomson, considerably improved the technique of Thomson and produced in 1919 the first of an extremely refined apparatus, known as the "*mass spectrograph*" for positive ray analysis and study of isotopes. For years he was the unique worker in the field, until Dempster, Cambridge and others brought in new improvements to the mass spectrograph, so that today the isotopic constitution of elements as well as the masses of single isotopes can be determined with extremely great precision.

The mass spectrograph, which derives its name from the optical analogue, is a veritable atomic balance designed to separate the different isotopes of an element and measure their individual masses as well as their relative abundances. By the use of electric and magnetic fields this instrument separates ions of different values, where *e* is the fundamental charge and M the mass of the ion. Positive ions having a common but differing in direction and velocity within certain limits are brought to a common focus. A variety of focusing schemes has been utilised, some resulting in velocity focusing, others in *direction* focusing; still others aimed directly at achieving a *linear mass scale*. So much success has attended these efforts that the conquests yet to be made are few. Aston has had a giant share in this laborious and delicate work till his death in 1946. As a result, it is now established that the different isotopes of any element have different masses, though identical in other properties, and enter into the composition of the element in different proportions and practically in every case the masses of single isotopes deviate slightly from the whole number rule, which fact has led to fundamental discoveries of great importance concerning the intimate structure of the atom,

9.2 SOURCES OF POSITIVE RAYS

Several methods are employed to secure positive rays. The method depends upon the nature of the substance under test.

Aston used spherical glass bulb discharge tube, to get positive rays.

The ordinary *spherical glass bulb discharge tube* was first used by Aston as a very decent source of positive rays, chiefly in the case of elements which could be conveniently had in the gaseous or *vapour form*. It had a concave aluminium plate perforated at the centre for the cathode, an aluminium wire in a side tube for the anode and a silica bulb mounted at the end opposite to the cathode in order to prevent the intense beam of cathode rays from striking the walls of the discharge tube and melting them away. The gas under investigation was introduced into the discharge tube through a narrow annular space near the cathode. The discharge was maintained by a powerful induction coil, the special shape and position of the electrodes making the bulb act as its own rectifier. The arrangement worked smoothly at current values of 0.5 to 1 milliampere and potentials of 20,000 to 50,000 volts.

The spherical discharge tube was perfected later by Aston by the following modifications. The cathode C was a solid hemisphere of aluminium, one inch in diameter, drilled with a hole of about 3mms. and shaped to a special curved form in front as shown in Fig. 9.1. This form although slightly less efficient was preferred to the concave plate type of cathode, because it was found to be less sensitive to change in position and did not require a special anti-cathode to dissipate the energy of the cathode rays. For, these rays were given off from the specially shaped cathode in two groups, a diffuse wide cone striking the walls at ED and a much more intense beem emerging from the cathode cavity as a cone FG, the energy of which was dissipated in the long cylindrical portion of the bulb which tapered and terminated in the anode A. The substance under test, if gaseous or of high vapour pressure, was admitted into the tube through a fine glass leak behind the cathode. Materials of low vapour pressure were introduced through a small stop-cock. Solids with high boiling points were placed in a glass bucket B and lowered through a vertical tube T into the anode neck, where they were vaporized by the heat of the cathode rays. These improvements made the spherical discharge tube produce a highly concentrated beam of positive rays at sufficiently low pressures, provided the cathode was properly set. The arrangement could be worked at 1 milliampere and 40,000 volts for a long period of time, without any artificial cooling of the bulb being required. Aston used this type of source of positive rays for practically the whole of his researches on mass spectra, with excellent results.

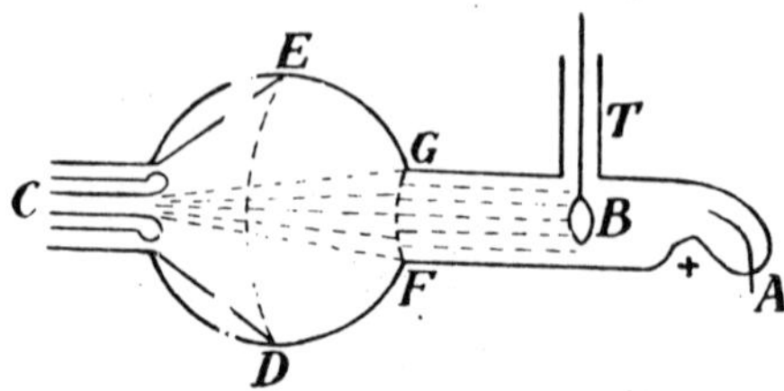

Fig. 9.1 : Source of Positive Rays.

Cylindrical discharge tubes with the end opposite to the cathode tapering (meant to dissipate the energy of the cathode rays) and the anode in a side tube, have also been used in the case of gaseous substances, chiefly in the early researches, by Goldstein, Thomson and others. They give a very good concentration of positive rays at the central hole of the cathode, much better than the spherical bulb type, chiefly at higher pressures and so is very effective for many kinds of research. In the actual working, however, instead of a solid beam, only a hollow cone of positive rays is obtained. Further, the adjustment of the tube for optimum working is very laborious and difficult, but when once this is secured, the arrangement functions very satisfactorily for a considerable time.

When the substances under investigation can be obtained only in *solid form not readily vaporisable*, as the majority of metallic elements, the technique described above is unsuitable. Other devices have therefore been employed where the substance is placed in some form or other in the anode, which itself produces the required positive rays. On account of this fact, the positive rays obtained are known as *anode rays*. Two distinct techniques, known as the hot anode and the *composite anode*, are in use.

The "*hot anode*" method consists in using for the anode a platinum strip which is coated with a salt of the metal under study and electrically heated. This anode is mounted close to the cathode about 1 cm. away, just in front of the perforation in the cathode. Heating the anode to a dull red heat and establishing the discharge with a large induction coil and a rectifying valve with a steady current value of 1 to 2 milliamperes at 20,000 volts, with the tube highly exhausted, a good stream of anode rays is produced. Owing to the very low pressure the rays are not visible, but could be inferred from the scintillations they produce at the cathode. Study of the mechanism of the production of positive rays by the hot anode has revealed that they originate not from the surface of the salt but from that of the platinum strip. A great drawback of this method is that the salt disappears very rapidly, so that it has to be replaced frequently. This method has been used with some success by G.P. Thomson and Dempster for Obtaining the anode rays of alkali metals.

In the "*composite anode*" method a paste made with graphite and a salt of the element under study is pressed into a hole drilled in the end of a small steel cylinder and the composite surface thus formed is used as the anode, which is automatically heated by the discharge itself, a concentrated beam of cathode rays from a concave-shaped-cathode falling on it. This arrangement produces a highly intense beam of anode rays. A speciality of this method is that the rays can be accelerated after production by a high P.D. with we aid of a subsidiary cathode and a kenotron. The mechanism of the production of positive rays in this device is not easily understood, since the actual working is often capricious and unaccountable. The method appears to be capable of wide application to metallic elements, even when the latter are not quite pure or could be obtained only in very small quantities such as the rare earths.

9.3 ANALYSIS OF THE POSITIVE RAYS

J.J. Thomson was the first man to determine, the specific charge of the positive rays.

The discharge tube in which the positive rays are produced is a large flask A, so as to get a steady discharge. The cathode B is placed in the neck of the flask. Its face is made of aluminium and is pierced at the centre by a brass tube of narrow bore of about 0.1 mm. as a continuation of the cathode. The cathode is kept cool by means of a water jacket C. The anode D is an aluminium rod, generally placed for convenience in a side tube. In order to ensure the supply of the gas under test, a steady stream of it is allowed to flow in through a capillary tube E and after circulating in A is pumped off at F. By adjusting the speed of the pump, the pressure in the discharge tube is maintained at any desired value, usually arranged so that the discharge potential is 30,000 to 50,000 volts. Under this high potential positive rays fly towards the cathode and those falling axially pass right through the fine tube, emerging as narrow beam.

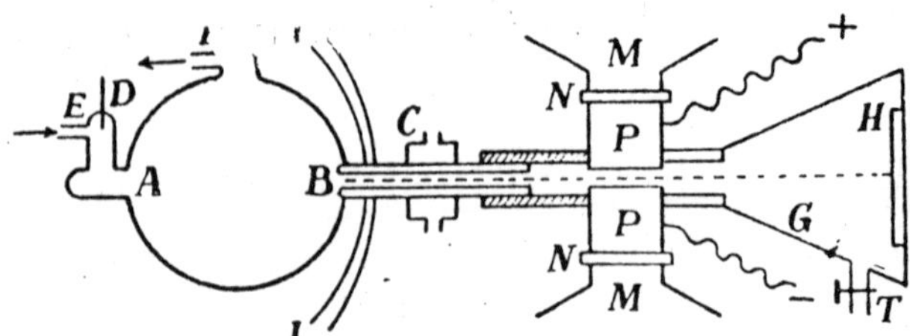

Fig. 9.2 : Thomson's Apparatus for Positive Ray Analysis.

This beam is subjected to parallel electric and magnetic fields simultaneously by the following device. Two soft iron pieces PP placed between the poles MM of a powerful electromagnet and electrically insulated from it by thin sheets of mica NN constitute at the same time the pole pieces of the magnet and the means of establishing an electrostatic field. By raising PP to any desired potential difference with the leads shown in the figure the deflecting electric field is obtained in the narrow gap on the path of positive rays. The deflecting magnetic field can be established in the same gap since PP form also the pole-pieces of the electromagnet. Thus electric and magnetic fields, parallel to each other, and perpendicular to the path of the rays can be made to act on the beam of positive ions. After passing through these fields the beam enters a highly evacuated camera G and is received on a photographic plate H. Owing to the very fine bore and length of the brass tube at the cathode, diffusion of gas is greatly prevented from the discharge tube to the camera, and the latter can therefore be maintained at a very low pressure by continuous pumping through charcoal immersed in liquid air at T, in spite of the relatively high pressure in the discharge tube A. This precaution is necessary, since the positive beam should not be disturbed by collisions with the molecules of the residual gases in the camera. In order that stray magnetic fields may not interfere with the discharge, shields of soft iron II are interposed between the magnet and the discharge tube. The photographic A plate when developed shows a series of parabolas as in the above photo, from which the masses of the positive ions that traced them can be determined.

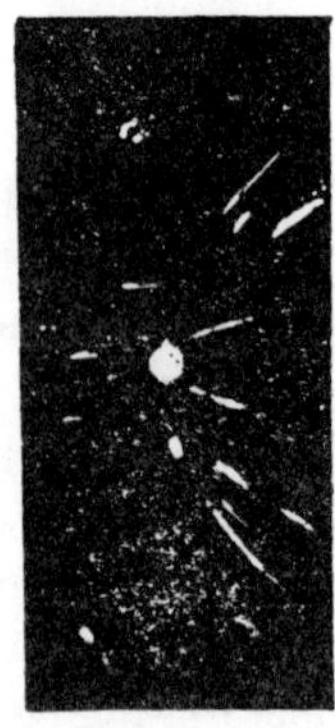

Fig. 9.3 : Positive Rays Parabolas (J.J. Thomson).

In the method of analysis of the parabolic traces, sufficient amount of hydrogen is always present to give a parabola on the plate. The analysis are calculated, interms of mass of hydrogen.

But the parabola method has number of limitations. In this method, many rays are lost by collision in the narrow tube. The traces on the photographic plate are blurred with no definite edges, so it becomes difficult to make accurate measurements. The influence of secondary rays makes analysis difficult. Some time the positive under test may make collisions with each other.

9.4 MASS SPECTROGRAPH

In 1918, Dempster built a mass spectrograph, suitable for analysis of substances, available in the solid state. After some years, that is in 1930, Brainbridge who was working, in the laboratory of the Bartel research foundation, designed a mass spectrograph.

Let us study the spectrographs designed by Dempster as well as Brainbridge.

1. Dempster's Mass Spectrograph

A metal cylinder A with its front surface C coated with a salt of the element under test and electrically heated serves as the anode. A filament *ff* electrically heated by the battery B_1 emits electrons. Maintaining the filament at a P.D. of 30 to 60 volts with respect to A by means of another battery B_2 the electrons bombard the heated salt with the result that the anode emits positively charged ions of the element. These ions are collimated into a fine pencil by the slit S_1 and then accelerated towards the slit S_2 by means of a high adjustable P. D. (800 to 2,000 volts) maintained between S_1 and S_2. On emerging from S_2 the ions accelerated *pradically to the sane velocity* enter a uniform and strong magnetic field acting perpendicular to the plane of the paper, produced between two semicircular iron pole-pieces of an electromagnet. Under the influence of this magnetic field the ions are swung round in a semi-circle towards the slit S_4 defined by the three slits S_2, S_3 and S_4. After passing through 84 the ions fall upon tbe-metal plate P which thus acquires a positive charge at a rate which can be determined by connecting P to a quadrant electrometer E. In the later design the electrometer was replaced by a Wilson tilted gold leaf electroscope and a compensating arrangement. The positive current carried by the ions to the plate P was balanced by the negative ionisation current obtained from old radium emanation tubes and whose strength could be accurately determined, the electroscope serving merely to indicate when balance was obtained.

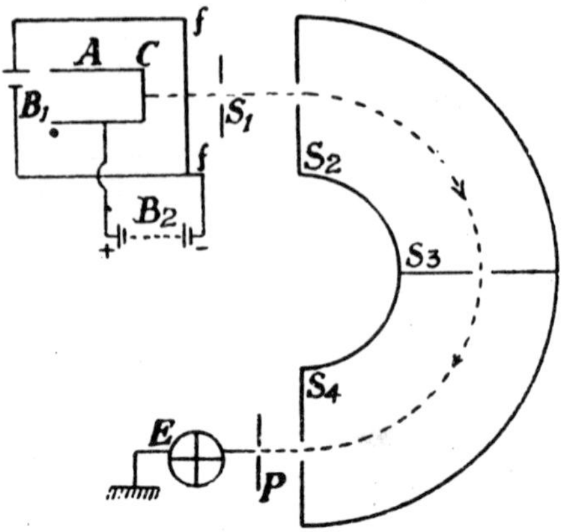

Fig. 9.4 : Dempster's Mass Spectrograph.

In 1930 Brain bridge built a mass spectrograph, based on the model of Dempster's apparatus. In 1933, he modified his apparatus.

Brainbridge used a special device called, the velocity filter. Using this device he gave the required velocity to the ions.

2. Brain Bridge's Mass Spectrogrpah

The apparatus is shown diagrammatically in Fig. 9.5. The beam of ions produced in a discharge tube in collimated by two slits S_1, and S_2 and then passed through a "velocity selector", which consists of a transverse electric field and a magnetic field perpendicular to the plane of the paper. The former is produced by maintaining the plates P_1 and P_2 at a suitable P.D., while the latter by an electromagnet represented by the dotted circle. The two fields are so arranged that the deflections they produce are in opposite senses. If X be the intensity of the electric field and H that of the magnetic field in the space between S, and then only those ions will pass through the selector undeviated which possess the same velocity v given by the relation $Xe = Hev$ or $v = X/H$. All the others will be prevented from reaching the slit S_3, since they will be bent away from the rectilinear path on account of the unbalanced effect of one of the two opposing fields. The ions that emerge from the slit S_3 at the exit of the selector with the same velocity are introduced into a uniform magnetic field of intensity H' acting at right angles to the plane

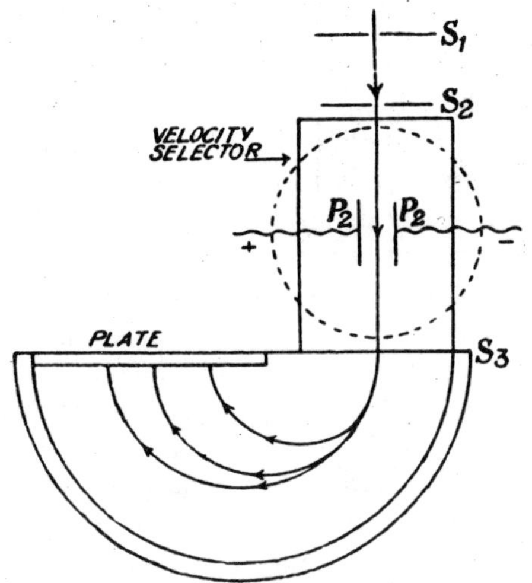

Fig. 9.5 : Mass Spectrograph Brain Bridge.

of the paper. Under the influence of this field they travel along circular paths which are governed by the usual relation $H'ev = Mv^2/r$, where M is the mass of an ion whose circular path has a radius r. The above relation gives

$$\frac{e}{M} = \frac{v}{H'r} = \frac{X}{H.H'.r}$$

Since v and H' are constant quantities, e/M is directly proportional to 1/r.

9.5 ISOTOPES

The existence of more than one type of atoms having same chemical properties but different atomic masses for the same chemical element, was discovered by, F. Soddy, while experimenting, with the naturally occurring radio active elements.

The isotopes of a given element have precisely the same chemical properties and cannot be separated by chemical processes. They are therefore considered as atoms of the same element but are not identical in all respects since they have different weights.

The notation usually adopted to designate a given isotope is $X^A{}_z$ where X is the chemical symbol of the element to which the isotope belongs, A the mass number, *i.e.*, the integer nearest to the atomic mass as determined by the mass spectrograph and Z the atomic number which gives either the total positive charge on the atom or the total number of electrons in its normal neutral state. For instance, the two isotopes of hydrogen are symbolically represented as H^1_1 (proton) and H^2_1 (deuteron) and the three isotopes of oxygen as O_8^{16}, O_8^{17}, O_8^{18}. Hence isotopes are ordinarily defined as atoms which have the same atomic number Z but different mass numbers A. With the actual dethroning of the chemical element from its pristine glory, the chemical atomic weight also has lost much of its former importance and is now considered to be merely a statistical mean value of the weights of the different isotopes in an element, though from the point of view of chemical analysis it is just as useful as ever.

We know at present about 280 different isotopes that occur in nature which include the 40 radioactive isotopes. Besides these, about 200 unstable radioactive isotopes have been produced in artificial disintegrations. The list is not, however, exhausted. The distribution of isotopes among the elements is varied and complex. For instance, some elements like arsenic, fluorine, iodine, gold, etc., have only a single isotope, while others have several, like xenon and mercury possessing 9 isotopes each and tin the highest number of 11 isotopes. In general, elements of odd atomic number almost never possess more than two stable isotopes, while those of even atomic number usually have a much larger number, *e.g.*, $LI_3^{6,7}$, $B_5^{10,11}$, $N_7^{14,16}$, $Cl_{17}^{35,37}$. When an element has numerous isotopes, it is always found that these occur in a fairly regular manner, *e.g.*, the isotopes of iron are 54, 56, 57, 58 or those of zinc 64, 66, 67, 68, 70. These regular sequences, which rarely exceed two mass units for successive numbers in going up the series, may be of importance in the intrinsic constitution of matter.

Accurate measurement of the relative abundances of isotopes also is of great importance, since it readily offers a means of checking chemical atomic weights by purely physical methods, of explaining anomalies found in the order of elements in the periodic table according to the chemical atomic weights, etc. Mass spectrographs have been used, as we have already indicated, to measure the abundance ratios of isotopes to a high degree of precision.

By measuring the total charges due to the different isotopes during a given interval of time Mass spectrographs of Dempster's or Nier's type is most suited for this purpose. The relative heights of maxima in the curves obtained by plotting the ionic currents against the accelerating potentials directly measure the relative quantities of the different isotopes of the substance analysed.

By measuring the relative intensities of the different lines traced on the photographic plate in the mass spectrograph of the type of Aston or Bainbridge. Special photographic plates are used in order to bring out the contrast of the different intensities. Comparisons are made only between lines on the same plate. The actual measurement of the intensities is done with a microphotometer by balancing the opacities of the lines against a standard wedge by a null method. The term "microphotometer" is applied to instruments by means of which the darkness or density of the lines on the negative is determined as a measure of the intensity of the rays that made them. These instruments trace automatically the density curves of the lines, photoelectric cells being used to convert the light effect into mechanical effect through the intermediary of valve amplifiers and sensitive galvanometers or electrometers. This method is beset with a great difficulty of inferring from the darkness of the line the number of particles which produce it. The charges being exactly equal and the masses being nearly the same for the isotopes of a heavy element, one may pretty safely suppose that equal number of atoms of such isotopes produce equal effect, but with very light elements this is not so sure. Aston has, however, used this method with some success. But the more accurate one, chiefly for light elements, is the previous electrical method which also eliminates the fundamental difficulty of realising a constant source, since' the relative currents carried can be compared by a null method. Further, due to the improvements brought to it by devices such as those introduced by Nier abundances as low as 1 in 100,000 can be measured accurately.

More than 250 abundance ratios of isotopes in various elements have been so far measured. There is a wide variation in abundances, as in the case of the number of isotopes in different elements.

While studying isotopes it is to be noted that, isotopes of even atomic numbers are much more abundant than those of odd atomic numbers.

The relative quantities of the different isotopes are found to be the same for every sample of a given element from whatever source it comes.

On account of the ever increasing practical applications of pure isotopes, the process of separation of isotopes, has got great importance.

Some of the methods employed for the separation of isotopes can be studied as follows:

Mass Spectrograph Method

The principle of this method is very simple. The photographic plate in the mass spectrograph of Aston's type is replaced by different vessels where the different isotopes can be separately-collected. The method enables the most complete separation, but the main difficulty is that extremely high currents of ions or long intervals of time are required to separate any appreciable amount of an element into its isotopes. The first complete separation of the isotopes of lithium was achieved in 1934 by this method in sufficient amount to be used in artificial disintegration experiments. During the next few-years the isotopes of potassium, boron and carbon were separated with success, but the yields were very low. But in 1941 under the urge of producing atom bombs with the lighter isotope of uranium (U^{235}) which is only 1 part in 140 of the ordinary element, this method was so perfected that appreciable amounts of U^{236} were produced. The "giant" mass spectrographs constructed for this purpose bear little resemblance to Aston's original apparatus.

Diffusion Method

The subject of the separation of a mixture of two gases by the method of diffusion through porous material has been thoroughly investigated by Lord Rayleigh. Theory shows that the rate of diffusion is inversely proportional to the square root of the masses of the molecules. Hence different isotopes of a gas will diffuse with slightly different velocities. The conditions under which maximum separation is to be obtained according to the theory are :

(a) That the "mixing" be perfect so that there is no accumulation of the less diffusible gas at the surface of the porous material through which diffusion takes place, and

(b) That the apertures through which the gases must pass are very small compared with the mean free path of the molecules. Hertz, in 1932, made use of this diffusion principle in a most ingenious way in the separation of the isotopes of neon. 'Since the differences in mass of the isotopes of an moment are usually very small, only slight separations are obtained in a single process, but by using a number of porous-walled diffusion units in series, the gases being circulated in a closed system, hotter results are obtained. Hertz used 48 tubes of diffusion containing special porous material at low pressure and 24 mercury diffusion pumps to circulate the gas, arranged in cascade, so that the process was repeated continuously, and obtained almost a complete separation of neon into Ne^{20} and Ne^{22}. Hertz later modified the apparatus so that the streaming mercury vapour itself of the pumps acted as the diffusion barrier. Isolation of argon and bromine have been rapidly separated by this process. Isolation of deuterium in a pure state from a 10% mixture with ordinary hydrogen has been achieved within about 8 hours. Partial separations of carbon, nitrogen and chlorine isotopes have also been obtained.

Thermal Diffusion

Enskog and Chapman have shown that if a mixture of two gases of different molecular weights is allowed to diffuse freely in a vessel whose ends are maintained at different temperatures, until equilibrium conditions are reached, there will be a slight excess of the heavier gas at the cold end and of the lighter gas at the hot end. In 1938 Clausius and Dickel employed this principle in the following simple technique which has since then become a method of isotope separation of great practical importance. If the mixture of gases is placed between 2 vertical parallel walls, one of which is hot and the other cold, thermal diffusion assisted by convection current results in two counter-current streams of different concentrations in the two constituents and produce a very rapid and marked separation of the lighter from the heavier molecules, the former concentrating near the hot wall and the latter near the cold wall. The two vertical parallel walls at different temperature are readily constituted by a vertical cooled glass tube with a coaxial wire electrically heated. As is to be expected, separation by thermal diffusion increases with the length of the diffusion column, so that a complete separation is attainable by the mere increase of the length of the cooled glass tube and the hot coaxial wire. Further, this length need not be in one stretch, but may be composed of a number of shorter columna, suitably coupled. It is found that a tube of one metre in length with the central wire heated to about 1650°C is equivalent to twelve Hertz diffusion units. The first outstanding success of the method was an almost complete separation of the isotopes of chlorine produced in 24 hours with a tube 36 metres long. The two isotopes of neon were separated in a practically pure state. Considerable enrichments of the rare isotopes of hydrogen, oxygen, carbon, nitrogen and sulphur have also been reported. This method is by far the simplest and most effective yet devised. It can also be employed for substances in the liquid state which bring in the following additional advantages that cannot be had when gases are used, *viz.,* the use of smaller column length and the larger bulk of material and no need of pumping apparatus. It is however limited to substances which do not react with the hot wire and are not affected by the high temperature used.

Pressure Diffusion or Centrifugal Method

If a heterogeneous fluid is subjected to a gravitational field, its heavier particles tend to concentrate in the direction of the field and if there is no mixing to counteract this, a certain amount of separation must take place. If, therefore, we have a mixture of isotopes in the gaseous or liquid state, partial separation should be possible by the application of a pseudo-gravitational field produced by a high speed centrifuge. Mullikan has given a complete theory of the separation of isotopes by thermal and pressure diffusions and shown that the latter method is likely to be of more value in the case of isotopes of high atomic weights, since the separation factor in the centrifugal method depends on the ratio of the masses of the molecules and not, as in other methods, on the square root of the masses. A special method named *evaporative centrifuging* has been developed successfully in America. The gas condensed at the periphery of a centrifuge at high speed is allowed to evaporate very slowly, the light fractions being drawn off very gradually at low pressure from the centre of the apparatus. It is suggested that this method ought to yield a separation 10 to 15 times as great in one operation as would diffusion or evaporation.

Evaporation Method

If a liquid consisting of isotopes is allowed to evaporate, the lighter isotopes escaping from the surface in a given time is greater than the heavier one. If the pressure above the surface is kept so low that none of the lighter atoms return to the liquid, the concentration of the heavier atoms in the liquid will steadily increase. Bronsted and Hevesy using this method have been able to separate a very small quantity of the isotopes of mercury, but did not obtain a complete separation. They were also able to announce the first separation of isotopes of chlorine by free evaporation of a solution of HCl in water at a low temperature of –50°C.

Separation by Electrolysis

If a large amount of water is electrolysed until only a very small quantity is left, the residue is found to consist largely of molecules of heavy water, *i.e.,* water in which the hydrogen atoms are those of the heavier isotope of mass 2. The concentration of heavy hydrogen in the residual water is due to the different ionic mobilities of the two isotopes of hydrogen. When electrolysis reduces the original volume of water to 1 part in 25,000 the concentration reaches almost cent per cent and heavy water is obtained in the pure state. A large quantity of electricity of the order of 30,000 ampere-hours is required to produce 1 gm. of heavy water. This heavy water is now manufactured on an industrial scale and is being widely used. The method is effective only in the case of hydrogen. It is worth while noting that heavy water has physical properties different from those of ordinary water. The maximum density of heavy water occurs at 11°C. Its freezing and boiling points at normal pressure are 3-8°C and 101.4°C; latent heat of steam 796 calories. It has a larger viscosity but smaller refractive index than ordinary water.

10

PASSAGE OF ELECTRICITY THROUGH GASES

10.1 INTRODUCTION

The laws of electrolysis tell us that there is an intimate connection between the elementary charge and the atomic view of matter.

We know well that at the time of lighting or thunder clap, very high electric current passes through the air for a short duration.

The basic arrangement for conducting experiments on the passage of electric current through a gas, can be explained as follows.

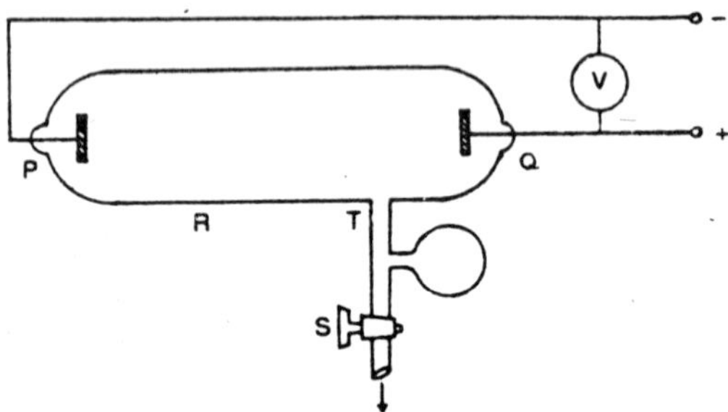

Fig. 10.1 : Experimental Arrangement for the Study of Electrical Discharge in a Gas.

All gases, however, become conductors allowing the passage of electricity under certain conditions. Thus, for instance, the combustion gases drawn from names are found to be conducting. Many agencies such as ultraviolet light, X-rays, electric sparks, radioactive substances, etc., also render gases comparatively good conductors. The conductivity persists for a time after the exciting agent is removed but does not last indefinitely.

The cause of the induced conductivity in gases has been found to be due to the presence of charged particles called ions. Under ordinary circumstances gases contain very few free ions and hence are excellent non-conductors; but when they are subjected to the influence of a suitable agent, a large number of free ions are produced and then they acquire a relatively free but true conductivity. The process of producing these ions is called the ionisation of gases. It consists in the detachment of one or more electrons from the neutral atom or molecule of the gas. The residue of the atom or molecule which is, of course, positively charged and consists practically of the whole mass of the original atom or molecule, constitutes the positive ion. The detached electron soon attaches itself to a neutral atom or molecule and constitutes the *negative ion.* As the two kinds of ions are formed from an electrically neutral atom or molecule, the charge on the negative ion must be numerically equal to that on the positive ion. Under the action of the electric field the positive ions will be drawn towards the negative terminal and the negative ions towards the positive terminal. Such a movement constitutes a current through the gas which is known as the *ionisation current.* Since the two kinds of ions have practically the same mass, their mobilities and diffusion coefficients are of the same order of magnitude at ordinary pressures. But as the pressure is reduced the negative ion throws off its attendant atom or molecule and the resultant negatively charged light particle, the bare electron, travels free and faster than the positive ion.

The process of conductivity in gases is, in some respects, strikingly analogous to electrolytic conductivity of a solution which according to the theory of Arrhenius is due to the presence of the two kinds of ions, positive and negative, produced by the dissociation of the electrolyte in solution.

The rate of production of ions depends upon the nature of the ionising agent, but the number of ions in existence does not increase indefinitely with time as some of the positive and negative ions are constantly recombining, the rate of recombination being proportional to the square of the number present. Hence also it is that the conductivity persists for a time, but does not last indefinitely after the removal of the ionising agent.

Townsend has shown that even with the strung ionisation produced by radium only one molecule out of a 100 million is ionised per second. In most cases of ionisation only ions electron is ejected from an atom, but positive ray analysis by Thomson and others indicates that the slow-moving positive ions may detach several electrons from certain atoms, *e.g.,* the ionisation of the mercury atom in this way may consist in the ejection of eight electrons from the atom.

10.2 DISCHARGE TUBE

When an electricity passed through gases at low pressure, an interesting phenomena can be observed. The simple experimental arrangement required to observe this phenomena is called the discharge tube.

The arrangement of a simple discharge tube can be explained as follows.

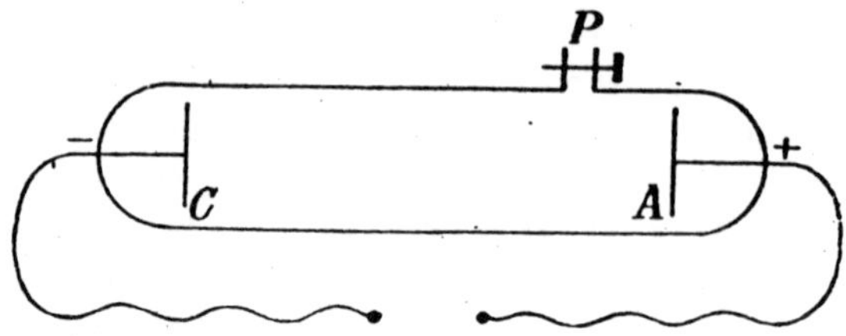

Fig. 10.2 : Discharge Tube.

It consists of a strong closed glass tube about 30 cms. long and 4 cms. in diameter, provided with two metal electrons C and A and a side tap P meant for connecting the tube with a high vacuum pump and a low pressure gauge. To obtain very low pressures a molecular pump and charcoal cooled by liquid air that has the property of absorbing most of the residual gas are used. The two electrodes are connected to the secondary of a powerful induction coil, capable of giving a high P.D. of the order of 50,000 volts and more. The electrode connected to the negative terminal of the secondary is the cathode G, while the one connected to the positive is the anode A.

As the pressure of the gas in the discharge tube is gradually reduced by working the exhaust pump, the following series of phenomena is found to take place in succession. When the pressure is about 10cms. of mercury irregular streaks of light appear accompanied by a crackling noise. As the pressure is diminished to the order of 1 cm. of mercury the crackling streaks broaden out into a luminous column with a continuous buzzing sound, extending from the anode A almost to the cathode C, its colour depending on the nature of the gas—reddish for air, bright red for neon, bluish for CO_2, etc. This luminous column is known as the positive column. As the exhaustion proceeds further and the pressure comes to about 3 or 4 mms. a difference between the two ends of the discharge becomes discernible, with a blue luminosity around the cathode and a discontinuity near it from the positive column. The glow on the cathode is called the *negative* or *cathode glow*, and the discontinuous dark space between the cathode glow and positive column is called the *Faraday dark space*. When the pressure reaches the order of 1 mm., the positive column gets shortened, the Faraday dark space extending to a greater length. Further reduction of pressure results in the detachment of the cathode glow from the cathode and a second dark space known as *Crookes dark space* is formed between the cathode glow and the cathode. When the pressure gets reduced to about 0.1 mm. the Crookes dark space increases in length and dark bands called *striations*.

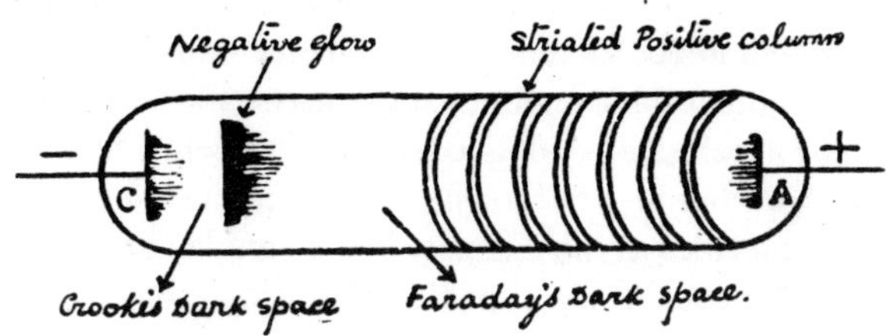

Fig. 10.3 : Appearance of a Discharge Tube at About 0.1 mm.

On further reduction of pressure the tube begins to decrease in brilliancy, the striations recede towards the anode and finally disappear. The Crookes dark space and the cathode glow expand. Then the cathode glow becomes ill-defined and finally vanishes. At a pressure of about 0.01 mm. the Crookes dark space practically fills the whole tube when a new phenomenon gradually comes into existence. The walls of the tube begin to glow with a bluish or greenish light, depending on the nature of the composition of the glass of which the tube is made. This is known as "fluorescence" caused by the action of a new type of radiation, called X-rays, resulting from the impact, of some invisible agent in the Crookes dark space. If the exhaustion is continued further the discharge has an increasing difficulty in passing, until finally the tube refuses to conduct.

When the pressure is reduced in the discharge tube the following changes can be observed.

1. Ionisation by collision.
2. Excited atoms or molecules, emit light.

Free electrons which abound in a low pressure discharge tube are mainly responsible for the different phenomena.

The accelerated positive ions in the discharge tube, strike the cathode, and cause it to liberate a great number of electrons. These electrons travel some distance accelerated through the strong electric field near the cathode, before they acquire sufficient energy to ionise gas molecules by collision.

As moving electrons do not emit light by themselves the space they travel before colliding with molecules will be dark, and this constitutes the Crookes dark space.

When the accelerated electrons, after traversing the Crookes dark space strike molecules and ionise them, the latter emit light which gives rise to the cathode glow. The positive ions resulting from this collision process tend to slow down the electrons, which in consequence lose their power to ionise molecules temporarily. Hence emission of light ceases and we have another dark region, the Faraday dark space. The electrons continue to be accelerated owing to the field gradient, although they have to move much farther than before in order to acquire the energy necessary for ionising molecules as the intensity in the Faraday dark space is lower. Once again the electrons are capable of ionising molecules by collision with the corresponding appearance of glow in the gas. This constitutes the beginning of the positive column, with the production of both positive and negative ions. The electrons thus created move rapidly towards the anode, while the heavier residual positive ions accumulate, once again retard the progress of the electrons and render them incapable of ionising molecules. Hence once more a dark space is produced in the positive column, which corresponds to a striation. The formation of alternate bright and dark bands in the positive column extending up to the anode is due to the repetition of the above-mentioned process.

It is to be noted that the extent of the dark space depends on the distance the electron can travel without collision, *i.e.,* the *mean free path* of the electron. This mean free path increases as the pressure of the gas is decreased. Hence the different dark regions in the discharge tube will increase in length as the pressure is reduced more and more. When the pressure is so low that the mean free path is larger than the dimensions of the tube, the whole tube appears dark as there is no further possibility of collision with molecules of the gas in the tube and consequent emission of light. But collisions take place with the walls of the tube even at this stage, giving rise to light emission due to fluorescence.

10.3 DIFFERENT RAYS IN THE DISCHARGE TUBE

More detailed analysis of the phenomena in the discharge tube at the different stages of exhaustion has brought to light the existence of three important radiations, called the cathode, positive, and X-rays.

In 1859 Pluecker observed the different types of rays in the discharge tube.

J.J Thomson, Perrin and some other scientists, established the important properties of these rays.

1. The cathode rays are shot out normally from the cathode, their direction being independent of the position of an anode.
2. A light body placed in the path of the rays experiences a force tending to move it away from the cathode.
3. The rays produce heat when falling upon matter.
4. A very large number of crystals, minerals and salts phosphoresce brilliantly under the impact of cathode rays.
5. The rays ionise gases like strong electric fields, and they can penetrate thin sheets of matter.
6. The rays are deflected by electric and magnetic fields, the direction, of deflection being that in which a stream of negatively charged particles would be deflected.

These electrodes are "preheated" by means of a mechanical press button arrangement or, better still, by a more convenient "starter" circuit for about 2 or 3 secs., during which time a current of about 1.5 amperes flows through them and makes them red-hot. Under these conditions a discharge current of 0.8 ampere is sufficient to maintain the main discharge that is struck. The preheating starter S_2 is of glow-type, and consists of two metallic strips inside glass bulb filled with helium. On closing the main switch S_1 of the 230 volts A.C. almost the full voltage is applied across the strips of the starter and a glow discharge results. This discharge heats up the strips which in consequence bend towards each other and finally touch, which enable the lamp electrodes to be heated. But when the strips are in contact the voltage between them drops to zero, the glow discharge is destroyed and the strips get cooled and separate again. Meanwhile the hot electrodes have the main discharge started between themselves. The condenser C, included in the starter circuit is meant to by-pass radio interferences and the resistance B in series with C, (about 100 ohms) prevents the starter strips from welding together. When the tube operates with the main discharge, it requires a choke Ch to limit the current to the required value. A condenser is connected across the mains to correct the power factor to a value of 0.9 lagging. The capacity of this condenser is of the order of 8 mfds.

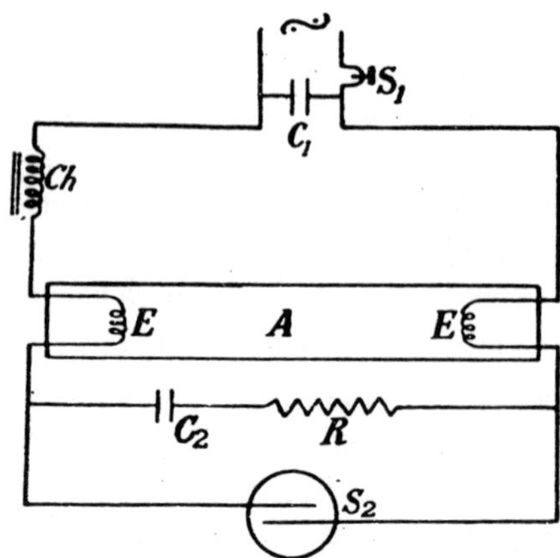

Fig. 10.4 : Fluorescent Lamp.

The amount of light obtained from these fluorescent lamps is more than three times when compared with the ordinary electric lamps for the same amount of energy consumed. The electric lamp does not give the same mixture of different wavelengths as daylight from the sun, while with these new lamps close imitation of daylight is possible by a suitable choice of the fluorescent powder. Not only daylight but any desired colour can be produced by using the proper fluorescent material. The manufacture of these fluorescent powders involves great technical skill and care as all impurities are to be excluded. More economic use of electric energy is made in the fluorescent lamps than in the electric lamps. In the latter a very great portion of the energy is wasted as heat, while in the former much less energy is turned into heat. Moreover, the fluorescent lamp gives all the extra light for nothing, since it uses no extra electricity for the fluorescent material to absorb the unused and even harmful intra-violet rays and turn them into useful and beneficent visible light. There is economy also as regards the average life of the lamp. The lifetime of an ordinary electric lamp is about 1,000 hours, while for the fluorescent lamp it is more than 3,000 hours. This great difference results from the fact that the electrodes of the fluorescent lamp are made of a narrow coil of thick tungsten wire which is not easily damaged, whereas the filament of the electric lamp is flimsy and delicate, which may therefore be snapped by any little accident. The greater cost of these new lamps is well compensated by the higher luminosity and longer life. Finally, since most of these lamps are built as tubes several feet long, the fluorescent light comes from the whole length of the tube. In consequence the shadows cast by them are much less dense and the light more restful to the eyes than in the case of the electrical lamp where the light coming from a narrow piece of wire causes glare and dense shadows. The minimising of glare and thick shadows and producing uniformly diffused light are of great advantage for certain purposes, such as work in the mines, delicate mechanical manipulation in factories, surgical operations, etc., which require almost shadowless but strong illumination.

If a magnetic field is applied by placing the discharge tube between the pole pieces of an electromagnet, the direction of the field being at right angles to the path of the cathode rays, the north pole in front and the south pole at the back, the luminous spot on the fluorescent screen is shifted vertically downwards. If the poles are reversed, the spot shifts vertically upwards. This is in accordance with the electromagnetic law represented by Fleming's left-hand rule. The ray should bend in a direction perpendicular to both its original direction and the direction of the magnetic field, hence in the vertical direction, either downwards or upwards according to the disposition of the poles. A direct evidence for the negative charge carried by the cathode rays was furnished by Perrin's experiment (1895) in which the cathode rays were deflected by means of a magnet into a metal cylinder sealed into the side of the discharge tube and connected to an electrometer. As soon as the cathode rays entered the cylinder a large negative charge was registered by the electrometer.

It is found that even very small magnetic fields are sufficient to deflect a beam of cathode rays. Further, the deflection, under like conditions, has been shown to be independent of the nature of the gas in the discharge tube and of the material of the cathode.

All these properties, considered together, lead to the important conclusion that the cathode rays are streams of negatively charged light particles, which acquire a very high velocity in the electric field maintaining the discharge, and fly away from the negative electrode. J.J. Thomson called them "streams of negative corpuscles", while Johnstone Stoney, who having found from electrolysis that electricity was atomic in nature, suggested the name "electrons", which name has been universally adopted today. Owing to the special conditions of ionisation at very low pressure in the discharge tube the electrons expelled from neutral atoms of the gas are practically isolated, without getting attached to neutral atoms and constitute the cathode rays. Thus the electron was discovered. It is also seen that these electrons are of very wide occurrence, entering into the constitution of all matter, as the stream of electrons is always produced in the discharge tube independently of the nature of the gas and of the material of the cathode. Subsequent investigation has confirmed that the electron is a universal negative carrier of electricity, whose nature is the same regardless of source or mode of liberation from matter. It may be noted that though the electrons are probably not "matter" as ordinarily under- stood, their mass being purely electrical in origin, they are to be considered as independent particles of electricity, having a mass, the lightest of all known masses, carrying a charge, the smallest quantity of electricity, and moving with a velocity, exceedingly high.

Goldstein, in 1886, working with a discharge tube where the cathode was perforated and the pressure of the gas was about 1 mm., noticed streams of luminosity at the back of the cathode, each stream proceeding from a perforation in the cathode. He called them *canal rays*, as they proceeded canalised by the holes in the cathode. Farther investigations showed that these rays produce phosphorescence, affect photographic plates, disintegrate metals upon which they fall ("sputtering"), penetrate thin aluminium foils, are deflected by electric and magnetic fields, the sense of deflection indicating that they are positively charged, and consist of a stream of positively charged particles of about the same mass as an ordinary gas atom or molecule. Hence they are known as positive rays or a beam of positive ions. The existence of such rays is readily understood with the general mechanism of discharge ionisation of the gas in the discharge tube, as we have seen, causes the expulsion of an electron from a neutral atom or molecule and the consequent formation of a positive ion. The electric charges on the electron and the positive ion are equal in magnitude but opposite in sign. The electrons resulting from ionisation being much lighter than the positive ions acquire much greater velocity in the electric field and move away rapidly from the cathode, while the heavier positive ions are less mobile and accumulate near the cathode moving towards it. These positive ions moving with a certain speed, that depends on the strength of the field and pressure of the gas, pass through the perforations in the cathode and become a beam of positive rays. If the cathode is not perforated, they simply strike the cathode and give rise to a faint glow at its surface. The condition for the development of positive rays is *ionisation at low pressure in a strong electric field*, the low pressure preventing the positive ions from colliding too frequently with other atoms or molecules and the strong field imparting the required high speed to them in spite of their relatively heavy mass.

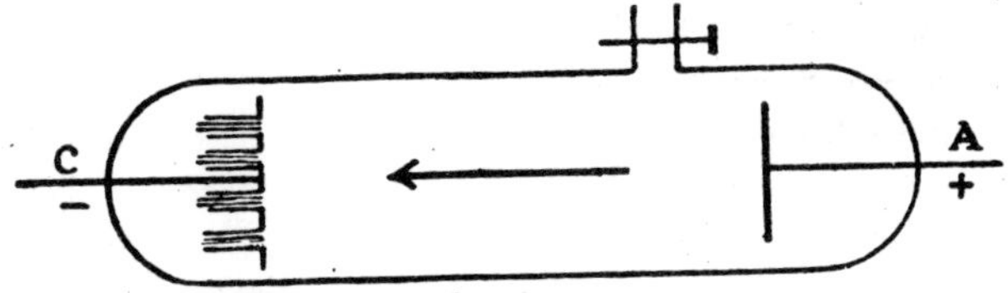

Fig. 10.5 : Canal Rays.

Positive rays can be formed from molecules as well as atoms so that they will be able to give direct information as to the masses of atoms and molecules, which refer to the atoms or molecules *individually*, and-not as in chemistry to the mean of a large aggregate. It in on this account that an accurate positive ray analysis is of very real importance. We shall come back to this point later.

In 1895, Roentgen, a German scientist, experimenting with a discharge tube of high vacuum of the order of 0.01 mm. and more, found, to his great surprise, that a fluorescent, screen in the neighbourhood of the tube became luminescent as though exposed to ordinary light. Likewise a photographic plate lying nearby was affected, although it was wrapped up completely in black paper, Interposing a thick plate between the tube and the fluorescent screen a dark shadow was produced on the screen. Light substances, such as thin aluminium sheet, paper, wood, etc., cast less distinct shadows. Roentgen was able to trace the origin of the emission which produced these effects to the walls of the vacuum tube upon which the cathode rays fell. On account of the unknown nature of the new type of radiation, he called them X-rays.

(1) They arise at the point where the cathode rays strike a solid obstacle, and travel in straight lines, obstacles placed on their path casting shadows.

(2) They excite fluorescence in many substances and affect photographic plates. They also ionise gases.

(3) A very remarkable characteristic of X-rays is their high power of penetration even through solid substances, which are opaque to ordinary light. Their penetrability depends oh the density of the substances. Flesh is much more transparent to them than

bones; hence their use in surgery, that followed almost immediately after the discovery, to examine bones, detect fractures, foreign bodies, etc. In this connection, it may be noted that the rays arising from different vacuum tubes differ in character: the higher the vacuum, the greater their penetrating power. Since the voltage necessary to force a discharge through the tube increases as the vacuum increases, in which case the tube is said to become *harder*, it is usual to say that a hard tube emits highly penetrating *hard* X-rays, while a tube of not so high a vacuum emits relatively less penetrating *soft* X-rays.

(4) When they fall upon substances, they give rise to other rays, to which the name of "secondary radiation" has been given. The secondary radiation is very complex, consisting of three different types, scattered X-rays, characteristic X-rays and corpuscular rays.

(5) They are not affected by electric or magnetic fields, which fact differentiates them from the cathode and positive rays and makes them resemble ordinary light.

Subsequent researches proved beyond doubt that X-rays are a train of electromagnetic impulses, hence essentially of the same nature as ordinary light but invisible and of very short wavelength, produced by the sudden stoppage of the electrons of the cathode rays by material obstacles situated on their path.

These rays invisible though they are form an important branch in modern science, in the study of atomic and crystal structures, in surgery and therapeutics, etc.

10.4 APPLICATIONS OF THE DISCHARGE TUBE

The discharge tube has many number of uses. The major applications can be studied as follows.

1. The Geisster tube
2. The modern lamps of great luminosity.

Usually, the Geisster tube is made in two shapes. In one type, a straight glass tube consists capillary portion about 7 or 8 cms, long with a bore of about 1mm.

The electrodes are made of either aluminium or platinum. The tube is filled with the gas. A high potential exceeding 2,000 volts is applied between the electrodes by means of an induction coil.

The second type of discharge tube is made in H shape. The shorter capillarly portion is about 4 cms. long.

These two types of discharge tubes have been used for many years in spectra of substances which can be obtained in the gaseous or vapour form.

Since these tubes require, very less amount of current, they are known as cold cathode tubes.

The sodium and mercury vapour lamps are classified as hot cathode discharge tubes since, though they do not require a very large potential like the Geisster tubes. They consume, high current, so electrodes get heated.

The sodium vapour lamp is considered as a low pressure lamp, while the mercury vapour lamp as a high-pressure one.

In the sodium vapour lamp the actual discharge tube A is bent in the form of a U-tube with the electrodes EE fused at the two ends and droplets of pure sodium metal are distributed uniformly on the inner walls of the tube. A very small quantity of neon gas is also introduced into it. The lamp ordinarily functions with 230 volts A.C.B. at to start the discharge, a voltage higher than that is momentarily required, which is supplied by a leak transformer introduced in the circuit. The lamp requires an operating temperature of about 300°C to produce a high luminosity. Even at this temperature the pressure of sodium vapour is low, about 0.01 mm. On account of the high temperature, heat insulation becomes necessary, which is secured by enclosing the discharge tube in a double-walled vacuum tube B. Due to the presence of neon, the light is at first red until the lamp temperature rises to a stage at which sodium volatilizes. Then the discharge is almost entirely maintained by the sodium vapour, when the light turns yellow. It takes some minutes for the lamp to bum at its full intensity, due to the gradual vaporisation of the solid sodium. The lamp is worked at relatively low currents, as the efficiency decreases when the current density is increased above a certain optimum value. The spectrum of the light produced consists chiefly of the two sodium lines 5,890 and 5,896 Å, so that the light may be used as a pretty good monochromatic source. The lamp made by Philips & Co. and patented under the name "Philora" is guaranteed to an average life of 2,500 hours. It is used for lighting up

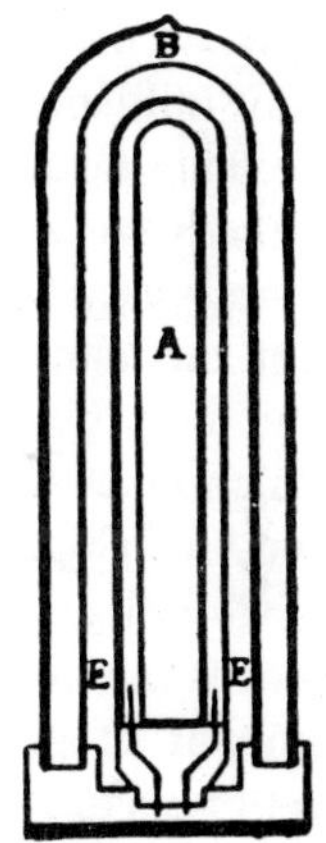

Fig. 10.6 : Sodium Vapour Lamp.

showcases in shops and public squares, but objects appear somewhat pale in its yellow light.

The mercury vapour lamp consists of a straight discharge tube T, mounted vertically in a closed outer bulb. The vertical mounting is to prevent the discharge touching the walls of the tube, which would thereby get overheated. The discharge tube contains a carefully controlled quantity of mercury with a trace of argon, the function of the latter being to help the discharge to grow at start until the mercury gets completely vaporised and can maintain the discharge by itself. When started the lamp works at low pressures, but the pressure builds up to a specific maximum value as the discharge increases. The lamp operates ordinarily on the 230 volts A.C. When toe mains are switched on, the voltage is sent across the main electrodes E_1 and E_2, which is, however, insufficient to start the discharge as the electrodes are not yet heated. An auxiliary electrode E_3, which is very near E_2 and is connected through a high resistance R (concealed in the lamp itself) to the electrode E_2 at the other end, is used to start the discharge. This glow discharge between E_1 and E_3 heats the electrodes, while consuming only a very little current of a few milliamperes due to the presence of the high resistance. When once the main discharge begins to work, practically all the current will pass through the tube between E_1 and E_2. The choke Ch in series with the lamp is used to limit the current. Usually a condenser C is put in parallel with the mains in order to raise the power factor, but it is not essential. The lamp gives out a pale bluish light which takes some minutes to attain its maximum high intensity. If the spectrum of this light is examined, there is practically no red, but there is a strong amount of green and some blue and yellow. These colours are visible, but there is another substantial amount of radiation, not visible to the eye, *viz.*, the ultra-violet rays, which play a prominent part in the construction and working of the fluorescent lamps as we shall see presently. The mercury vapour lamps are produced by the G.E.C. under the patent name "Osira" and are widely used for street lighting.

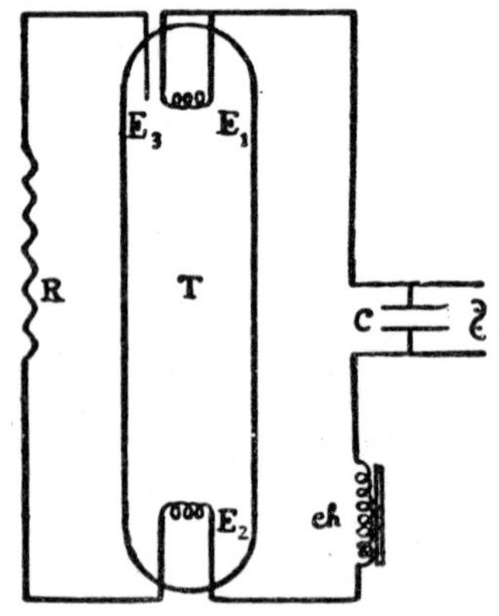

Fig. 10.7 : Mcreury Vapour Lamp.

10.5 THE FLUORESCENT LAMPS

In the fluorescent lamps, invisible ultra-violet rays are utilised to produce brilliant light. Unused and wasted energy is converted into useful light.

The inner surface of the discharge tube is coated with suitable fluorescent material.

A little more or little less of one or the other of these fluorescent materials enables one to control the spectrum of the light produces, with in narrow limits, so that, a light which approximates very closely to natural daylight results as in the case of the widely used fluorescent day light lamps.

The ultraviolet rays produced by collisions of the electrons with the molecules of the mercury vapour in the discharge tube falling on the fluorescent coating on the inner walls, of the tube cause the emission of light radiations.

The fluorescent lamp consists of a long and wide glass tube A, the inner surface of whose walls are coated with the fluorescent material.

A small quantity of mercury is introduced in the tube, which is then exhausted to a low pressure of about 1 mm, it is seated at the two ends, which consists two electrodes.

11

RADIOACTIVITY

11.1 INTRODUCTION

The French physist Henri Becquerel was the first to discover the radioactivity. The discovery was quite accidental. When he was investigating an interesting point in connection with X-rays, he found the H phosphorescence of certain fluorescent salts, activated by sunlight. At that time he was experimenting on, the double sulphate of uranium and potassium.

After his experiment Becquerel came to the conclusion that the uranium salt, even in the dark, might be emitting radiations which penetrated the paper wrappers and affected the photographic plates.

Becquerel, continuing his researches with different compounds of uranium, in solid form and solutions, found that the newly discovered radiations were emitted by uranium, irrespective of its state of chemical combination and physical conditions of temperature and that there was no connection between this phenomenon and phosphorescence. Further experiments proved that these radiations caused the discharge of electroscopes by ionising the air through which they passed. The name of "*Becquerel rays*" was first given to the radiations emitted by uranium. It was later changed into *radioactivity* to designate all such rays emitted not only by uranium but by several other inhabitance which were soon discovered, such as thorium, radium, polonium and actinium.

The most significant discovery among these was radium, made by Madam Curie and Pierre Curie, her husband, in 1898. In a careful search to discover the phenomenon, of radioactivity among a number of minerals, Mme. Curie found that pitchblende, a uranium mineral, was many times more active than a pure uranium salt. By fractional crystallisation the Curies succeeded in separating from the mineral two new highly active elements which were named radium and polonium. The activity of polonium is many times that of uranium but that of radium enormously greater—more than a million times that of uranium for equal weights.

Madam Curie **Pierre Curie**

Fig. 11.1

In the same year, 1898, Schmidt discovered radioactivity in thorium which was more or less of the same strength as in uranium. Debieme in 1900 was able to separate from pitchblende in association with the iron group another radioactive element which was named actinium. Its activity is comparable with that of radium. From amongst the most important of the remaining radioactive elements, *radiothorium* and *mesothoriun* were discovered by Hahn in 1905 and 1907 respectively; *ionium* in 1907 by Boltwood and Hahn independently. In the same year Campbell succeeded in establishing the radioactivity in *potassium* and *rubidium*. The discovery of *protactinium* was made by Hahn and Meitner in 1918. Hevesy and Paul in 1932 showed that *samarium* was also a radioactive substance. The process of separation of the radioactive substances from the original mineral ores is an exceedingly difficult task, a ton of uranium ore, for instance, yielding only a few decigrams of radium.

From 1898 onwani there was a tremendous rush of investigators into the now field, exploring almost every conceivable aspect of radioactivity—chemists ascertaining the chemical properties of the radioactive elements and physicists observing their physical properties. From these researches it soon became clear that there is nothing abnormal about these radioactive elements as regards their physical and chemical properties and that they all have their places in the periodic table Thus, for instance, radium, except for the special feature of activity, fits in at the end of the second group of the periodic table having chemical and spectral properties similar to calcium, strontium and barium, the other members of the same group. Hence the ordinary physical and chemical properties of the radioactive elements will not help one very much in the study of the phenomenon except indirectly in so far as the pure radioactive elements are separated from the parent mineral ores by the use of chemical and physical methods.

The various effects produced by radioactivity, such as photographic action, penetrative power, ionisation of gases, production of scintillation on fluorescent screens, development of heat, chemical effect, etc. were also studied extensively in the initial researches. These investigations not only paved the way to some very important applications in science and industry, but also were extremely helpful in the progressive understanding of the phenomenon. Thus, for instance, the very discovery of radioactivity was first made by the photographic action and the photographic plate still remains a very suitable piece of apparatus for some important experiments on radioactivity; the working of the electroscope which forms an essential part of the technique of radioactive measurements is based on the ionising property of the rays; their penetrative power has been utilised in the original distinction of the three kinds of rays that constitute radioactivity.

The researches conducted by Lord Rutherford, Mme. Curie and William Crookes at the initial stages of the study of radioactivity may be mentioned on account of the very important conclusions that have been drawn from them.

Soon after the discovery by Becquerel, Lord Rutherford and his collaborators at Cambridge investigated the penetrating power of the radiations from uranium, using sheets of tin-foil as absorber and the gold leaf electroscope to measure the intensity of ionisation produced by the transmitted radiations and showed that the original radiations emitted by uranium were of two distinct kinds: the one, very soft, easily absorbed and capable of producing intense ionisation, which Rutherford called the a-rays, and the other, much more penetrating than the first and much less effective as an ionising agent, which he called the β-rays. With this finding it became clear that the radiation which affected the photographic plate in Becquerel's experiment was the β-rays. It was later discovered by Villard that there was also a third kind of rays still more penetrating than the above-mentioned two and it was named the γ-rays.

Mme. Curie and her collaborators, using the electrical method based on the ionising property of radioactivity, similar to that employed by Rutherford, were also able to show that the activity of any uranium salt was directly proportional to the quantity of uranium in the salt, thus proving that radioactivity is an atomic property independent of the state of chemical combination of the atom.

Sir William Crookes in 1900 showed that if uranium salt is precipitated from solution by means of ammonium carbonate and then redissolved in excess of the carbonate reagent, a slight precipitate remained which was several hundred times more active in photographic action than the original uranium. This active residue he called uranium X (UX) and its occurrence at first appeared to contradict the view that radioactivity is an atomic property. But the correct explanation was found in the formation of a new radioactive substance during the emission of radiations by the original substance. The intense photographic action being due to β-rays, the new radioactive UX was emitting α-rays. Had the examination been done by means of the ionising property which is mainly due to the α-rays, it would Rave been found that the original uranium still possessed this power and not the UX.

11.2 TYPES OF RADIOACTIVE RADIATIONS: THE ALPHA-RAYS

The α-rays consist of discrete particles. The determination of the ratio to the charge 'E' to the mass M of α-particles, is the important task.

While determining the e and M of α-particles Rutherford used the method, and this was same, which was made by J.J. Thomson.

To find out e and M of α-particles, electric field can be used.

Two parallel metallic plates, 35 cms. long and 4 mms. apart arranged immediately above the source of α-particles. A strong electric field of the order of 20,000 volts/cm, is maintained between these plates. The a-particles, in passing through this field, are deflected at right angles to their path, and by reversing the field, two traces D_1 and D_2 are obtained on the photographic plate. From the geometry of the apparatus and the distance between the traces it is possible to determine the deflection produced by the field and thus get a numerical value of $(M/E)v^2$. By combining this with the result of the magnetic experiment both E/M and v are obtained, as in the case of the cathode rays.

Rutherford found that for the α-particles from RaC the value of E/M = 4,820 *e.m.u.* per gram. The value of E/M of the hydrogen atom is about double that obtained here, *viz.*, 9650 *e.m.u.* To determine the mass M of the α-particle from the value of E/M it is necessary to measure the charge E carried by the α-particle. This was also done by Rutherford in collaboration with Geiger as follows.

Determination of the Charge E of the α-particle

A known quantity of RaC is deposited in a shallow dish R which is placed in a highly evacuated vessel V. The dish is covered with a very thin aluminium foil to keep back recoil atoms. The α-particles after passing through a window W of known area and covered with very thin aluminium foil fall upon the collecting plate P connected to an electrometer. The charge acquired by the plate P in a known time is determined by observations on the electrometer. The whole apparatus is maintained in a strong magnetic field, to bend away the β-particles given off by the source preventing them from reaching the collecting plate as well as to make the secondary electrons liberated from the surface of the collecting plate by the impact of α-particles return back to the plate.

To obtain the charge per particle, it is necessary to know the number of α-particles falling on the plate. This number was counted by Rutherford and Geiger by means of an ingenious device known as the *Geiger α-particle counter.*

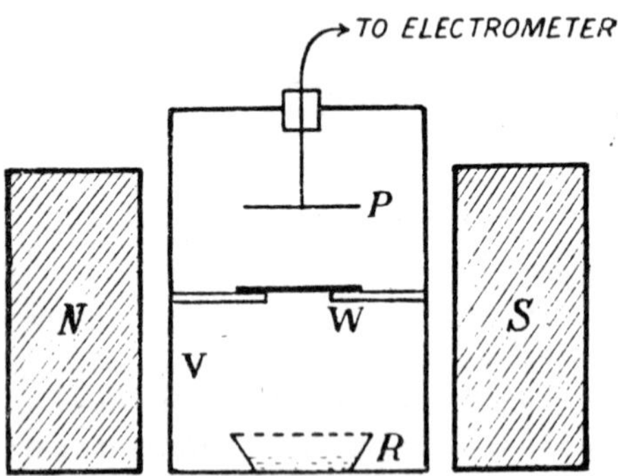

Fig. 11.2 : Apparatus for Measuring the Charge E on the α-Particle.

The mass M of the α-particle is deduced from the experimentally found values of E/M and E. It turns out to be 6.64×10^{-24} gm., a mass four times that of the hydrogen atom and hence equal to the mass of a helium atom. Thus, the α-particle is found to have the mass of a helium atom and to carry two positive elementary charges, which reveals the fact that the α-particle is to be identified-with a helium atom that has lost both its electrons, *i.e.,* with the bare helium nucleus.

This conclusion was further confirmed by Rutherford in the following manner. The α-particles emitted by a large quantity of radon enclosed in a thin-walled glass tube were allowed to penetrate into a highly evacuated discharge tube. After two days a spectroscopic examination of the discharge tube showed that helium had appeared in it. The spectral lines of helium became brighter with increasing time, indicating that more α-particles had penetrated into the discharge tube.

Rutherford calculated also the amount of helium that would be evolved from a given amount of radium in a given time if α-particles were identified with helium nuclei and found it to be in substantial agreement with the volume of helium emitted actually in laboratory experiments.

The evidence is quite complete, therefore, that the *α-particles are helium nuclei.*

The magnetic and electrostatic deflection experiments give also the velocity of the α-particles from RaC as 2×10^9 cms./sec. and hence very high, nearly 1/15 of the velocity of light. The velocity of α-particles varies from 1.45 to 2.2×10^{-9} cms./sec. with different radioactive sources. It was first believed that all the α-particles from a given radioelement had the same velocity and hence the same energy except for slight variations due to "straggling" effect caused by impacts of the fast moving α-particles with other atoms. But later, it was established from experimentally observed facts, such as the "long-range" α-particles and ''α-ray spectra", that the first surmise was not correct but that α-particles even from a single source had different velocities. This discovery led to very important conclusions about the internal structure and state of the nucleus, as we shall see later.

The scintillation method or spinthariscope, first devised by Crookes, which has proved to be a sensitive method of great value. When a screen is coated with ZnS crystals and exposed to α-rays, it becomes luminous due to fluorescence effect produced by the rays. When viewed with a low-power microscope the luminosity is found to consist of a number of scintillating points, which appear like shooting stars. Each flash is due to the impact of a single α-particle upon a crystal. This offers a delicate means of counting individual α-particles.

Wilson's cloud chamber is another very powerful method of studying the individual particles. As α-particles pass through the expansion chamber at the right moment, they ionise the gas along their paths and leave tricks which can be photographed, to be studied at leisure, as seen from the adjacent photo. Each track corresponds to the passage of a single α-particle. The use of the cloud chamber has furnished a number of very important data concerning not only the α-particles but also atomic nuclei.

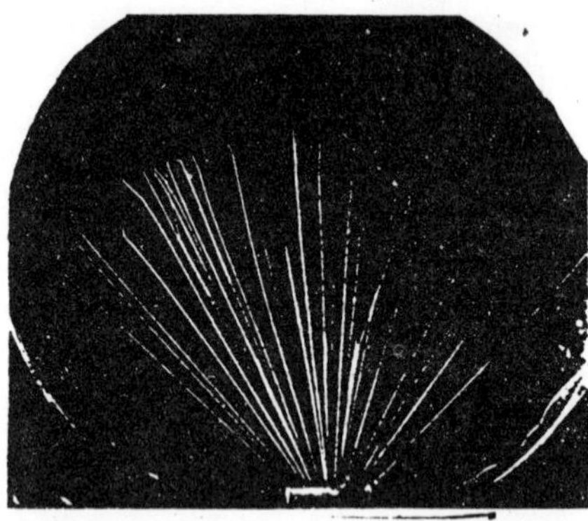

Fig. 11.3 : Cloud Chamber Photograph of α-Particle Tracks.

When α-particles are passed through matter they cause ionisation by knocking electrons out of the atoms lying along their path.

α-particles leave a trial of positive and negative ions. During ionisation process α-particles lose energy continuously and their velocity is correspondingly diminished. At last their speed reaches a certain critical value. Below this value they lose their power to ionise.

The velocity of α-particles rises only when the velocity has falls to very near the critical velocity and then suddenly it falls to zero.

The mass of the α-particles being very great as compared with that of the electrons, the (α-particles knock electrons out of their way without being deviated to any great extent. On rare occasions, however, a collision with the relatively small but heavy nucleus of the atom takes place in which case the α-particle suffers a large deviation, sometimes even greater than 90°. This phenomenon has thrown great light on the internal structure of the atom as we shall see. These events can be readily observed in cloud chamber photographs of particles, as seen from the photo taken when α-particles passed through a cloud chamber containing oxygen. The abrupt bends which occur at the ends of the tracks are caused by the "straggling" effect. The forked tracks represent a close collision between an α-particle and an oxygen nucleus, resulting in a single large angle scattering.

If thin sheets of absorbing material path of the α-particles, the range of the α-particles is reduced. This is because the α-particles in passing through the material knock out electrons from the atoms in the material and so lose energy. The stopping power of any material is defined as the equivalent cms. of air path by which the normal range has been reduced. It depends upon the initial velocity of the α-particle and the thickness, atomic weight and density of the absorbing material. Bragg and Kleeman found, as a result of their researches, that the stopping power is proportional to the square root of the atomic weight of the absorber.

Geiger found the important relation between the range of an α-particle and its velocity.

The range 'R' is proportional to the cube of the velocity 'V'.

or, $R = av^3$ (a = constant)

By measuring the velocities of a-particles, Geiger derived the above formula.

The velocity and range of the α-particles from one radioactive substance, being known, the velocity of emission of the α-particles. By knowing the range one can find the radioactive substance. This becomes easier by Geigers law.

11.3 THE BETA-RAYS

The β-rays consist of negatively charged particles. So the carry negative charge. β-particles have the same e/m value, as the electrons is applied. But the e/m value for the β-particles is not constant, as with the electrons of the cathode rays.

β-rays, are much more deviated in a magnetic field than the α-rays. This clearly indicates, that, β-particles are much lighter than the α-particles.

The effect of the electric field on the β-particles is to deflect them towards the positive plate, causing them to traverse a parabolic path, while under its influence. The path after leaving the field becomes rectilinear. The effect of the magnetic field is to deflect the β-particles along circular paths in a plane perpendicular to the field, hence perpendicular also to the deflection produced by the electrostatic field. The electrostatic deflection is proportional to $1/mv^2$, while the magnetic deflection to $1/mv$. These two deflections being at right angles to each other when the two fields are applied simultaneously, if the velocity v of the β-particles varies continuously; a parabolic trace will be obtained on the photographic plate, each point on it corresponding to (β-particles of a certain velocity. By measuring the coordinates of any point on the trace the corresponding values of v and e/m can be calculated exactly in the same way as in Thomson's determination of e/M for positive rays.

In the actual experiment, in order to make accurate measurement, the electric field or magnetic field was reversed, when a second parabolic trace which was the exact image of the previous one was obtained parabolic traces actually obtained were, however, found not to coincide with the parabolic curves, shown by dotted lines in the figure, that would be expected on the assumption that e/m was the same for all velocities. This discrepancy between theoretical prediction and actual result could be interpreted only on the basis that e/m was not constant in the case of β-particles moving with velocities of the same order as that of light (the velocities measured in the experiment varying from 2.36 to 2.83×10^{10} cms./sec.), but diminished as v increased. If it could by assumed that e the charge on the β-particle was independent of its velocity, the observed result must be regarded as indicating that the mass m of the β-particle increased with the velocity v.

This fact had already been predicted by Lorentz from the following considerations. J.J. Thomson treating the electron as a uniformly charged sphere of radius a moving with a uniform velocity v along a straight line, had shown, as we have already seen, that the mass of the electron was entirely due to its charge e, and obtained the expression $2e^2/3a$ far this electromagnetic mass, on the basis of the classical electromagnetic theory, in which the charge is regarded as throwing out tubes of force into space, the tubes having energy associated with them. The derivation of the electromagnetic mass) is valid only uncondition that the velocity v of the electron is small compared with the velocity c of light. Heaviside pointed out that as the velocity v became great, the tubes of force representing the charge crowded together tending to set themselves at right angles to the line of motion.

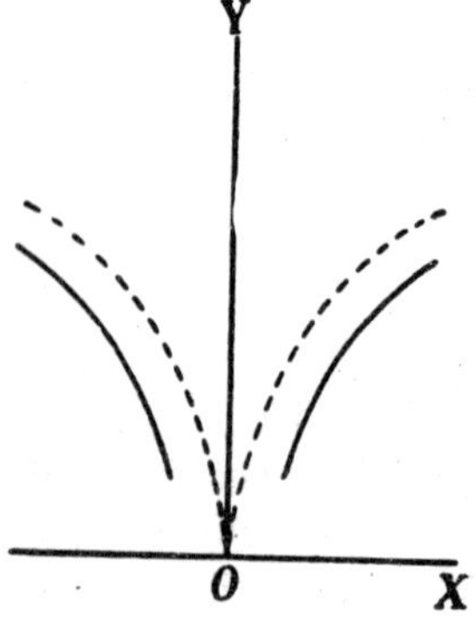

Fig. 11.4

The velocity of β-particle varies considerably. For example the velocity of β-rays emitted by *v*X, is about 0.6C.

The β particles emitted by any one radio element have also widely different velocities. The upper limit which gives the maximum velocity could be fixed exactly.

As the α-particles, in their passage through matter, knock electrons but of atoms and produce ionisation, their speed is slowly reduced without, however, being deviated to any great extent from their straight path, except on rare occasions when they collide with nuclei. But when the β-particles knock electrons out of atoms and produce ionisation, they are not only reduced in speed but also deviated very much from their paths *by impacts with electrons or nuclei* unless they possess of extremely high velocities. These peculiarities of the passage of β-particles through matter can be easily observed in a cloud chamber. The straggling effect becomes so large that β-particles do not exhibit a well-defined range as in the case of α-particles. For the same reason, the absorption and scattering phenomena interfere with each other in the case of β-particles. Another great difficulty is the want of homogeneity of β-rays, having a complex energy spectrum. A further complication arises from the fact that the β-particles change in mass with change in velocity. Several researches, however, have been made as regards absorption and scattering of β-particles, into the details of which we cannot enter here. We may just mention some of the results obtained.

It is interesting to note that the highly inhomogeneous β-particles from radioactive substances show an absorption which is almost exactly *exponential* up to a certain thickness of matter. This is, however, only an accidental result of the initial energy distribution and of the scattering effect. Experiments made with a beam of homogenous β-particles show that the absorption is not really exponential. After traversing a certain thickness of the absorber the energy of the β-rays is no longer large enough to be detected. Hence, one may speak of a *range*, for β-rays, as in the case of α-rays, although ill-defined on account of the great preponderance of the straggling effect. The ranges of β-particles are large, about 100 times those of α-particles and are found to increase rapidly with the speed of the particles, roughly independent of the nature of the absorber.

The ionising power of the β-particles, which is much less than in the case of α-particles is lost below a certain critical velocity; but above this, the ionisation rapidly increases to a maximum value. After reaching the maximum the degree of ionisation falls with increasing speed to a constant value. The maximum ionisation occurs at a velocity corresponding to an acceleration of the particles by a P.D. of 1000 volts, more than 1000 ion-pairs being formed per cm. in air at N.T.P. Very high velocity β-particles produce only about 50 pairs per cm, which is about 200 times less than in the case of α-particles of the same speed.

Scattering of β-rays, experiments show that multiple scattering plays a prominent part. With gold foils, for instance, as much as 50% of a β-ray beam is diffusely reflected by "*multiple Scattering*". As the angle through which a particle is scattered is inversely proportional to the square of its energy, β-rays, being much less energetic, are scattered much more than α-particles. "*Single scattering*" can take place also by impact both with electrons and with nuclei, as demonstrated by the cloud chamber tracks of β-particles, where abrupt changes through more than 90° are observed.

11.4 THE GAMMA-RAYS

γ-rays can be easily identified by their, property of, high penetrative power.

γ-rays have more or less same nature of X-rays but, they have extremely short wavelength. When the γ-rays are passed through thick absorbing material, one can find the true exponential absorption. Most γ-rays from radio active substances are heterogeneous.

The interaction of γ-rays with matter is a very complex phenomenon. From the experimental results so far obtained we may distinguish three different types of effects, *viz., the protoelectric effect, the scattering effect and the positron-electron pair production.* The first two are met with X-rays while the third only with γ-rays. The relative importance of the three processes depends both upon the energy of γ-rays and the nature of absorber.

The Photoelectric Effect of Gamma-Rays

When comparatively soft γ-rays interact with heavy elements almost all the absorption is due to the photoelectric effect. In this process, the γ-rays knock out electrons from the atoms of the absorbing material, which results in the ionisation of the atoms and the emission of characteristic fluorescent radiations. The energy of the photoelectrons produced by γ-rays is very high, since there takes place a complete absorption of the total energy of the incident γ-rays by the electrons. The γ-rays do not ionise directly, but through the intermediary of the electrons which they eject from matter. That no direct ionisation takes place follows from the fact that the penetrating power of the secondary electron liberated by the γ-rays independent of the path previously traversed by the γ-rays. Had the γ-rays given up energy by direct ionisation along the path covered before they ejected the secondary electrons, the energy and consequently the penetrative power of the latter would depend upon the length of that path, which, however, is not found to be the case. The inability of γ-rays to ionise directly is due to the fact that they carry no electric charge. For the same reason they *cannot be directly detected or measured* by the ordinary methods used in the case of charged particles, such as α- and β-rays. All the devices used to detect or measure the intensity or energy of γ-rays are, in fact, sensitive not to the γ-rays as such, but to the secondary radiations they produce by interacting with matter. The *fluorescent radiations* produced by γ-rays have the same features as characteristic X-rays, dependent on the nature of the substance that emits them.

As in the case of X-rays, so also with γ-rays, there occur' coherent scattering without change of wavelength and incoherent scattering or *Compton effect* with change of wavelength. This second type of scattering plays a prominent part when *hard* γ-rays interact with *light* elements, ejecting electrons from the atoms of the absorber. These electrons are called recoil or *Compton electrons* in order to distinguish them from photoelectrons. The energy absorbed by these Compton electrons is only a small fraction of the total energy of the incident γ-rays, unlike in the case of the photoelectrons.

The Production of Positron-Electron Pairs by Gamma-Rays

The γ-rays interact with matter, also in a third and novel manner, a peculiar photoelectric effect which has no classical analogy and is not met within ordinary X-rays, that is,, the creation of electron-positron pairs. As the story of the experimental discovery of this phenomenon coincides with that of the discovery of the positron in cosmic ray researches, we shall postpone it to a later chapter. We may mention here briefly the main features of this effect of γ-rays. The pair production process assumes a great importance when *high energy* γ-rays interact with *heavy elements*. The energy of the γ-rays required to produce this effect must be at least that which is equivalent to the rest-masses of the electron and the positron produced. For less energy, the phenomenon does not occur, while for greater energy, the excess, which is left over the amount spent in the creation of the two material particles, the electron and the positron, is communicated to the formed particles as kinetic energy of motion. Hence, in this case also there is complete absorption of the total energy of the γ-ray which simply disappears, as in the photoelectric effect. The positron or positive electron has been shown to be identical in nature with the electron having the same mass and same fundamental charge, except for the sign of the charge, *i.e.,* the charge it carries is positive while that of the electron is negative. The phenomenon is an excellent illustration of a vertiable *materialisation of radiant energy*, undreamt of in Classical Physics. It also offers a very good experimental confirmation of the theory of relativity.

To measure the wavelength of γ-rays is not an easy task. The wavelengths can be measured successfully, by the use of crystals as diffraction gratings.

Rutherford and Andrade first time used this method, for the measurement of wavelength of γ-rays.

Thibaud and Frilley used the rotating crystal method. They used strong source of γ-rays. During the experiment they kept the photographic plate at a large distance.

Now let us study about the important applications of the radio activity.

Radioactive elements are used in the preparation of, Luminescent paints. These paints are, coated, on the points and, figures of clocks and watches

Today Radium therapy has got much more importance. Various skin diseases can be cured by this method.

Radioactivity has got many number of medical application. Artificially produced radioelements are used in biological researches.

The organic constituents of living body, Metabolism in animal and plant systems, can be easily understood by Radioactivity.

12

THE THEORY OF RELATIVITY

12.1 INTRODUCTION

Einstein the great scientist formulated the theory of relativity for the first time. So, we can say the term relativity, applies to Einstein's theory.

Space, time and mass, can be explained thoroughly, with the theory of relativity.

Now let us deal with space.

Physics which is directly concerned only with the *measurement of space* reduces the study of space to the measurement of a length, *i.e.*, of the distance between two points in space. In Classical Physics space thus understood is *absolute*, "by its very nature remaining the same and immobile without relation to anything external," as Newton has declared. This means to say that the length of an object is independent of the conditions under which it is measured, such as the position or motion of the object or the experimenter. The classical notion of absolute space led to two useful results, *viz.*, *first,* a fixed frame of reference by which the position or motion of any object in the universe can be determined, and *second*, the constancy of the geometrical form (aubical spheiwair etc.) of an object, independent of the position or state of motion of the object or observer.

Fig. 12.1 : Issac Newton.

While dealing with the measurement of time Physics deals with, the method of comparing, different phenomena. For example clucks measuring time are compared to and regulated by the periodic rotation of the earth.

Newton's idea about time is quite different, and absolute, by its very nature, flowing uniformly without reference to anything external.

After, time now let us try to understand some thing about mass.

While dealing with mass one must remember about, the volume and density of the body whose mass is to be calculated.

Mechanics, mass is given by the ratio between the force acting on a body and the acceleration thereby acquired. Newton's Second Law of motion which states that the force is proportional to the change of momentum it produces, implies that the *mass of a body is absolute* and constant, independent of the motion of the body or the observer.

Einstein's theory rejects this absolute nature of the fundamental quantities, space, time and mass, postulated by classical mechanics, by denying their independence from the position or motion of bodies or observers.

It must be noted, however, that Einstein's theory does not break off completely from everything contained in Newtonian mechanics, since the latter has clearly enunciated a certain principle of relativity which is preserved in the new theory also as one of its fundamental postulates. *The Newtonian principle of relativity* may be stated in the words of Newton as follows: "Absolute motion which is the translation of a body from one absolute place to another absolute place can never be detected; for, translatory motion can be perceived only in the form of *motion relative to other material bodies*." From the context the absolute motion referred to by Newton is simply uniform rectilinear motion of a body between two points in absolute space which is considered as a fixed frame of reference. What Newton meant by his principle of relativity is that the laws of mechanics are unaffected by the uniform rectilinear motion of the system of reference. Hence, it is impossible by mechanical experiments performed within the system to detect the uniform rectilinear motion of the system. We can measure therefore uniform motion only relatively with observations made from outside the moving system and not with those confined to the system. This is the reason why Newton law away with the distinction between a state of rest and that of uniform rectilinear motion, as is seen from his first law of motion.

Let us consider a person inside a train. At a station, when the train at rest, to thrown a ball vertically upwards and finds it to fall vertically downwards. Now, let the train be moving at a uniform speed along a straight track. The person in the moving train repeats his experiment and finds the same result as when he along with the train was at rest at the station, *viz.*, the ball thrown up vertically comes down vertically. Hence, he realises that from the behaviour of the ball or similar mechanical experiments performed within the train it is not possible for him to say whether he and the train are at rest or in uniform motion; he cannot therefore deduce anything about his motion.

If another person, stationed outside-the moving train and at rest, watches the same experiment conducted within the moving train, he finds that the ball thrown up vertically does not fall vertically down but travels along a parabolic path from which he can readily detect the motion of the train relative to himself. Hence, all rectilinear uniform motion can be measured only relatively. It may be noted that if the motion of the train is not uniform or rectilinear, as happens when the train turns round a curve or its speed is accelerated, the ball thrown up vertically will not fall vertically down even to the person inside the train so that non-uniform motion can be detected by observations made within the system itself. Hence, the principle of relativity of Newton applies only to uniform rectilinear motion.

Einstein has been able to enrich this principle of relativity and make it produce wonderful results by removing from it the classical notions of absolute space and absolute time and by introducing a new and original idea, *relativity of simultaneity,* in contradiction to the classical absolute simultaneity.

12.2 EINSTEIN'S THEORY OF RELATIVITY

Einstein realised that motion relative to material bodies has more physical significance, we know well that, in classical mechanics, time was considered as absolute, and universal.

According to Einstein there is no absolute time, as time may be variable, from one to another.

For Einstein, physical event was, never merely a fact.

Einstein, while explaining his ideas upholded two things. Let us study one after another.

Special or Restricted Relativity and General Relativity. The special theory was formulated in 1905 to deal with problems of uniform rectilinear motion, *i.e.,* where one observer and his measuring instruments are moving with constant velocity with respect to a second observer and his instruments. It chiefly aimed at solving the riddle of absolute motion, so much disputed in connection with optical phenomena. In 1915 the theory was farther extended to interpret more general cases of non-uniform motion, *i.e.,* where one frame of reference moves with an accelerated velocity relative to the other. This extension is called 'general relativity' because it is applicable to bodies with complicated accelerated motions, as in the case of heavenly bodies, including the earth; its chief achievement is the explanation of the law of gravitation on a more refined and accurate basis than that given by Newton. We shall now outline the main features of the theory of Einstein under both aspects.

The two fundamental postulates used in the development of the special theory are :

(i) The laws of physical phenomena are the same when stated in terms of two systems of reference in uniform translatory motion relative to each other.

(ii) The velocity of light in vacuum is a constant, independent not only of the direction of propagation but also of the relative velocity of the source and the observer.

The first, which is a *principle of equivalence*, is a generalisation from a wide range of physical experience. It appears as a natural extension to all physical laws of the Newtonian principle of relativity which applies directly only to mechanical laws. All physical processes are based on the fundamental equations, of mechanics. Hence, if the latter are the same with regard to all frames of reference that move uniformly in a straight line, the equation of all physical phenomena must follow the same principle of equivalence. There is no implication, however, that this postulate is self-evident; it has to, be tested for its validity like the assumptions made in any physical theory by comparing predictions made by it with experimental facts.

The second postulate affirming the constancy of the velocity of light as well as its independence from the relative velocity of source and observer appears to represent a simple experimental fact. Of course, the absolute constancy of the velocity of light; cannot be directly verified by experiment, but it is implied in the very definition of simultaneity. As regards its independence from the relative motion of the source and observer the different experimental measurements of the velocity of light, such as Michelson's rotating mirror method, Romer's astronomical observations on the satellites of Jupiter and De Sitter's work on variable stars, offer a physical justification for its acceptance.

Of the two postulates, the second is the more important and more basic to the theory than the first. For, the theory departs from the classical ideas not through the first postulate of relativity but through the second, the constancy of the velocity of light. To pass from the coordinates which describe a physical event in one system to the coordinates which describe the same event in another system moving uniformly and rectilinearly with respect to the first, without violating in any way the principle of equivalence, the universal constancy of light in both systems must be used. From such a combination of the two postulates there follows a series of very important deductions, such as the futility of ether, the intrinsic quantum nature of radiations, the variation of the mass of a moving body with its velocity and the universal mass-energy relation.

While dealing with the special theory of relativity it is needed to study about futility of ether.

The existence of ether has for long remained a necessary axiomatic assumption in the explanation of all natural phenomena, such as gravitation, propagation of energy in vacuum, electric and magnetic fields, etc. in which there is apparently *action at a distance*. But it had to be defined by the exigencies of the circumstances and in consequence often resulted in glaring contradictions. Thus, ether was supposed to be *rarer than the most rarefied gas*, which penetrated all matter and did not offer any resistance to the propagation of light waves and motion of material bodies. At the same time, it had to be the *most rigid solid* to account for the transverse nature of the light waves demanded by the phenomenon of polarisation. Hence it appears to be endowed with intrinsic properties which are contradictory. Such an 'ether' medium was the life-centre of Fresnel's elastic solid theory of light waves. Maxwell's electromagnetic theory of radiations, dispensing with the elastic solid idea and considering light as a result of quick alternate electric polarisation in the medium, minimised the difficulty to some extent.

Einstein, however, was impressed by the fact that Maxwellian electrodynamics, as originally interpreted and applied, led to asymmetrics which did not show themselves in any observation. Reflections such as these, combined with the insoluble problem concerning motion in ether, led Einstein to first reduce ether to a mere shadow and ultimately ignore it, adopting as a fundamental principle that relative motion alone determines the nature of all the phenomena observed.

With the omission of ether in the theory of relativity, another important conclusion followed concerning the intrinsic nature of radiations. The classical view of considering them as perturbations in ether had to be remodelled on new lines. Radiations came to be considered as particles of energy, which propagate in absolute vacuum with a velocity that, depends not on the physical properties of the medium but upon their own intrinsic nature, *i.e.,* upon their energy content, which is, however, a function of the constant standard the velocity of light. This new conception of radiations is forced upon us not only by the theory of relativity, but also by the observed facts of black body radiation, photoelectric, Compton and Raman effects and their theoretical explanation by Planck's quantum or corpuscular theory. It may be noted here that if radiations are to be considered as particles of energy the commonly accepted *classical distinction between matter or mass and radiation or energy has to be given up.*

Although the theory of relativity leads indirectly to the new quantum conception of radiation by its ignoring ether, yet it does net deny the validity of the laws of electromagnetism. In fact, it leaves electromagnetic field laws just as they stand, unlike in the case of mechanics. This is to be expected, since the theory of relativity actually developed out of experiments, on light which forms part of electromagnetic phenomena. What is of great interest is that the new theory envisages together both the corpuscular and wave aspects of electromagnetic radiations. As regards the wave aspect, it introduces a further refinement by considering the distinction between the electric and magnetic fields as a relative one, depending upon the frame of reference employed. This conclusion can be readily established by means of the following simple consideration. An electric charge at rest in a system S will produce, as observed in that system, only an electrostatic field. But to an observer in another system S', in motion relative to S, the same charge will constitute a current-element accompanied by a magnetic field.

We know well that a body which is not subjected to force, remains at rest or in uniform motion, which can be explained as inertial systems.

According to Einsteins theory, in all such systems the physical laws have the same form.

But classical mechanics, deals in different way. According to this, the difference between a body moving with uniform velocity and a body having accelerated motion lies in the fact that, the forces of inertia come into play.

Thus when a body is started or stopped suddenly, a jerk is felt; this is because, owing to inertia, a body at rest tends to remain at rest and a body in motion continues to be in motion, so that such bodies oppose any change in their actual state. Similarly the centrifugal force due to rotation is also an inertial force. In these cases where inertial forces arc acting the motion can be detected by experiments performed in the system. Hence, acceleration was considered as absolute. But Einstein generalised the principle of relativity so that it applies even to accelerated systems, and acceleration, in consequence, need not be considered absolute.

Coming to the particular case of acceleration due to gravity, the classical conception of gravitation presupposes *action at a distance* and postulates the existence of a force of attraction between material bodies. For instance, a mass, such as the earth or the sun, is surrounded by a space in which there is a latent force ready to act upon and produce movement of a body that penetrates into that apace. In consequence, the body falls towards the mass, as if attracted by it. This power of attraction exists permanently in the space near the mass even if there is no body manifesting its existence. The region of space surrounding a mass thus constitutes the gravitational field winch imparts a uniform acceleration known as the acceleration due to gravity to all bodies located in that field. The mass seems to act at a distance sinur the attraction is exerted even through vacuum where there is no ponderable matter, as for example, in the case of the sun attracting the planets.

With the introduction of ether, the difficulty arising from this action at a distance was minimised but not wholly removed, since ether brought in ifs wake other more serious difficulties which could not be easily explained. General relativity gives altogether a new interpretation of the phenomenon of gravitation, where action at a distance completely falls out of the picture.

Einstein started with the revolutionary idea that the classical, picture of a gravitational field is artificial, since it is possible for ass-observer not to detect such a field at all by choosing a suitable franie of reference. For instance, consider a stone attached to the end of a string and whirled round, the stone moving along a circular path. We know that a tension is developed in the string called the centripetal force. To an observer on the stone there is a repulsion called centrifugal force equal and opposite to the centripetal force, while to the one who is whirling the stone there is no such centrifugal force. Einstein saw that this apparent repulsive force is due to the curvilinear motion of the stone and he extended the idea of such curved motions to a sort of curved space in order to explain gravitational force. He then generalised these facts into what is known as the principle of equivalence, according to which the effects of a gravitational field are precisely the same as those due to uniform acceleration of the material frame of reference relative to which the phenomena are observed, this acceleration being equal and opposite to the acceleration which the gravitational field would give to a particle in the frame. Hence, there can be no difference between a stationary gravitational field due to attracting matter and the field due to uniform acceleration of a frame of reference moving outside all gravitational field, provided its acceleration is equal and opposite to the acceleration due to gravity, so that one can be replaced by the other.

This principle may be illustrated by the following simple example. Imagine an enclosure, from which nothing can be known of what goes on outside it, falling *freely under the action of gravity*. To an observer within it bodies put anywhere and left to themselves would float just at that place; a horizontal projectile would trace a straight line; bodies would have no weight; thus all traces of gravitation would have disappeared for him. A gravitational field can therefore be suppressed by a system such as the enclosure having a proper acceleration. Conversely, suppose the same enclosure is *outside all gravitational field*, at an infinite distance from any mass, pulled up by a rope and kept in uniformly accelerated motion, the acceleration being equal to that of gravity. For an observer within it everything would take place as if ho were in a gravitational field; bodies would fall with respect to the enclosure in accelerated motion, the acceleration being the same for all and equal to the acceleration doe to gravity; bodies would therefore have weight; a horizontal projectile would no more travel in a straight line but along a parabola. A gravitational field can therefore be imitated by giving to a system of reference an acceleration equal and opposite to that of the field. It must be noted, however, that only a uniform field can be replaced fully by a single system of reference. This ia possible only in a small region of space. Considerations over larger regions lead to varying fields and no single reference system will now be sufficient; both accelerated and non-accelerated systems would have to be used and the problem gets extremely complicated.

The principle of equivalence explains why the inertial mass of a body is equal to its gravitational mass. The *inertial mass* is a coefficient which measures the resistance of inertia of the body opposing the action of a force. If the acceleration given to a body by a force F is α then its inertial mass is the ratio $F/\alpha = m$. The gravitational mass is a coefficient which determines the force of attraction experienced by a body in a gravitational field. The weight of a body of mass m is $w = mg$. It is found experimentally that $w/g = m$. The inertial mass is therefore equal to the gravitational mass. The same property of a body manifests itself according to circumstances as inertia or as gravity. The force of gravity is ultimately a force of inertia. For a body in space, either there is a state of accelerated motion and no gravitational field or the body is at rest in a gravitational field. It is impossible to distinguish between a field of force due to inertia and a field of force due to gravity.

The principle of equivalence is found to hold good for electrical and optical phenomena as well. Thus, for instance, in the case of a ray of light propagated rectilinearly with respect to a system S of uniform motion it is possible to show that the path of the same ray of light is no longer rectilinear, considered in relation to a second system S' of accelerated motion.

12.3 GALILEAN TRANSFORMATION AND ELECTROMAGNETIC THEORY

Before the advent of the electromagnetic theory it was believed that light waves were of the same type as the waves in an elastic medium ex. sound waves.

Such waves have velocities determined by the elasticity and density of the medium. Since light waves are transverse in nature as the polarisation experiments show, the medium needed for their propagation must have the properties of a solid. Further, the medium must be infinitely rigid in order that there might be no longitudinal light waves. Since the velocity of light is very high, the density of the medium must be very low while its elastically must be very high. The medium must pervade all space including the interstices of matter in order to account for the propagation of light through matter. However it must not hinder the movement of celestial bodies which are known to move in non-viscous medium. This hypothetical solid medium, endowed with such contradictory properties as very high elasticity, very low density and infinite rigidity was called the *ether*. Many scientists believed that the absolute frame of reference about which Newton had talked and in which his laws of motion were supposed to be strictly valid, was nothing but *ether*.

The formulation of the electromagnetic theory of light by Maxwell in 1864 and its experimental confirmation through the series of experiments by Heinrich Hertz, led to the realization that light was a kind of electromagnetic wave which is a combination of very rapidly oscillating and mutually perpendicular electric and magnetic fields propagated through space with the velocity $c = 3 \times 10^8$ m/s. Later researches have established that not only light waves, but x-rays and γ-rays on the short wavelength side and radiowaves, radar, microwaves etc. on the long wavelength side are examples of electromagnetic waves. After the formulation of the electromagnetic theory of light, the view that light waves are a kind of elastic waves in the hypothetical medium ether was given up.

Maxwell, Hertz and others still held the view that the electromagnetic waves required a medium for propagation and ether was supposed to be this medium. However the expected properties of ether were so strange that serious doubts were expressed about its existence. So it became necessary to investigate very carefully the reality of the existence of ether. The most important of these experimental investigations was carried out by the two American physicists A.A. Michelson and E.W. Morley in 1887. This along with some other important experiments having a bearing upon the question of the existence of ether are discussed below.

Aberration of Light

It is well-known that the observed positions of the stars vary slightly throughout the year. There are two reasons for such variation, *viz.*, parallax and *aberration*. Parallax is important for the nearer stars and is due to the finite distances of these stars from the solar system. On the other hand, the aberration of star light 5s due to the fact that the apparent direction of a star in the sky is different from its true direction. This causes the apparent position of the star in the sky to describe a circle or ellipse during the whole year and was first studied by J. Bradley in 1728. The shift in the observed direction of a star due to aberration which is much greater than that caused by parallax depends on the relative velocity of the light source (*i.e.*, the star) and the observer (*i.e.*, the earth). Since this is the same for all stars, the shift due to aberration of light has the same magnitude for all stars.

The earth moves in its orbit with a velocity $\upsilon = 30$ km/s. As will be evident to see a star the telescope should be tilted towards the direction of the earth's motion from the true direction of the star by an angle a given by.

$$\alpha \approx \tan\alpha = \frac{\upsilon}{c} = \frac{30\times10^3}{3\times10^8} = 10^{-4} \text{ radian} \neq 20.5'' \text{ of arc.}$$

Six months later, the telescope has to be tilted by the same angle in the opposite direction.

The conclusions drawn from the observations of stellar aberration are that if there is such a medium as ether through which light waves are propagated, then the stars must be at rest with respect to it while *the earth moves through ether with its observed velocity.*

Trouton and Noble's Experiment

If the earth moves through ether, there should be an ether wind like the wind felt by a passenger in a railway compartment blowing in a direction opposite to the motion of the train. A number of experiments were performed to detect this ether wind. The most famous of these experiments was the Michelson-Morley experiment to be described in the next section. Another such experiment was performed by Trouton and Noble (1903) who tried to detect the torque on a charged capacitor due to the motion of the earth through ether.

The charge on any one plate of the capacitor moving with the earth behaves like an electric current which produces a magnetic field at the position of the opposite plate. Since the charge on this second plate also moves with the same velocity it is equivalent to an electric current flowing in the opposite direction so that the magnetic field exerts a force on it, determined by Biot-Savart's Law. An equal and opposite force acts on the first plate for the same reason. The two together give rise to a torque on the suspended condenser which should thus be turned parallel to the *ether wind*. By measuring the angle through which the condenser is turned, the turning moment can be measured.

Very careful measurements by Trouton and Noble and later by others failed to detect any such turning moment. Thus the expected ether wind could not be detected.

12.4 IMPORTANCE OF THE THEORY OF RELATIVITY

According to many scientists and mathematicians, the theory is a marvelous synthesis, from the physical point of view. At present the theory is admitted by all scientists.

Certain postulates of the theory such as the constancy of the velocity of light as well as its being the superior limit of all velocities attainable by material bodies, the relativity of simultaneity, etc. appear not to be finally demonstrated in all cases without exception. But the theory as a whole is to be considered as a *product of genius*, which perceives the complicated relations of things by intuition rather than by the pain-process of experience and verification. Further, its epoch-making conclusions, chiefly as regards mass and energy and the law of gravitation, have been experimentally confirmed, while the mathematical treatment involved proceeds so logically that it has not been seriously called into question so far.

By a new definition of simultaneity and by adoption of the constancy of the velocity of light, Einstein has been able to show that physical phenomena can be interpreted without the intervention of ether. Old anomalies of mechanics and electrodynamics vanish into thin air; classical physics is cleansed of its wrong absolute presuppositions, while its laws are comprehended within the framework of a new wholesale scheme. Physics dealing with the actual measurement of space, time and mass in one form or other must necessarily make use of concrete events of concrete bodies even when the problem is treated in an abstract manner. Classical physicists following the lead of Newton seem to have made the mistake of measuring abstract space, abstract time and abstract mass which, not being referred to any concrete case, are in a sense universal and absolute, but as such are only mental concepts. Physics, however, clearly does not deal with the measurement of mental abstractions. Herein lies the merit of the theory of relativity in having put things in their right places. Concrete physical space, time, mass and motion cannot be measured absolutely, since the material bodies to which they have necessarily to be referred are always dependent on other material bodies in all their activities and passivities. This becomes self-evident when one considers the motion of celestial bodies. There is no absolute motion as such in the material universe : the earth and other planets are moving round the sun and the solar system as a whole about other constellations and so on.

The theory does not deny real and intrinsic properties belonging to objects such as dimensions or intervals of time; it clearly admits that the proper dimensions and times measured by observers at rest in a system are absolute, *i.e.,* the same for all. What the theory claims to have proved is that these properties cannot be directly measured ai easily as ½ as supposed, at least in the case of an observer in motion with respect to the objects. Under certain conditions two observers using the same units find different values for the dimensions of the same object or for the same interval of time. Hence, objective reality, which is more complicated than any isolated single observer could have thought, cannot be defined by space or time taken separately, since they vary with observers. A combination of both, however, is found to be invariant; this forms the space-time continuum in which world events are to be observed and measured. Against this new background mass and energy become mere aspects of one and the same entity; the different principles governing physical phenomena merge into one single universal principle of mass-energy equivalence; motions of astronomical bodies and the more familiar law of gravitation receive a more accurate explanation, while the old interpretations of these phenomena are retained as first approximations which require correction.

13

WAVE MECHANICS

13.1 INTRODUCTION

In 1926 Schröedinger developed a mathematical theory, later which has received the name of wave mechanics.

When old quantum theory was failed, Heisenberg thought about new ideas. He came to the conclusion that one should try to develop an atomic mechanics on the basis of quantities which are actually observable in the atomic domain.

It was on this basis that Heisenberg (1925), for the first time, invented the *matrix mechanics* which was further developed by Born and Jordan (1925). Soon afterwards, Erwin Schrödinger (1926) independently developed *wave mechanics* on the basis of de Broglie's hypothesis of wave-particle duality and proposed a wave equation for describing the motion of atomic systems. The Schrödinger wave equation has become the foundation stone of non-relativistic quantum mechanics or wave mechanics.

Later Schröedinger (1926) showed that though the matrix mechanics of Heisenberg and his wave mechanics were entirely different in form, in content they were identical.

Subsequently P.A.M. Dirac (1930) proposed a general formalism of quantum mechanics which is a unifying concept of the above two formulations. Besides, the theory of interaction of the radiation field with atomic particles was developed by Dirac (1927), Jordan and Pauli (1928) and others. The foundations of relativistic quantum mechanics were also laid about this time by Dirac (1928).

The space time behaviour of an atomic system is determined by the laws of probability. Just as the intensity of light at different points in interference or diffraction experiments is determined by the amplitude squared of light wave, so also in the case of an atomic electron, the probability of its being timed at a particular point is determined by the square of the amplitude of the corresponding wave function.

Thus, we may say that for any quantum system the wave function ψ (**r**, t) determines the entire space-time behaviour of the system. Out first step is to develop a wave equation satisfied by ψ (**r**, t).

This equation cannot be 'deduced' on theoretical considerations. We have to guess the equation. Its success will depend on its ability to explain various observed phenomena.

Some fundamental characteristics of the equation can be inferred:

(a) The differential equation satisfied by the wave function must be *first order in time*, so that if 'ψ is known at some point **r** at the initial instantly, then at any subsequent instant t, the function ψ (**r**, t) at the same point will be uniquely determined. This may be regarded as the basic postulate of quantum mechanics.

(b) The wave equation must be linear so that if ψ_1 (**r**, t) and ψ (**r**, t) are two solutions of the equation, then any linear combination (C_1 ψ_1 + C_2 ψ_2) w_1s0 a solution. This ensures the validity of the superposition principle discussed earlier.

(c) The wave equation must be consistent with de Broglie's hypothesis and correspondence principle, *i.e.*, the conclusions of the new theory should be valid even in those cases for which classical mechanics is applicable. Just as the laws of geometrical optics are obtained as approximations from the wave theory of light, so also the laws governing the motion of the wave-packets as a whole in de Broglie's theory should be derivable as approximations from the new wave mechanics. This may be termed "correspondence principle".

13.2 DIRAC'S WAVE MECHANICS

Dirace has succeeded to bring the non-relativistic theory into harmony with the principles of relativity by this relativistic wave equation of the electron.

Dirac succeeded in formulating a wave equation for an electron moving in a potential fields in such a way as to make it relativistically, invariant.

Although, the rather complicated theory of Dirac cannot be given here in detail, it must be emphasized that the solution of this new relativistic wave equation not only led to a complete interpretation of the fine structure of spectral lines, but also to a rational explanation of the spin and magnetic moment of the electron itself.

In Dirac's theory, in addition to the solutions corresponding to the normal electronic levels found experimentally, there were others which seemed to represent no observed facts. These solutions predicted the existence of certain states, in which the electrons possessed a negative kinetic energy and hence did not correspond to particles in any usual sense. These states, however, could not be ignored, since transitions must theoretically occur between them and the normal states corresponding to positive kinetic energy.

Dirac suggested that such a difficulty might be avoided if it were supposed that *all the negative energy states are normally occupied and, further, that the totality of electrons in such states produces no external field and hence not physically observable*. For, in such a case, the process of an incident electron passing 'into the negative state, which is already filled normally, is impossible. Further, since there are an infinite number of states of negative energy even in a finite volume, there should be an infinite density of electrons in negative energy states in all space. If it now be assumed that the number of electrons existing in the universe is slightly greater than the number of available negative energy states, then some electrons, finding no place in these negative energy states will have to occupy states of positive energy; these are the ordinary electrons which are observed.

Let it be further supposed that from the sea of electrons occupying the negative energy states, electromagnetic fields, say in an act of absorption in the atom, can raise an electron and bring it into the region of positive energy states; it then takes its place as an ordinary observable electron. There now remains, however, an empty place in the midst of the negative energy states, called the *Dirac, hole*, which has entirely the same character as a positive charge. Theoretical formulae show that this hole in the sea of negative states behaves as an exact counterpart of the electrons, *i.e.*, a particle identical in mass and magnitude of charge with the ordinary electron, but with a positive sign for charge.

Since at the time when the theory was first proposed the only fundamental positively charged particle known was the proton, Dirac thought at first that the above-stated holes corresponded to protons, although he was quite aware of the difficulties arising from the difference in masses of the proton and the electron and from the very stable existence of the proton. Bůt after the experimental discovery of the positron by Anderson in cosmic ray researches, this particle was identified M ith the Dirac hole. Hence in *Dirac's theory the positron is considered as an unoccupied state of negative energy, a perfect image of the negative electron*; its actual discovery argued to the essential correctness of the theory.

The most important consequence of Dirac's theory is the prediction of two novel processes, *viz.*,:

(i) *Materialisation of Energy*, and

(ii) *Annihilation of Matter*.

In terms of the theory, the first of these consists in a photon of energy greater than $2m_0c^2 \approx$ MeV raising an electron from a state of negative energy to a state of positive energy, which results in an *electron-positron pair production*, since the electron in the final positive energy state will be observed as an ordinary negative electron while the Dirac hole in the negative state produced in the process corresponds to a positron. This process cannot take place in empty space, because energy and momentum cannot be conserved; but can occur in the electric field in the neighbourhood of a nucleus which can take up the extra momentum. The second process, *viz., annihilation of matter*, corresponds to the reverse case of the fall of an electron from a high positive state into a hole in the negative state. The law of conservation of momentum demands that if this process were to take place in empty space, two light quanta, each of energy $m_0c^2 = \frac{1}{2}$MeV would originate. Both the phenomena have been experimentally observed. Recent discoveries (1956) of the anti-proton and the anti-neutron, which bear the same relation to the proton and neutron respectively as the positron does to the electron, have added further support to the basic ideas of the theory. In spite of these successes, Dirac's theory cannot be considered as perfect in every respect, on account of the several arbitrary and not readily understood assumptions involved, (such as the existence of an infinite density of electrons in space), as well as other very serious difficulties, with which it is-confronted when the interaction of the negative- energy electrons with the radiation field is considered.

13.3 INDETERMINACY PRINCIPLE

In 1927 Heisenber draw a very interesting idea, later which was known as principle of indeterminacy.

Wave mechanics says, it is the wave associated with a particle that represents all that, about the particle.

By making the wave packet very small (*i.e.*, ψ practically zero except within a very small region) the position of the particle can be more or less fixed, but since it can be shown that the velocity spread of such a packet is large, the velocity of the particle becomes indeterminate. On the other hand, if we consider a large wave packet with many crests, the velocity spread is so very small that the particle velocity can be fairly accurately determined, but the position of the particle becomes very uncertain.

Extending these results to the limiting cases of infinitely large and infinitely small wave packets it is readily seen that certainty about velocity involves complete uncertainty about position and *vice-versa. Hence it is impossible to determine simultaneously both position and velocity of a particle with accuracy.* Experiment can never enable us to say precisely that a given particle occupies such and such a position in space and that its velocity has such and such a magnitude and direction. All that experiment can tell us is that the position and velocity of the particle lie in between certain limits; in other words, there is only a *calculable probability* that it has a certain position and velocity.

Heisenberg, who was the first to realise these consequences of wave mechanics, has expressed them mathematically by means of the following equation known as the uncertainty relation : $\Delta x . \Delta p \approx$ h, where Δx; and Δp are the respective uncertainties involved in the simultaneous measurement of the position and momentum of a particle and A the Planck's constant. It states *that the product of the uncertainties in determining the position and momentum of a particle is approximately equal to Planck's constant.* According to it, the smaller the value of Δx, *i.e.,* the more exactly we determine the position, the larger the value of Δp, *i.e.,* the less exactly we determine the momentum. The converse is equally true.

The relation also brings out the fact that it is the existence of the constant h which prevents us from knowing simultaneously both the position and motion of a particle with accuracy, while if h were zero such simultaneous accurate knowledge would be possible. The constant h therefore represents an absolute limit to the accurate and simultaneous measurement of position and momentum, a limit which in the most favourable cases we may reach, but which we can never get beneath. The essential point insisted upon in the uncertainty principle is *not the simultaneity of determination of the two quantities but the accuracy involved in the simultaneous determination.* As a matter of fact, position and momentum can be simultaneously measured in the macroscopic bulk state as is done in ordinary physics, but the accuracy obtained is much less than that stated in the uncertainty relation. Since h, the ultimate limit of possible accuracy, is a small quantity (6.55×10^{-27}) the indeterminacy is completely masked by the experimental errors in the macroscopic state, while it becomes evident in the microscopic atomic state.

Although, the principle of indeterminacy was first formulated in connection with wave mechanics, it was soon realised that It governs all Nature and fundamental law. In every physical phenomenon there remains a margin of uncertainty which enshrouds it with a fundamental indefiniteness and discontinuity that cannot be overcome by any instrument of measurement, even of the highest precision. This generalised indeterminacy can be illustrated by one or two examples.

Determination of the Position of α-Particle by a Microscope

Suppose to attempt to determine accurately the position of an electron using some instrument, such as a microscope of very high resolving power. Since the lower limit of resolution depends upon the wavelength of the light employed to illumine the particle, it follows that we must use radiation of the shortest wavelength, such as γ-rays, if we wish to determine the position as accurately as possible. But the employment of γ-rays involves the Compton effect so that the electron experiences a recoil. Thus, at the instant at which we locate the electron by irradiating it with γ-rays and observing the scattered γ-rays, its momentum undergoes a discontinuous change. Furthermore, there is indeterminateness about the magnitude of this change, for it will vary according to the direction in which the scattered γ-ray leaves the point of impact. We cannot limit closely the range of possible directions of the scattered γ-rays without a serious loss of resolving power which would involve less definition of the position of the electron.

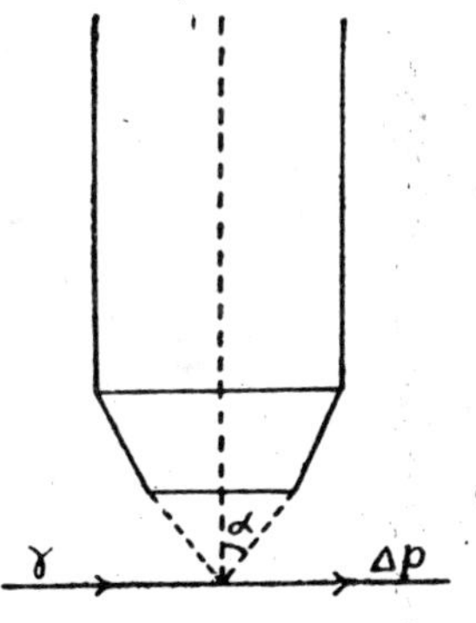

Fig. 13.1

From classical optical theory it can be shown that the resolving power of a microscope is given by $\Delta x = 1/2 \sin \alpha$, where Δx represents the distance between two points which can just be resolved by the microscope and hence the possible error that is involved in the determination of the position of the electron, λ the wavelength of the light used and α the angular aperture of the microscope.

Now, according to the corpuscular view, the minimum amount of light that could be used for irradiation is a single quantum $h\nu$. The electron can be actually seen only when it scatters this quantum into the microscope. But in this process the electron suffers a Compton recoil of the order of magnitude $h\nu/c$, the direction of which is indeterminate to the same extent as the direction of the scattered quantum. Since the scattered quantum can enter the microscope anywhere within the angle α, the uncertainty in direction of both the scattered quantum and recoil electron is given by α. Hence the component of momentum of the electron in a, direction perpendicular to the axis of the microscope is uncertain to the amount Δp given by

$$\Delta p \approx 2\frac{h\nu}{c}\sin\alpha \approx \frac{2h}{\lambda}\sin\alpha$$

The product of the uncertainties in the simultaneous determination of the position and momentum is therefore

$$\Delta x \times \Delta p \approx \frac{\lambda}{2\sin\alpha} \times \frac{2h}{\lambda}\sin\alpha$$

i.e., $\Delta x.\Delta p \approx h$ (Heisenberg's uncertainty relation).

Diffraction of a beam of electrons through a narrow slit, from the corpuscular standpoint we have to regard the phenomenon of diffraction of electrons as occurring due to the deflection of the individual electrons at the slit, either upwards or downwards and the diffraction pattern recorded on a photographic plate as arising from the statistical superposition of the electrons in the beam. Every electron which is registered on the plate must have passed through the slit, but at what

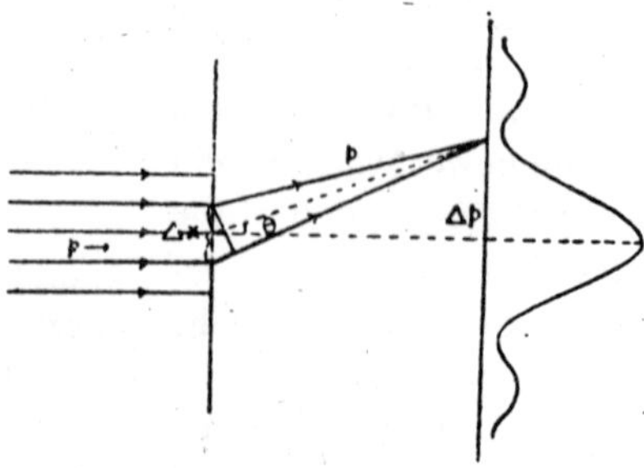

Fig. 13.2

place in the slit is indefinite. If Δx is the width of the slit, I the uncertainty in the specification of position of the electron perpendicular to the direction of flight is given by Δx (Fig 13.2). Since the electron is deflected at the slit it acquires an additional component momentum perpendicular to the original direction of flight. If p is the momentum of the electron, the component perpendicular to the initial direction is $p \sin \theta$, where θ is the mean angle of deflection. Since the electron may be anywhere within the diffraction pattern the uncertainty in the knowledge of the above-stated component is

$$\Delta x \approx \Delta p \; 2p \sin \theta$$

On the basis of the wave theory of diffraction

$$\sin \theta = \frac{\lambda / 2}{\Delta x}$$

$$\Delta x = \frac{\lambda}{2 \sin \theta}.$$

13.4 THE WAVE NATURE OF MATTER

In 1924 the French physicist Louis de Broglie, put forward his suggestion that, matter which is ordinarily considered as made up of discrete particles, atoms, protons, electrons... . But in the meanwhile scientists have admitted about the dual nature of matter (that is radiation nature.)

This dual nature of matter gave encouragement to those scientists who were attempting to, reconcile existing concepts with the double aspect of radiation.

De Broglie's concept of "matter waves" was able to bridge over the almost insuperable gulf which Classical Physics had created between the two fundamental forms in which Nature manifests herself, *viz.,* matter and radiation, and arrive at a certain unity between them, since both matter and radiation could be conceived to possess a dual nature of the same type. But the synthesis was not to be as easy as it appeared at first sight.

At any rate, scientists, encouraged by the important suggestion of De Broglie, set themselves to elaborate it by means of precise mathematical theories as well as check it by delicate experimental researches. Thus arose a very interesting and important but difficult chapter in Modem Physics which comprises the phenomenon of *diffraction of material particles*, such as electrons, protons, etc., forming a solid experimental proof of the existence of waves associated with matter and the theories worked out along different lines such as the quantum, mechanics of Heisenberg and the *wave mechanics* of Schröedinger, wherein the "matter wave" concept not only received a firm mathematical basis and thereby became extremely useful in the interpretation of many a difficult problem in Atomic Physics but also led to a strange conclusion known as the *Principle of Uncertainty* that shook Classical Physics to its very foundations.

The aim of the present chapter is to give a brief sketch of the main features of this important branch of Atomic Physics. After stating De Broglie's concept of matter waves we shall describe some of the experiments carried out on the diffraction of electrons and other material particles. We shall then study the theoretical treatment of the problem given by Heisenberg and Schröedinger, without entering into all the complicated mathematical details involved. Finally, we shall consider the revolutionary Principle of Uncertanity that has resulted from the analysis of the wave nature of matter.

Louis De Broglie, attempting to develop a theory of radiation in terms of light quanta or photons, was led to the new conception of matter waves by the following considerations:

According to this principle, the two fundamental forms, matter and energy, in which Nature manifests herself, must be mutually symmetrical. Since radiant energy has been shown to possess a dual nature, wave and particle, matter also must possess the same dual nature, particle and wave.

For, the *Maupertian principle of least action* in mechanics and *Fermat's principle of least time* in optics implied similar conditions which were very suggestive. According to the, former, a moving particle chooses always that path for which the action is minimum, *i.e.,* the integral of the momentum over the path is a minimum.

The close analogy of the two principles belonging to two different branches of Physics argued to the probability of the behaviour of matter as a wavelike entity under suitable circumstances. Just as radiation, ordinarily treated as a wave, has to be supplemented with a particle characteristic for a satisfactory explanation of observed optical phenomena, so also material particles, considered in mechanics as having a corpuscular structure alone, might have to be supplemented with a wave aspect for a full understanding of their behaviour.

In 1913 Bohr basing himself on the quantum theory was able to give a very satisfactory explanation of the structure of electronic orbits in the atom and of the origin of spectral lines. According to this interpretation, an electron was shown to remain in an orbit of definite size for a considerable time without radiating energy, further the orbits available to the electron were rigorously selected by quantum rules from many orbits permitted by classical mechanics and it turned out that their radii were proportional to the square of integral numbers. Thus, the stable non-radiating orbits of the electron in the atom are governed by integer rules. De Broglie wishing to find a satisfactory explanation for these restrictions of the quantum theory argued that since the only phenomena involving integers in Physics were those of interference and modes of vibration of stretched strings, both of which imply wave motion, the electrons in the privileged orbits could not be regarded simply as material particles but a certain intrinsic periodicity should also be assigned to them.

These reflections led Louis de Broglie to make bold to suggest in his doctorate thesis the new idea of matter waves, for which he was later (in 1929) honoured by the scientific world with the award of the Nobel Prize. In his thesis he wrote that there is an intimate connection between waves and corpuscles not only in the case of radiation, but also in the case of matter. A moving particle of matter has always got a wave associated with it and the particle is controlled by the wave in a manner similar to that in which a photon is controlled by waves. To study the path of a beam of monochromatic radiation we use the wave theory, while to calculate the amount of energy transactions of the same beam we have recourse to the photon or quantum of energy $h\nu$. In a similar way, the electrons are particles, that is to say, their charge, mass and energy are observable in particle form; but if we want to find the path of a beam of electrons and whether and how it is reflected by objects, we must treat it as though it were a beam of waves. It is to be noted that the energy is carried by the electrons and not by the waves associated with them. In other words, whatever might vibrate in the matter wave it is not a strain in a real medium possessing energy.

The matter waves conceived by De Broglie are called "De Broglie waves" for obvious reasons. Since the essential feature of any wave is its wavelength, De Broglie next directed his attention on this quantity and derived an expression for the wavelength of the matter waves using the general equation of a standing wave system and the principles of the theory of relativity.

13.5 STUDY OF MATTER WAVES

One must remember that the electrons are the material particles which readily lend themselves to experimental investigation.

In 1927 the electron waves are experimentally detected, by two American physicists. Davission and Germer.

Davisson and Germer were studying the reflection of electrons from a nickel target. After, some time the target was transformed into a group of large crystals. Later the group of large crystals. Later the reflection of electrons became anomalous.

Both the physicists suspected that a beam of electrons might be diffracted from crystals like X-rays, which would mean that electrons behave like waves under certain circumstances.

The electron beam is produced from what is known as an *electron gun* G. This contains a tungsten filament F heated to dull red, when electrons-are emitted by thermionic action. The electrons are then accelerated in an electric field of known P.D. and collimated by suitable slits so that a fine parallel pencil emerges. This pencil is directed to fall on a large single crystal of nickel, known as the *target* T which is capable of rotation about α-axis parallel to the axis of the incident beam, the purpose of which will be made clear presently. The electrons are *reflected* from the crystal in different directions, the angular distribution being measured with a Faraday cylinder called the *collector* C, which is connected to a sensitive galvanometer and can be moved along, a

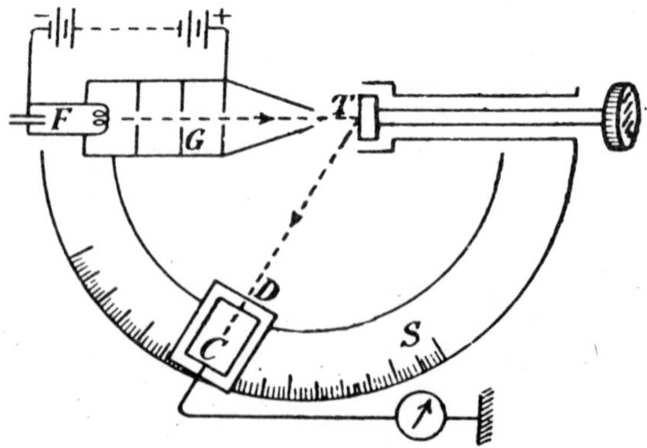

Fig. 13.3 : Apparatus of Davisson and Germer used for Diffracting Electrons.

graduated circular scale S, so that it is able to receive the reflected electrons at all angles between 20° and 90°. The collector has two walls insulated from each other. A retarding potential is applied between, the inner wall C and the outer D so that only the fastest electrons, *i.e.,* those possessing nearly the incident velocity, but not the secondary slow electrons excited by collisions with atoms, may enter the collector and be detected by the galvanometer. The accelerating potential used ranged from about 30 to 600 volts and the retarding potential was nine-tenths of the accelerating voltage. The whole apparatus was completely enclosed-highly evacuated and degassed.

The nickel crystal which is known to be of the face-centred cubic type is cut so as to have a smooth reflecting surface parallel to the lattice plane, *i.e.,* perpendicular to one of the diagonals of the cube. By rotating the crystal about the axis stated above, any azimuth of the crystal can be presented to the plane defined, by the incident beam and the beam entering the collector.

With the beam of electrons at 'normal' incidence on the surface of the crystal, when a diffraction effect from the surface layer acting as a plane grating was produced. For each azimuth of the crystal, a beam of low voltage electrons was made to fall normally on the surface of a crystal, the collector was moved to various positions on the scale S and the galvanometer current at each position noted. The current, which was a measure of the intensity of the diffracted beam of electrons, was plotted against the angle between the incident beam and the beam entering the collector, known as the "*colatitude*". The observations were repeated for different voltage electrons and several curves were drawn. It is seen that a "bump" begins to appear in the curve for 44 volt-electrons. With increasing voltage the bump moves upward and attains its greatest development in the curve for 54 volts at a colatitude of 60°. At higher voltages the bump gradually diminishes, there being hardly any trace of it at about 68 volts.

The bump in its most prominent state of development offers a convincing evidence for the existence of the electron waves.

Now according to experiment we have a diffracted beam at a colatitude of 50°. Applying the well known relation of a plane reflection grating, $n_1 = d \sin \theta$, n referring here to the first order, d being equal to 2.16 Å, as given by crystallographic analysis,

$$\lambda = 2.15 \sin 50^\circ = 1.66 \text{ Å}.$$

This excellent quantitative agreement is very important, showing as it does, that a beam of electrons does really possess wavelike characteristics.

With the beam of electrons falling 'obliquely' upon the crystal, when diffraction effect from a space-lattice, *i.e.*, from successive parallel layers of atoms in the crystal, analogous to Bragg's X-ray diffraction, was produced, as was indicated by the existence of *regular selective reflections* depending upon the velocity of the incident electrons. If the electrons were simple corpuscles one cannot explain these selective reflections at definite electron velocities. On the other hand, if it is assumed that electrons have waves associated with them, their wavelengths can be calculated using Bragg's formula $n_1 = 2d \sin \theta$, appropriate to the case. The values so found agree with those calculated by the De Broglie relation, $\lambda = (150/V)$.

If the electron gun and the collector are fixed, the glancing angles of incidence and reflection can be kept constant. Then varying the electron velocity, the galvanometer current is measured for each value of velocity. Plotting the current values against the corresponding electron velocities or accelerating voltages, a curve with several sharp maxima are obtained. The different maxima correspond to the various orders. From the curve the accelerating voltages which produce maximum reflection in the different orders can be found and hence λ calculated using De Broglie's relation. On the other hand, with the known values of the glancing angle θ and the grating space *d*, using Bragg's relation, λ is found.

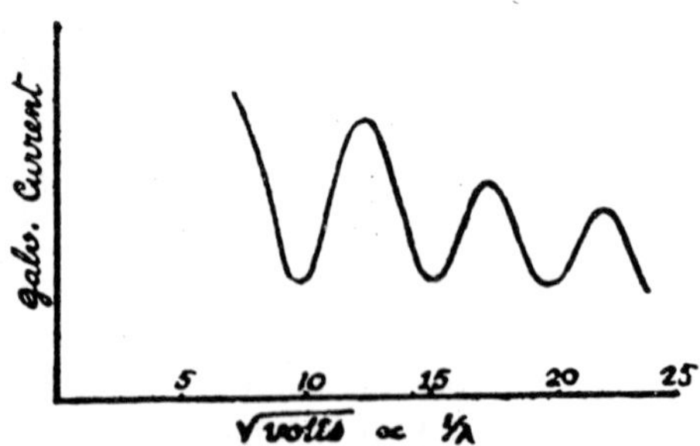

Fig. 13.4

13.6 ELECTRON WAVES

In 1934, J.V. Hughes was able to verify De Broglies's law for electrons with velocities comparable with that of light.

But before this, Rupp, a German was succeeded in diffracting glow electrons by means of a ruled grating with 1300 lines per cm.

Today the electron diffraction, being used to extend the methods of X-ray diffraction, to the study of the inner structure of materials, particularly to study the complex molecules.

In the study of surface structure, electrons have two inherent advantages over X-rays, *viz.,* they penetrate less deeply and their interaction with the atoms of the body under test is more intimate than in the case of X-rays. A further practical advantage of electron diffraction is the high intensity of the diffraction pattern so that only very short exposure times are necessary for photographic purposes, whereas much longer times are required when X-rays are used. G. P. Thomson has used electron diffraction to investigate surface layers of unknown composition. The nature of the patterns produced enables the crystalline structure of the surface to be determined. Buhl and Rupp have shown that reproducible patterns are obtained when electrons are reflected from oil surfaces. Dauvillier has used the transmission method of diffraction in the structural analysis of organic materials prepared in thin films. Jenkins, Morison, Langmuir and others have used electron diffraction in the study of lubrication, by graphite, grease and oils. Modern electron microscopes are usually fitted with an electron diffraction unit as a very useful complement.

The close analogy between the electron waves and X-rays has naturally led investigators to search for polarisation in these waves, chief among whom were Davisson, Germer and Bupp. The method used is similar to that applied in the case of X-rays, *viz.,* to produce a double reflection of the electron beam on two similar surfaces and examine whether the intensity of the second reflected ray depends upon the azimuth of the second reflector with respect to the first. It was found that there was no variation in intensity greater than 0.5 per cent. On the whole, the polarisation of the electron waves appears doubtful, but investigations are still in their initial stages. From the theoretical point of view the results obtained are more promising. For, Dirac in 1928 by an ingenious application of the principles of relativity to wave mechanics, established that the electron possessed a magnetic moment and spin. From this finding, it could be argued to the possibility of polarisation of electron waves, since just as the polarisation of light waves can be attributed to the spin of the photon, so also the spin of the electron should cause polarisation of electron waves.

Since De Broglie's theory applies to any material particle, experiments designed to show the wave characteristics of free atoms and molecules have been tried by several workers. Special difficulties have to be overcome in producing diffraction effects with atoms and molecules, which do not arise in working with electrons. First of all, as the atoms and molecules are much heavier than electrons the De Broglie wavelength will be much smaller for them and it will therefore be more difficult to detect diffraction effects with them. Secondly, the free atoms or molecules possess a Maxwellian distribution of velocity, whereas for diffraction experiments a beam of uniform velocity is desirable. Thirdly, neutral atoms or molecules are very much harder to detect than are charged particles. However, Dempster in 1927 and Estermann and Stem in 1930 succeeded in obtaining clear diffraction effects with hydrogen and helium by suitably reflecting them from crystal gratings, thus lending-full support to De Broglie's general concept of matter waves. It may be noted here that for bodies in bulk .such as those we deal with in daily life, a cricket ball for instance, the wavelengths are far too small to be observable, being of the order of 10^{-24}Å But there is no reason to doubt that even these have a De Broglie wavelength associated with them.

It is clear from the experiments described above that some sort of wave must be associated with all material particles whatever their nature. This fundamental feet leads to a new form of mechanics different from the classical one when the fine structure details of matter are to be considered. This new mechanics must be able to combine valid aspects of the wave theory of light, the quantum theory of radiation and the dynamics of particles in motion; it should also be able to throw more light on the nature of the matter waves which, according to the simple analogical conception of De Brogue, can just be conjectured as a kind of wave packet, not of the same type as that of radiation, but without further physical significance. We shall, therefore, now pass on to the study of the new type of mechanics which the matter waves called for.

One of the reasons which made De Broglie have recourse to the idea of matter waves was the concept of the stable and non-radiating electronic orbits restricted by quantum rules to integral numbers in the Bohr theory of atomic structure as we have already mentioned. In the same theory, on the assumption that energy is radiated as spectral lines when the electron jumps from one quantum orbit to another and that the frequency of the emitted radiation is strictly governed by a law known as Bohr's frequency condition, great advance was made in the analysis of optical and X-ray spectra. But after a time it was realised that Bohr's picture of the atom was not complete. Spectral lines were observed which could not be made to correspond to any known electronic orbits and special orbits had to be provided, often with very little reason other than that the spectrum seemed to demand them. With the experimental analysis of fine structure of spectral lines and anomalous Zeeman effect etc., the Bohr atom model was being more and Bohr frequently disregarded because of its increasing inability to account satisfactorily for the observed facts.

In 1925, therefore, even before the wave nature of the electron had been discovered, Heisenberg and Schröedinger taking up the clue of matter waves of De Broglie as a possible solution that might successfully overcome the difficulties met with in experimental spectroscopy developed independently and from two different directions a new mechanics. The one of Heisenberg is known as *quantum mechanics* and that of Schröedinger *wave mechanics*. The former involves a very difficult branch of mathematics known as matrix calculus and it is well nigh impossible to describe it in simple terms. Hence, within the scope of this book, we can only indicate the main line of argument of Heisenberg and that too qualitatively. The wave mechanics of Schröedinger is more readily understood and lends itself to a more or less pictorial representation. We shall therefore deal with it somewhat more in detail. But here also, applications to particular cases involve fairly complex mathematical solutions and hence only the results of such solutions will be made use of, wherever necessary, in the study of Atomic Physics.

13.7 MERIT OF THE THEORY OF MATTER WAVES

The theory of matter waves has solved all the problems, for which it was designed. The theory leads to equations whose solutions correspond to definite energy values, comparable to the energy terms of Bohr's atom model. This is how the quantum numbers of atomic structure and spectroscopy are inherent in the theory and are given in fundamental terms.

Wave mechanics also accounts for facts which are unexplicable by the older theories, such as half integral quantum numbers, zero-point energy, etc. It has even led to predictions which later have been verified experimentally, as diffraction of electrons, resonance phenomena in spectroscopy, etc.

Furthermore, events impossible to the classical theory are found, when treated by wave mechanics, to have a very small but finite probability of taking place. On the basis of this prediction, natural radioactive process and artificial disintegration have received a very satisfactory explanation. Application of wave mechanics to these specifically nuclear phenomena has resulted in some understanding of the nucleus itself including its structure and energy content.

More than all these special problems, the general behaviour of matter in the atomic state, as the Dr. Jekyll and Mr. Hyde, has been interpreted to a high degree of satisfaction by wave mechanics. It has shown that not only radiation but also matter exhibits sometimes wave-like properties and sometimes corpuscular properties; it has further led the way to treat the two concepts, wave and corpuscle, as complementary aspects of viewing one and the same objective process, which only in definite limiting cases admits of complete pictorial representation and is therefore nearly always subject to certain indefinitenesses, given by the principle of uncertainty. Herein lies the great advance made by the new mechanics over the absolute determinism of classical mechanics.

But this final conclusion of wave mechanics, *viz.*, the principle of uncertainty, *must be regarded not as a complete, surrender of determinism in physics, as many seem to think, but as a further refinement in the right direction showing that the deterministic conception of Nature is neither complete nor universal.* For, the whole of the material universe, both in the macroscopic and microscopic states, appears to be subject simultaneously to determinism and indeterminacy. In other words, no physical phenomenon, whether large or small, is wholly deterministic or wholly indeterminate. It is true that in the large-scale phenomena dealt with in classical mechanics, as in astronomy, for instance, determinism predominates, while in the small-scale atomic phenomena indeterminacy is clearly perceived. But from this it does not follow that there is no determinism at all but only absolute indeterminacy in the microscopic state, while apparent determinism, *i.e.*, an admixture of determinism and negligible indeterminacy is restricted to the macroscopic state alone.

In general, if *the doctrine of determinism* were to be rejected as completely wrong, there would be no order or regularity in physical phenomena and no scientific knowledge of them could exist. But physics, even atomic physics, does exist : it is an actual fact and has proved its value by its progress arid the number of its practical applications. We have a convincing proof of this in the modern art of artificial transmutation of elements, which is possible only if there exists a certain amount of determinism coupled with indeterminacy in material bodies, even in the atomic state.

Planck's constant 'h' which is considered to be the limiting barrier of determinism and a measure of indeterminacy, is really an enigma of Modern Physics; it still remains the unknown syllable in Nature's crossword puzzle. Although, its finite value shows that the microscopic state, where indeterminacy becomes apparent, is not merely a small scale representation of the greater macroscopic state of glaring determinism, yet the former tends asymptotically towards the latter. Such a tendency is possible only when with the states have common features, *viz.,* that they are partly deterministic and partly indeterminate.

Again, when one considers the strange way in which the constant *h* enters into Atomic Physics, *viz.,* through the corpuscular aspect in the case of radiation, but through the wave aspect in the case of matter, it becomes possible to surmise that every physical reality, big or small, is complex, containing within itself, *i.e.,* independent from and prior to all experimental investigation, a double complementary aspect of determinism and indeterminacy. For, on this assumption alone one could understand why an obviously wave form radiation acts as corpuscles and corpuscular matter behaves like a wave.

The marked element of paradox, that appears in the wave mechanical conception of physical reality leading to the principle of indeterminacy, also seems to favour the opinion proposed." For, waves of probability, pure abstractions as they are nevertheless propagated "in space" just as though they were continuous and homogeneous elastic waves. Likewise, the particles whose existence is intermittent cannot be localised and defined with exactness. Thus the wave and particle of the new mechanics, are, each of them, a complex of indeterminacy and determinism. According to the principle of indeterminacy, both the position and velocity of a particle cannot be determined simultaneously with accuracy.

14

THE CONVERSION OF THE ELEMENTS (TRANSMUTATION)

14.1 INTRODUCTION

The conversion of the one element into another is named as transmutation. Transmutation is often called as, disintegration. Because during the process the transmuted element disintegrates, forming a different new element.

The conversion of the elements into one another, was the dream of human race.

In the middle ages it occupied a very prominent place, under the name of *alchemy* whose main purpose was to arrive at what was known as the *philosopher's stone* which would on touch convert base metals Into noble i ones. The alchemist's heroic search for this wonderful agent of transmutation resulted in failure, chiefly due to the meagre data then available about the ultimate constituents of elements.

In fact, one cannot aspire to transform element into element, if one does not know what elements really are. Hence, before the problem of transmutation could be seriously tackled, the ground had to be prepared and this took a long time.

Several centuries of painstaking labour at last brought the scientists close to the fundamental constituents of matter. All matter was first classified into *92 simple elements* and the smallest unit of any one of these, *the atom*, which completely defined the individual nature and properties of that element, was fully recognised. It was next established that the atom was itself composed of two distinct entities, *viz.*, a central positively charged *nucleus* of extremely small dimension, but wherein practically the whole mass of the atom was concentrated, and very light negatively charged particles, the *electrons*, gyrating round the *nucleus* at relatively great distances. These peripheral electrons were found to be highly organised and responsible for most of the chemical and physical properties, while the more wonderfully constituted nucleus gave the essential individuality to the element. It was therefore realised that to transform one element into another, the very atom kernel, the nucleus, should be attacked and altered.

The final and decisive stage in this long preparation came with the discovery of radioactivity. The study of the natural radioactive phenomena not only proved beyond any doubt the complexity of nuclear structure but also paved the way for artificial transmutation. For, the occurrence of spontaneous disintegration of the nucleus in the heavy radioactive elements readily suggested the idea whether it might not be possible to break up nuclei of other ordinarily stable elements artificially by bombarding them with high speed particles. Rutherford took up the hint and performed a series of experiments in which the atoms of different stable elements, such as nitrogen, aluminium, phosphorus. etc., were subjected to an intense bombardment by the high velocity α-particles from natural radioactive substances and finally, in 1919, succeeded in transmuting these elements, thus realising the old dream of the alchemists.

14.2 STUDY OF TRANSMUTATION

The experimental technique which is employed in artificial transmutation, has got complexity and many difficulties.

Since the dimension of the nucleus is of the order of 10^{-12} cm, small projectile, which is capable of penetrating sharply the electron groups, surrounding the nucleus is needed. So we can say, the transmutation will be a phenomenon of rare occurrence.

Another difficulty which lies in the process of transmutation is the presence of a positive electric field.

If the projectiles are positively charged, then many of them will be deflected away can penetrate through the potential barrier.

If the fragments of the broken nucleus are very heavy, they move slowly and remain lost in the mass of the element, and in consequence cannot be detected by any external device. If at least one of the fragments is a light particle, such as a proton, neutron, or α-particle, as is often the case, the process can be more easily checked, since the light particle moves fast enough to get out of the mass of the parent element and project itself into space.

But another great difficulty arises here, *viz., the possibility of mistaking particles which have nothing to do with the disrupted nucleus* (such as the incident projectiles themselves which might have been scattered, the recoiling nucleus, the protons from hydrogen which may be present as impurity in the element under study, the particles given off by radioactive impurities, etc.,) *as the true fragments of transmutation.* Hence, it must be established that at least some of the particles observed are not of the spurious kind, but really parts of the fractured nuclei. To determine this point with precision, the *nature* of the expelled corpuscles, their *direction* of motion and energy must be carefully studied. But this is no easy task, in practice, and a whole complex technique is required to detect accurately the results of the experiments.

Artificial transmutation of elements was first accomplished by Rutherford using the high speed α-particles from RaC' as projectiles and more than ten years passed before other particles were considered. At present, the projectiles that have been successfully applied to nuclear disintegration are the cosmic rays, α-particles, γ-rays, protons, deuterons, neutrons and electrons. The first three of these may be considered as *natural agents*, being derived from either the mysterious radiations which continuously bombard the earth (cosmic rays) or the natural radioactive substances (α-particles and γ-rays), while the rest as artificial, as they are either produced as a result of artificial transmutation (neutrons) or rendered intense and accelerated to a high speed by artificial devices (electrons, protons and deuterons).

Among the *natural agents*, the *cosmic rays* are the most, powerful, endowed as they are with fabulous energies of the order of billion electron volts. The *α-particles* are obtained from many radioactive substances. For quite a long time they were the sole agents of transmutation and even today they remain very important, capable of effecting very interesting and instructive transmutations. Of considerable importance has been the use of polonium (RaF), since this can be separated from parent materials which are β- and γ-active and so affords a pure source of α-particles of energy about 5 MeV. Higher energy α-particles from RaC' (about 8 MeV) are also frequently employed. The γ-rays obtained from radioactive substances have energies reaching up to 2 to 2.6 MeV. These have been used in a special type of transmutation known as photo-disintegration. Very high energy γ-rays, up to even about 17MeV, obtained as products of certain disintegration processes, have also been used as bombarding agents.

The general drawback of these natural projectiles is that they are beyond the control of man. It was the desire to control the conditions of his experiment that led the physicist to devise his own artillery. He wanted to be able to choose the kind of projectile and to regulate the speed and intensity with which the projectiles struck the target. Thus it is clear that the special techniques with which we are dealing here, are chiefly directed towards the production of artificial agents.

Of the *artificial agents, neutrons* which have proved to be by far the *best agents of transmutation*, must be considered apart and the artificial accelerators described below do not directly refer to them, on account of their peculiar property of having no electric charge. Per one thing, they cannot be accelerated like charged particles, and for another, they turn out to be the more effective agents of transmutation the lower their speed. Hence, in their case artificial means are to be applied, rather to slow down than accelerate them. But they may require the indirect use of artificial accelerators in so far as, for instance, deuterons, accelerated by these services to high speed and bombarding a lithium target, produce an intense supply of high velocity neutrons which might be needed for effecting certain transmutations. A *standard and efficient type of neutron source for transmutation purposes* is a small sealed capsule containing a mixture of powdered beryllium and radon gas. The atoms of Be are thus bathed in α-particles emitted by radon; many nuclei of beryllium are thereby disintegrated by the α-particles, producing a great abundance of neutrons. For many experiments, the small size of such a source makes observations exceedingly simple and effective owing to the largo solid angle-available for targets.

The techniques of producing artificial agents of transmutation are therefore applied chiefly to the two remaining positively charged particles, the protons and the deuterons, although they have been extended to α-particles in order to overcome the low intensity of the α-particles available from natural radioactive sources.

14.3 ANALYSIS OF TRANSMUTATION

While studying about the transmutation, it is needed to give more concentration towards the fragments which fly forth from the bombarded nuclei and studying their nature.

One must bear in mind that there is no one single technique which is capable of furnishing at one stroke all the required data. To get, sure conclusion many different methods are taken together.

The various techniques which are employed in the analysation of nuclear bombardment are as follows:

(1) Scintillation method.

(2) Chemical separation method.

(3) Magnetic spectrograph method.

(4) Ionisation chamber method.

(5) Counter method.

(6) Cloud chamber method.

(7) Photographic emulsion method.

As most of these methods were devised even before artificial transmutation was discovered, chiefly in connection with the study of natural radioactivity, we have already dealt with several of them in their appropriate places. But many of them have been modified and improved under the impetus of modern alchemy to such an extent that they bear little resemblance to their simple and primitive forerunners. Hence, in the present account of these methods, we shall limit ourselves to such of their special features as have reference to their employment in artificial transmutation.

METHOD OF SCINTILLATIONS

It is a simple and direct method which enables charged particles to be counted individually. It has also an historic interest, as the one by which the first proof of artificial transmutation was obtained. The fluorescent materials commonly used for the screen on which the scintillations are observed are barium platino-cyanide, calcium tungsten and zinc sulphide. The duration of a scintillation is of the order of 10^{-4} sec.

The method may be well suited for counting α-particles it can hardly be used in the case of protons and it is almost impossible to use it for counting electrons. If the fluorescent screen is bombarded with the γ-rays, often produced during transmutations, electrons are liberated from the fluorescent material and the surrounding matter, which make the screen shine all over with a feeble glow, against which the flashes made by the particles to be counted are difficult to discern. Such are the drawbacks of the method; yet, on account of the extreme simplicity of the technique which has also the unique advantage that no barrier, not even a gas, need intervene between the detecting screen and the source of the fragments, it is still used, but in a highly refined form, known as the *scintillation counter*, as we shall see presently.

Chemical Separations

The use of certain standard chemical separation techniques, however, has materially assisted in the interpretation of the results of transmutation, chiefly in the case of induced radioactivity. A good guess as to the nature of a nuclear reaction makes it possible to separate the possible products and by observing in which of the separated residues the activity lies, the nature of the element produced in transmutation can be determined. This is usually accomplished by employing a radioactively inert element of the kind to be separated, as a carrier, in order to have a sufficient quantity to work with.

This method was introduced in 1934 for α-induced radioactivity by Curie and Joliot and for neutron processes by Firmi. Positron activity obtained by deuteron bombardment has been studied in a like manner by McMillan and Livingstone.

A valuable modification was initiated by Szilard and Chalmers, when they realised that chemical bonds could be broken by the recoil of the activated atom and so the free element could be released from a chemical compound. If the element in question is bombarded in the form of a compound which, once dissociated by recoil action in the process, is not reproduced spontaneously, the activated atoms remain in a chemical state different from the bulk of the inactive substance and can be separated by ordinary chemical methods.

In the magnetic spectrograph method, three steps are taken into consideration.

The first one is photographic plate. β-rays given off by the radioactive substance under study are bent by a constant magnetic field.

In Geigger tube counter method, β-particles can be counted accurately.

By using Faraday cylinder the method can be made more successful.

The basic principle employed in the Ionisation chamber method is that, charged particles in motion produce in a gas ionisation varying with their nature and velocity. So, it is possible to know about the nature of the ionising particle as well as its energy.

The ionisation chamber, as regards its essential features. Mention will be made here, therefore, only of the additional special technique involved in its use with transmutation experiments.

Maintenance of a constant high p.d., the P.D. must be high enough, so that the electric field between the electrodes reaches approximately saturation, in which state most of the ions formed reach the electrode before they could recombine.

The dimensions of the chamber as well as the nature and pressure of the gas used in it are determined by the type of radiation to be measured. Thus, for example, in the case of α-particles which are completely absorbed by a few cms. of air, the chamber is usually filled with air at atmospheric pressure and its linear dimensions need not be greater than the range of the particles, since large size would not increase the ionisation produced. For β- and γ-rays, which are much more penetrating, larger chambers have to be used. In order increase the ionisation in the case of γ-rays, it is found useful to replace air with a more intensely absorbing gas, *e.g.*, methyl iodide, or to increase the pressure to 30 to 50 atmospheres.

The sensitivity or quick collective power of the chamber can, in general, be increased by the use of highly absorbing gases at high pressures. It must be noted, however, that it does not increase in proportion to the density, since at high pressures there is an increasing degree of recombination of the ions. This can be avoided to a large extent by the use of very pure rare gases, such as argon, in which the recombination effect is small up to very high pressures.

The sensitivity of the electrical device which measures the feeble ionisation current has been progressively improved so that at present it is possible to detect the primary ionisation due to even single particles.

Electroscopes and Electrometers

In the initial stages, the gold leaf electroscope of the tilted type of great sensitivity or the quadrant electrometer, whose sensitivity was increased to a high degree either by increasing its voltage sensitivity or by decreasing its capacity, was used. They were able to detect and even measure the total charge of a few thousands of ions without amplification.

String Electrometer

The next step in the development was a modification of the electroscope, designed by Lauritsen, where the gold leaf was replaced by a tiny metal-coated quartz fibre, about 5 microns in diameter and 6 mm, long, the movements of which could be observed by a microscope, or better still, recorded by photography on a moving film. This arrangement known as the string electrometer possesses the advantages of *small size* which enables the fibre to return to its normal position very quickly after each displacement, *high sensitivity* due to the low capacity of the small fibre and *special ruggedness* lending itself to any orientation. The primary ionisation produced by single particles can be readily measured with this device. Each ionising article is revealed by a jump on the photographic film, the size of the jump being a measure of the number of ions produced by a particle. But the *major difficulty* in this method, as in the older ones, though minimised to a good extent, is the *lack of quickness of response to particles that follow one another in rapid succession.*

With the advent of radio-technique of amplification by thermionic valves, the above difficulty was overcome to a great extent by a device, known as the linear amplifier, which augmented without distortion the feeble ionisation current from a single particle to a value which would easily register on an oscillograph or an electrical counter. Originally developed by Greimacher, this method has been brought to a high degree of perfection by Ward, Wynn Williams and Cave of the Cavendish Laboratory, by Dunning of Columbia and by Maurice de-Broglie of Paris.

Counter method is very useful one and efficient method which is used today. Today three types of counters are in use.

1. The Geiger Counter

It was first devised by Rutherford and Geiger in 1908. In 1928 it was developed further by Geiger and Müller.

The Geiger counter is essentially a simple ionisation chamber, very compact on account of its small dimensions and very efficient due to Two automatic amplification, inside the chamber itself, of the weak ionisation current produced by the passage of a single charged particle. This internal amplification is secured by the application of a carefully adjusted strong electric field between tile two electrodes of the chamber. In an ordinary ionisation chamber, the voltage on the electrodes is just large enough to collect the ions produced by the passage of a charged particle. Following, a suggestion made by Gorton, in 1905, for X-ray ionisation, Rutherford and Geiger increased the voltage on the chamber to such a high value that the electrodes in the chamber just did not spark over, so that even a single charged particle passing through the chamber was enough to set the chamber off and cause a discharge. Under these conditions, the relatively few ions produced by the particle, finding themselves in a strong electric field, are immediately and violently accelerated towards the electrodes. In their passage they produce more ions by collisions with neutral molecules, until the cumulative effect results in a discharge, large enough to be measurable.

It is essential that the discharge caused by the passage of the particle should not become permanent, but must be quickly and automatically extinguished. This is achieved either by the gas in the chamber or by a suitable adjustment of the accompanying circuit or by both. For, then only the system will be reset in proper condition for responding effectively to successive particles. The counter acts somewhat like an *automatic rifle* which resets itself every time it has been fired by the 'trigger' action of the passing particle.

There are a number of empirical rules (some partially understood, some not at all) about the size and the shape of the chamber, the proportioning and conditioning of the electrodes, the nature, purity and density of the gas and the intensity of the electric field, which makes the problem of the inner mechanism of the counter operation difficult for a complete solution. Given, however, the great value and the frequent use of these counters, we shall state briefly some of the important practical details about their construction and operation.

There are two kinds of Geiger counters in use, called the *point counter and the tube or Geiger-Müller counter*. They are differentiated chiefly by the shape of the electrodes.

In the *tube* or *Geiger-Müller counter*, the outer electrode is a metal tube T (brass or nickel), 1 to 5 cms. in diameter, 10 to 50 cms. long and the inner electrode a very fine wire W, 0.1 to, 0.5 mm. thick, usually of tungsten, stretched along the axis of the tube and well insulated from it by means of the definite plug PP. The electrodes are usually sealed in a thin-walled glass tube G. The gas contained in the counter may be either air, hydrogen, argon, or even a mixture, at a reduced pressure of 2 to 10 cms. of mercury. The counter is evacuated and filled with the desired gas at the chosen pressure with the help of a glass tube K, which is then sealed off. The introduction of a gas mixture in the counter not only lowers the values of the high voltage to be applied but also aids in quenching the discharge. A strong electric field is established between the two electrodes by the application of a steady high potential. The external circuit contains a high leak resistance R as in the previous case.

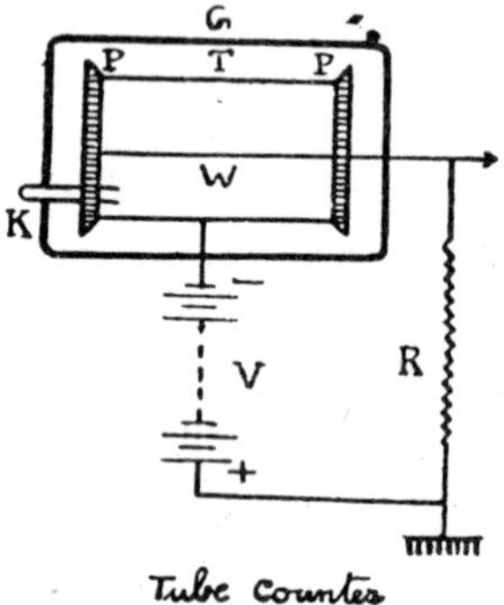

Fig. 14.1 : The Geiger Müller Counter.

The design of the enter tube electrode as well as the nature of the gas employed vary according to the different uses of the counter. Thus, for instance, to detect γ-rays, the tube has pretty thick walls (1 to 3 mm), although the practice is to keep the amount of material at a minimum to order to reduce the contamination background Secondary electrons released chiefly from the walls of the tube produce the discharge. When intended for the study of β-particles a thin aluminium window, about 0.1 mm. thick, is used to admit the radiation. Heavy particles such as *protons* or *α-particles* are readily detected with a thin-walled tube counter operating at atmospheric pressure, for the detection of *neutrons*, the counter is filled with hydrogen, the protons chased by the impact of neutrons with the hydrogen atoms constituting the 'triggering' particles; or better still, the tube is made of silver, easily activated by the neutrons, in which case, the electrons of disintegration of the radioactive substance artificially produced by the neutrons will operate the counter.

The tube counter has several advantages over the point counter *viz.,*:

(1) the duration of the discharge is reduced,

(2) the sensitive region extends throughout the total space between the electrodes,

(3) a wider range of voltage is obtained for the sensitive condition, and

(4) it is not so readily disturbed by discharge.

The Scintillation counter is really the super-speed counting device. The counter has following main parts.

1. A phosphor
2. A photo multiplier
3. Electronic valve equipment.

The organic phosphorous are highly transparent and they can often be produced as large single crystals.

A photomultiplier an electron multiplier. In the multiplier the two successive pulses are registered separately, and they are of the order of 10^{-10} sec. This is about 10 thousand times shorter than the resolving time of a gas filled counter.

Electronic equipment functions as an amplifier, scaler and register.

Choice of the Phosphor

This depends on the special uses of the counter. For example, when it is necessary to investigate the radioactive content of small weak sources, or of sources emitting radiation of low energy, a pretty large size crystal phosphor is taken, a small hole is drilled in it and the radioactive sample is placed in that hole. A more recent and better method is to dissolve the sample in a liquid phosphor, in which case practically all the emitted radiation is absorbed within the phosphor, thereby giving a high detection efficiency.

Mounting the phosphor demands the following conditions:

(i) Light other than that produced ia the phosphor has to be excluded;

(ii) The scintillations produced have to be efficiently, directed to the photocathode surface of the photomultiplier;

(iii) The phosphor must be protracted from mechanical damage and if hygroscopic, from moisture;

(iv) The phosphor must be screened from extraneous radiation. Numerous types of phosphor-holder have been developed to meet these requirements, each depending upon the purpose for which the counter is designed. The problem of maximum light-collection from the phosphor has been solved by *either* silvering *or* better still, frosting all the surfaces of the phosphor, except that through which light passes to the photomultiplier.

Light guide is a device to make the light produced by the phosphor reach the photomultiplier with as little loss as possible, chiefly when the photomultiplier has to be kept separated by a good distance from the phosphor, as in the case when a scintillation counter is used in conjunction with particle accelerators like the cyclotron, employing huge magnets. It is usually a glass tube containing water or a glass rod several feet long, at one end of which the phosphor is mounted and at the other the photomultiplier. The light from the phosphor is carried along the guide without much loss to the photomultiplier, thanks to the phenomenon of total internal reflection. The light guide is usually enclosed in a light-proof outer covering.

Voltage Supply to the Photomultiplier

A constant voltage produced by a stabilised 'power pack' has to be supplied to the chain of resistors which divides the P.D. into the necessary steps to supply each dynode with its correct tension.

Discriminator in the Amplifier-Sealer Circuit

When counting the pulses caused by the incident radiation, it is necessary to exclude the many smaller pulses due to the spontaneous emission of electrons from the dynode surfaces. For this purpose, a 'discriminator circuit' is introduced between the amplifier and sealer circuits. It is an electronic device for sorting out the required size of pulse from all those passed on by the amplifier. Such discriminators are used chiefly in the analysis of a heterogeneous incident radiation, when the components are to be separated and their relative intensities are to be measured.

Scaler and Recorder

The scaling circuit is usually so arranged that the sorted pulses received by it are passed on 'in powers of ten' to the recorder. Different kinds of recording are in use : the total number of pulses may be indicated visually or recorded graphically; the count rate, *i.e.*, the number of pulses in unit time may be directly given by the recorder; in some cases the cathode ray oscillograph is used as the recording device.

Scintillation counter has number of applications in various fields:

1. Scintillation counter can be used as, efficient detector of protons deuterons and α-particles.
2. Organic phosphor counter is used to detect β-particles.

Recently, many types of scintillation counters have been devised in the *study of cosmic rays* for the detection of mesons and other unstable particles of very high energy. This has been possible due to the ability of the scintillation counter to respond specifically to events separated by extremely short intervals of time. The scintillation telescope forms naturally an important technique in these investigations, since the axis along which the phosphors are mounted can be rotated, so that the particles arriving in any direction relative to that of the primary beam can be studied.

But the most interesting and efficient type is the *liquid scintillation counter*. With the advent of liquid organic phosphors, it has become possible to use large-volume liquid counters which are bound to throw light on problems not approachable by other methods. For example, such counters have been used in the measurement of the energy spectrum of extremely fast moving charged particles met with in cosmic rays or produced by the modern powerful particle accelerator, such as the synchro-cyclotron. The energy of the primary particles analysed may go up to several billion volts and the scintillation counter technique involving a novel type of radiation is quite ingenious.

In 1934, Cerenkov, a Russian scientist, discovered that very fast electrons moving through a transparent medium emitted light at a definite angle from the direction of motion of the electrons. Such an effect was called *Cerenkov radiation* and was interpreted as due to the electrons moving with velocities greater than that of light in the medium where the electric field of the electron is strongly perturbed and an optical "shock wave" is generated. Can a particle travel with a velocity greater than that of light? According to the theory of relativity no particle can travel with a velocity greater than that of light *in vacuum*. In the present case, we are dealing with particles travelling in a *medium* with velocities greater than light and giving rise to a new type of radiation. That such a phenomenon really occurs can be readily understood by an example from acoustic. When an aeroplane travels with velocities smaller than that of sound in air, the sound emitted is due only to the oscillatory motions of the engines and other moving party of the plane. But when supersonic velocities are reached, as in the case of modern jet planes, then even the uniform motion of the aeroplane produces an additional radiation of very intense sound waves called the Mach's waves. On this analogy one may call the Cerenkov effect as "singing electrons". Cerenkov used β-rays and Compton electrons. With electrons of 2 MeV fairly strong source of radiation could be had. The radiation was found to be bluish white in colour and was identical in appearance for all solids and liquids.

In 1934 crystals counter was discovered by Sir R. Robertson. In 1945 Van Heerden conducted serious investigation of crystals as radiation counters.

In 1947 E.W. Wooldrige, A.J. Abearn and J.A. Burton obtained positive results, with diamond as a-particle counter. Later crystal counter took good position in Text Book of Atomic Physics.

Crystal counter is found to be faster than the Geiger counter.

Cloud Chamber Method

For the study of artificial transmutation, the cloud chamber is used as a powerful tools. This apparatus provides, all the important information required in the analysis of transmutation.

In 1897 the fundamental principle of the cloud chamber was discovered, by C.T. R Wilson. In 1912 Wilson designed the first cloud chamber.

This prototype of all modern cloud chambers is shown diagrammatically in Fig. 14.2. A is a cylindrical expansion cloud chamber with walls and roof of glass, containing dust-free air saturated with water vapour. Directly below A is a movable piston P whose rapid descent produces the adisbatic expansion. A large evacuated vessel V is connected to the piston through a tube provided with a valve C. When C is opened, the air under the piston rushes into the vacuum chamber, thereby causing the piston to drop suddenly. The wooden blocks WW are to reduce the air space below the piston. Water at the bottom of the apparatus ensures saturation in the chamber. The expansion ratio can be adjusted by altering the height of the piston. An electric field E is maintained in A to sweep out the spontaneously produced ions that would create a diffuse fog. This electric field is usually cut off just before expansion.

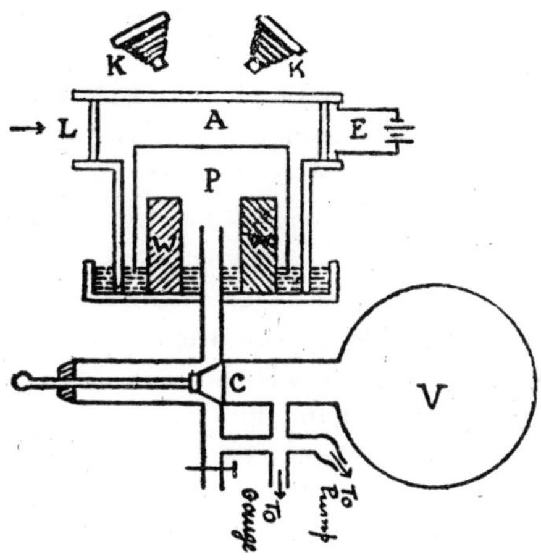

Fig. 14.2 : Wilson's Cloud Chamber.

Immediately after expansion, the particles are allowed to enter the chamber through a side window; droplets are formed on all the ions produced by the particle, so that they martyr the track of the particle. At this precise moment, if the chamber is strongly illumined with a horizontal beam of light I/from a mercury vapour discharge lamp, the track of the particle can be visually observed as a white line of fog against the blackened top surface of the piston. The tracks of individual particles can be photographed by a camera, or better still, by two cameras KK focused on the interior of the chamber through the glass roof. The various operations, *viz.,* expansion, admission of ionising particles into the chamber, illumination and photographing must be performed in very rapid succession to obtain good results. For this purpose, Wilson used an ingenious arrangement with a falling weight, which first opened the valve C and allowed the expansion to take place, then let the particles into the chamber and finally made a contact which flashed on the arc light and clicked the camera for taking the picture. As the chamber heats up very quickly, in about 1/50th of a second, the observations must be completed before the chamber has become so warm that the condensation droplets disappear.

With such a relatively simple apparatus, Wilson and others were able to obtain beautiful photographs of individual ionising particles, such as α-particles, β-particles, etc.

Diffusion Cloud Chamber

In 1939, Langsdorf, gave the idea about sensitive diffusion cloud chamber.

In the diffusion cloud chamber, the diffusion of a vapour is carried from the warm root of the chamber. Condensation takes place in a sensitive layer near the bottom.

The chamber, is usually made with glass rings about one foot in diameter and half a foot in height, strengthened by a brass cylinder with perspex windows. The bottom of the chamber is a metal plate (brass or aluminium) ¼ inch thick blackened with stove polish or black bakelite. When the bottom surface becomes covered with liquid, it forms an excellent background for photography. The roof of the chamber is a metallic plate, with a central glass window for observation and photography. The top and bottom plates are sealed to the glass ring with rubber gaskets to have the chamber gas-tight.

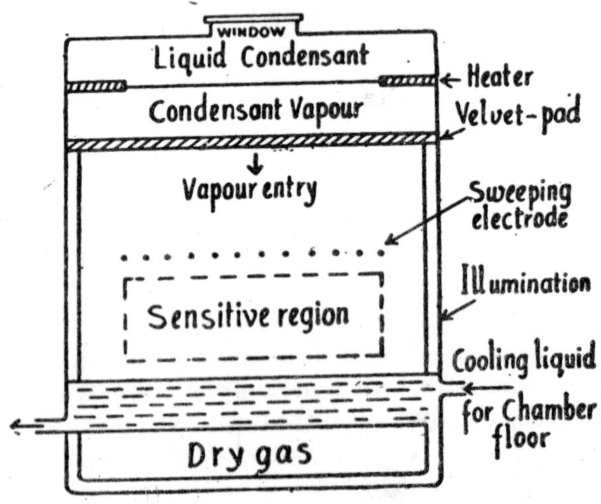

Fig. 14.3 : Diffusion Cloud Chamber.

The condensdnt liquid is contained in a ring-shaped metal trough just below the top-plate and an electric heater is used to promote evaporation. The bottom of the trough has small perforations and is covered with a velvet pad which thus gets uniformly soaked with the liquid. It is essential for the maintenance of stable operation that the condensant vapour should be light as compared with the gas used in the chamber. The deepest sensitive layers have been obtained with mixtures of methyl alcohol, ethyl alcohol and water. However, since the composition of the mixture changes with evaporation, pure methyl alcohol is usually employed.

The gas in the chamber, air may be used as the inert gas in the chamber, although argon appears to give better results. Methyl alcohol diffusing through air atmospheric pressure is found quite satisfactory. Lighter gases, like hydrogen and helium, have also been used under high pressures.

Temperature conditions of the bottom and top of the chamber, the bottom of the chamber is usually maintained at about –70°C by a block of dry ice pressed firmly against the metal base-plate. Another method of cooling is by partially submerging the base plate in a bath of alcohol and dry ice. Although the temperature of the floor of the chamber is not very critical, it is desirable that it should be below –40°C to obtain a suitable high degree of super saturation. The top of the chamber is maintained in the whereabouts of 40°C, depending on the size of the chamber and the nature of the gas and vapour used. As the temperature of the top is increased, the depth of the sensitive layer is increased until an optimum temperature is reached above which the sensitive region breaks up into different layers, which is evidently to be avoided.

14.4 PHOTOGRAPHIC EMULSION METHOD

Kinoshita and M. Reingmanns, have discovered the, Photographic emulsion method. Earlier days this method was the main tool for the detection of the charged particles.

The method is very simple charged particles when allowed to fall on a photographic plate, will have the same effect on the sensitive layer plate as would exposure to light.

Some technical details of the photographic Emulsion method can be explained as follows.

In the ordinary photographic plate, the number of grains that are "exposed" by the passage of a particle through the emulsion is not great enough to make interpretation of the plates easy. As early as 1939, Dr. Powell had been in touch with Ilford Co., with the hope of their preparing plates in which the number of sensitive grains would be greater than normal. The first attempts met with little success and it was only in 1947, that a series of new plates was produced with special emulsions which contained a *very high concentration of sensitive grains*, about ten times as much silver halide for a given quantity, of gelatine as an ordinary emulsion. This close packing of the grains would enable the track lengths of the particles to be measured accurately and hence energy determinations based on track lengths become more reliable.

In these special emulsions, the *size of the sensitive grains vary from* 0.5μ to 1μ according to which the different grades of plates are named A, B, C, D, etc. By varying the average size of the grains it is possible to control the sensitivity towards different types of radiation. Thus a plate can be prepared which will permit a somewhat selective detection of protons, α-particles or fission fragments.

These specially prepared plates have also *much thicker coatings of emulsion than, in ordinary photography*, a thickness commonly employed being 100m (1/10 mm.). This would be the longest distance of any track in a direction at right angles to the surface of the plate. Since the range of particles in the emulsion runs from 10μ to a full mm., many tracks could be followed only when they run parallel to the plate surface. The photographic record of a nuclear disintegration will thus be a *star-shaped explosion* within the emulsion, with the tracks of the resulting particles scattering in a variety of directions. Some might leave the plate before their course is completed and others come to rest within the emulsion, in which case their ranges can be measured. For the recording of particles which enter the emulsion at large angles with regard to the plane of the emulsion, very thickly coated plates up to 306μ have been used.

The Experimental Set-up

The arrangements of the specially prepared plates to record nuclear phenomena depends to a great extent on the particular problems to be solved; but two simple methods have been used so for, in connection with artificial transmutation.

One of them, applicable to cases where there are only two products of disintegration, *viz.*, the recoiling nucleus and a fast light particle, such as a proton The bombarding beam, say of deuterons, accelerated by a Cockcroft and Walton tension multiplier impinges upon a sample of the element, say lithium. The resulting particles, protons, may be stopped to a known extent by a mica window and then passed through a diaphragm into a camera and on to the photographic emulsion, the angular position of which with regard to the direction of the incident protons may be varied, and measured. If necessary, the camera chamber can be evacuated. The protons falling on the emulsion will produce their characteristic tracks in it, which, after having been developed and fixed, can be observed and studied.

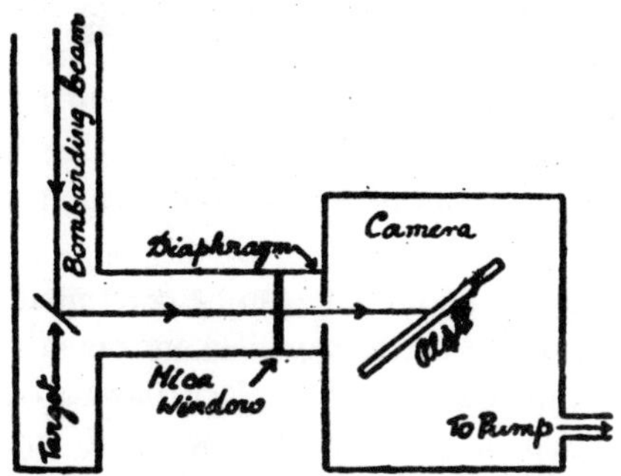

Fig. 14.4 : Appatus for Exposing Photographic Emulsion to Particles of Disintegration.

To study the minute tracks of the particles, they are magnified considerably by a microscope.

Later the magnified image of the track is projected on the screen and photographed. If the track dips into the emulsion it is necessary to focus on different parts of the track in turn and take a series of microphotographs.

Generally, speaking, the nature of the particle can be deduced from the appearance of its track, which depends upon the number of ionisations the particle causes per unit length. The chance that a silver halide grain is mad developable depends of the number of ionisations it receives. An α-particle with its twofold charge and mass of four units causes many more ionisations per unit-length of path than, for instance, a proton of the same energy with its single charge and unit mass. Hence the intensity of ionisation produced by an *α-particle* is more than sufficient to render developable all the silver halide grains it traverses. *A proton* having the same initial energy as an *α-particle* but a greater speed will cause so few ionisations in individual grains that not all the grains at the beginning of the track will be made developable. It is this characteristic which enables the investigator to distinguish between α-particle and proton tracks in the photographic emulsion. For the same reason, a *deuteron* track will be roughly intermediate between those of the proton and α-particle. Although *electrons* cause ionisation and undoubtedly bring about a blackening of the emulsion, actual electron tracks are not easily observed. This is partly due to their very light mass, which causes them to be scattered in zig-zag paths in the emulsion. In addition, the ionisation per unit length is much weaker than in the case of protons. Another difficulty is the background of fog grains, (*i.e.*, those which are developable without exposure) which makes it impossible to the observer to distinguish the grains made developable by electron. Recently, however, (1948) Kodak Ltd. has produced an emulsion mutable for the detection of electrons.

The energy of the particle can be determined from the *range of its track* trade. Bit in the connection than are a number of points which have to be carefully solved. In the first place, there is an appreciable shrinkage of the photographic emulsion in-processing, amounting to as much as 43%. Dr. Powell's greatest achievement, accomplished by long and patient investigation, is that he developed a method of measuring the particle tracks accurately in spite of this shrinkage. He was able to establish that the vertical shrinkage of the emulsion was not only uniform over the central area of the plate but was also constant for all plates of the same type.

Secondly, since the emulsion is a mixed material, it is of equal importance to obtain a direct and experimental comparison between the ranges of particles of different types and speeds passing through air and through a photographic emulsion. This comparison work is carried out as follows: Firing a given kind of particles, whose range in air as known, into the photographic emulsion a range calibration for the particular emulsion in use can be made. Thus, for instance, using the α-particles from polonium which is nearly a pure α-emitter, if the source is placed at increasing distances from the photographic plate and the corresponding lengths of the tracks recorded in the emulsion are measured, it is found that:

(a) The number of silver grains produced in a track is directly proportional to the length of the track, and

(b) The range (38 cms.) in air of α-particles emitted by P_0 corresponds roughly to 20 to 25μ in a photographic emulsion. The range in air of an unknown α-particle can therefore be easily determined from the length of its track in the emulsion using the calibration curve and consequently its energy calculated from the well-established range-energy relation in air.

14.5 TYPES OF NUCLEAR REACTIONS

The nuclear reactions can be observed with different kinds of projectiles like α-particles, proton deuterons and neutrons.

A very important feature in substitutional reactions is the evidence they furnish for the existence of quantum states of excitation in nuclei by the readily observable "groups" of outgoing particles. The incident particle, on entering the nucleus can be captured, not on the ground level, but on a level corresponding to an excited state of the nucleus. In such a case, the emitted particle will have an energy less than, normal by the amount of excitation and, if several excited states exist) one may observe several groups of emitted particles in complete analogy with, the phenomenon of "fine structure" of α-particles met with in radioactive substances.

It is to be expected also that transformations which are characterised by several groups of emitted particles will be accompanied by radiations corresponding to the transition of the product nucleus to the normal state after the reaction. The energies of the γ-rays should be equal to the differences in the energies of the various groups. It is generally assumed that the group of highest energy or responds to the product nucleus left in the ground state; the others of lower energies correspond to higher excited states. The differences between the highest energy group and the others thus give directly the energies of the excited levels of the product nucleus. The value of the reaction energy Q usually given in connection with the reaction equations corresponds to the highest energy in the groups of the emitted particles.

Radiative capture reaction is less probable than particle disintegration. This is because, the radiation width of nuclear levels is usually smaller than the particle width.

When particle disintegration is extremely small this type of reaction takes place.

For this the following reasons can be given:

1. Sufficient energy is not available.
2. The potential barrier is too high for particles.

The evidence for the radiative capture process cannot be had directly as there are no fast particles as products of disintegration, but only indirectly, either by detecting the γ-rays emitted in the process by their secondary effects, or by the product nucleus when it happens to be β-active, as in the case of many transmutations caused by the capture of neutrons.

Radiative Capture of Protons

Postponing the study of radiative capture of neutrons to the section on neutrons, we shall state here briefly the results obtained with the radiative capture of charged incident particles. The first experimental evidence of this process was furnished by Cookcroft (1934) who noticed that a carbon target bombarded by an intense beam of protons becomes positron-active, the activity being evidently due to the formation of radioactive N^{13} according to the reaction

$$(p, \gamma)\ {}_5Cl^{12} + {}_1H^1 \rightarrow *{}_7N^{13} + hv$$

$$*{}_5Cl^{13} \rightarrow {}_6H^{13} + e^+ \ (T = 10 \text{ mins.})$$

This process is evidently due to the soft that sufficient energy is not available to make up the masses of the possibly produced particles [condition (a)].

Other cases that have been discovered and studied extensively by Hafstad and Tuve, Crane, Delsasso, Lauritsen, Fowler and others are:

$$(p, \gamma)\ {}_3Li^7 + {}_1H^1 \rightarrow *{}_4Be^8 + hv$$

$${}_9F^{12} + {}_1H^1 \rightarrow *{}_{16}Ne^{20} + hv.$$

These reactions have been detected and analysed either by the emitted γ-rays or by the radioactivity of the product nucleus or by both.

In the first case of the *γ-rays from the capture of protons by Li*, experiments conducted by Delsasso and his associates, in 1937, have shown that there is only one strong γ-ray line at 17-1 MeV and probably one or more weak ones between 10 and 17 MeV. The first line corresponds to-the transition to the ground state of Be^8. The reaction energy Q calculated from the masses of the nuclei involved and the energy of the incident protons (440 KeV) is 17.4 MeV, in good agreement with the observed 17.1 MeV. A line at 14 MeV may be expected from a transition to the excited state of Be^8 of 2-8 MeV excitation energy known from other nuclear reactions.

The second case of γ-rays from the capture of protons by fluorine is interesting since it gives a good example of the working of selection rules [condition (d)]. There seems to be a single γ-ray line having an energy of 6 MeV about. The reaction energy Q deduced from nuclear masses and the energy of incident protons (0.33 MeV) is 13.23 MeV.

It appears, therefore, that the transition to the ground state is forbidden, evidently by selection rules.

Resonance disintegrations : It was early realised that in the (α, p) reactions certain groups of the emitted protons have a definite energy which does not vary with the incident α-particle energy. Furthermore, these groups are observed only for a definite energy of the α-particles. These facts indicate the existence of the phenomenon of resonance, first suggested by Gurney on wave-mechanical principle and experimentally observed in the bombardment of Al with α-particles by Pose. It has been subsequently found by many other observers that the resonance disintegration is a pretty common occurrence, not only in the case of bombardment of light elements by α-particles, but also in proton and neutron produced processes.

The principle underlying this resonance process is as follows : If the energy of the incident particle is such that the total energy of the system is close to one of the virtual levels of the compound nucleus, the probability of the formation of the compound nucleus and hence of the nuclear process will obviously be much greater than if the energy of the particle falls in the region between the levels. Therefore, one may expect to find characteristic fluctuations of the *yield* of a nuclear process with the energy, from high values at the resonance energies to low values between resonance levels.

In the exchange reactions the process will give rise to two product nuclei which will then product nuclei which will then move apart with an energy determined by the energy balance appropriate to the reactions.

Now let us deal with the photo disintegration. The disintegration of nuclei, the atomic photoelectric effect has been termed as photo disintegration or nuclear photo effect.

In most cases, the process results in the *emission of neutrons*. Consequently the product nucleus is always less in mass by unity than the bombarded nucleus. If the *product nucleus* does not exist in nature in a stable state, it will be *radioactive*, and in general, a *positron/emitter*. The reaction can therefore be detected by either observing the neutrons emitted or more conveniently by the radioactivity of the residual nucleus.

Photodisintegration can occur only when the energy of the incident γ-rays exceeds the binding energy of the particle to be ejected. As the energy of the incident γ-ray is steadily increased, the reaction begins at a certain "threshold" value of the photon energy, and the yield then increases more and more rapidly as the energy is increased. Theoretical considerations by Bethe and Peieris show that the greatest probability of disintegration occurs when the γ-ray energy is just twice that of the binding energy of the ejected particle. The cross-section for the process is of the order of $10^{-27}cm^2$. Although this is a small quantity, the great penetration of γ-rays exposes a large number of nuclei to bombardment, which results in an appreciable number of disintegrations.

In principle, all nuclei can be disintegrated by γ-rays of sufficiently high energy. It is to be expected also that particles other than neutrons, *e.g.*, protons, α-particles or deuterons can be ejected in this process. The relative probability of these various types of photodissociation will be determined by the respective "particle widths" of the levels of the compound nucleus, *i.e.*, primarily by the penetrability of the potential barrier for the particle to be emitted; the probability will therefore be smaller for heavier nuclei.

Photodisintegration was first discovered in 1934 by Chadwick and Goldhaber who observed that deuterium gas, when irradiated by the γ-rays of ThC' of energy 2.62 MeV, was disintegrated, the deuterium nucleus splitting up into a proton and a neutron according to the equation

$$_1D^2 + h\nu \rightarrow {}_1H^1 + {}_0n^1 + Q.$$

A calculation from the masses involved in this reaction shows a mass deficiency of 0.00234 m.u., which means that the reaction is *endoergic* (*i.e.*, Q is negative) ,and an energy Q equivalent to the mass defect or 2.178 MeV must be supplied. Hence to produce photo-disintegration with deuterium, the incident γ-radiation must possess an energy of at least 2.178MeV.

Chadwick and Goldhaber proved this to be true from experimental data as follows. The energy of the protons produced in the disintegration can be measured by means of a linear amplifier or a cloud chamber. Assuming that the neutron carries approximately the same amount of energy as the proton on account of their nearly equal masses, the total kinetic energy carried away by the two product particles can be taken as twice the measured value of the proton energy. The total energy thus estimated is 0.45 MeV. Subtracting this amount from the energy of the incident γ-rays, *viz.*, 2.62 MeV the energy absorbed in the process is 2-17 MeV, which agrees well with the values calculated from the masses of the nuclei appearing in the equation. This result is further confirmed by the fact that Photo-disintegration of deuterium is produced also by the γ-rays of RaC' of energy 2.198MeV, although with much reduced intensity.

Fig. 14.5 : Prof. R. Peierls.

14.6 INDUCED RADIOACTIVITY

In 1934, induced Radioactivity was discovered by Mme. Curie Joliot and her husband M. Frederick Joliot. Today one could actually produce about 500 radioactivity elements which do not occur normally in nature.

As soon as Curie and Joliot announced their discovery, the study of artificial radioactivity was taken up in many other laboratories as Cavendish in Cambridge and Pasadena in California, and new cases of it were announced at a rapid rate. It was soon found that the phenomenon occurred not only under α-particle bombardment, but also with other projectiles, as protons, deuterons and neutrons and that the artificially produced radioelements not only emitted positrons but also electrons in several cases.

Wilson's cloud chamber method was applied in the analysis of the activity of the artificial radioelements by Ellis and Henderson at the Cavendish Laboratory and by the workers at Pasadena and beautiful photographs of the tracks of the emitted particles curved by magnetic fields were obtained in confirmatory evidence of the discovery. One such photograph is reproduced here. It was obtained by Anderson in California, when a carbon target, after having been bombarded for several minutes with protons of 0.9 MeV energy, was placed right into the chamber. The tracks obtained have the specific aspect of the electron tracks and in the applied magnetic field of 800 gauss they are curved in a sense which proves the particles to be positive. The positron activity is evidently due to the formation of radionitrogen N^{13} in the "capture" (p. γ) reaction of proton with carbon, already described.

Fig. 14.6 : Induced Pomtron Activity from Carbon Bombarded by 0.9 MeV Protons. (Andersen).

The nature of the dependence of the induced radioactivity en the kinetic energy of the bombarding particles was also investigated. The Joliots were able to establish that the half period was independent of the kinetic energy of the projectile when the energy of the α-particles of P_0 was varied from 6.3 to 1MeV. This proved that a *single kind of unstable nucleus alone resulted from the reaction.*

14.7 NUCLEAR ISOMERS

The manifestation of different radioactive properties exhibits to certain differences in the internal structure of nuclei, which are identical in all respects. They are called nuclear isomers.

Many cases of the artificial radio elements reveal one and the same nucleus having the same mass number (A) and the same atomic number but they disintegrate in different ways with different decay periods. These isomers exhibit certain typical properties.

Soddy was the first, in 1917, to suggest that such isomers might exist among the natural radioactive elements, and Hahn was able, in 1921, to establish experimentally the isomeric property of UX_2 and UZ which were the only cases of nuclear isomerism known for a long time. In recent years, however, several isomeric pairs have been discovered among the artificial radioactive nuclei. The study of this phenomenon is of special interest to nuclear physics, as it argues to the existence of low lying metastable states in nuclei. A brief summary of the experimental study and theoretical interpretation of nuclear isomerism will now be given.

Nuclear isomerism has been experimentally investigated in the following three ways, *viz.*,

(i) a first identification of isomeric cases,

(ii) a closer examination of the internal mechanism of isomers, and

(iii) separation of isomers.

Identification of isomers consists chiefly in deciding whether the two different period β-activities experimentally observed in certain cases are isomeric or not. This is done by a simple "cross-checking" with different types of nuclear reactions, in which the element in question and its immediate neighbour are implicated, along with the knowledge of the masses and relative abundances of the corresponding stable isotopes. It must be emphasised that the simultaneous existence of two decay periods is, by itself, no sufficient criterion for the identification of isomers, but that "cross-checking" must necessarily be used to arrive at a safe conclusion. Let us illustrate by the isomerism of ${}_{35}Br^{80}$, which was the first case to be discovered among artificial radioelements.

The Isomers of Br^{80} : There are only two known stable isotopes of bromine of mass numbers 79 and 81 and relative abundances 50.7% and 49.3% respectively. When bromine is bombarded by slow neutrons, radioactive elements are formed which disintegrate by the emission of electrons with three different periods, 18 mins., 4.2 hrs. and 36 hrs. Chemical tests show that the radioactive products are isotopes of bromine.

When bromine is bombarded with γ-rays of very high energy (17MeV), radioactive elements chemically identified as isotopes of bromine are formed, which decay emitting β-rays, again with three different periods, 6.3 mins., 18 mins., and 4.2 hrs. The reaction is evidently a *photonuclear (γ, n) effect* represented by

$$\left.\begin{aligned} {}_{35}Br^{79} + h\nu &\rightarrow {}^{*}{}_{35}Br^{78} + {}_{0}n^{1} \\ {}^{*}{}_{35}Br^{78} &\rightarrow {}_{34}Se^{78} + e^{+} \end{aligned}\right\} \text{for mass 79}$$

and

$$\left.\begin{aligned} {}_{35}Br^{81} + h\nu &\rightarrow {}^{*}{}_{35}Br^{80} + {}_{0}n^{1} \\ {}^{*}{}_{35}Br^{80} &\rightarrow {}_{36}Kr^{80} + e^{-} \end{aligned}\right\} \text{for mass 81}$$

Here also there are only two radioactive isotopes for bromine formed, of masses 78 and 80, which are, however, responsible for the three observed periods, so that again to one of these must be attributed a double period.

Cross-checking these two different reactions, it is seen that two of the observed periods, *viz.,* 18 mins. and 4.2 hrs. are common to both which must therefore be assigned to that radioactive isotope which is also common to both reactions and hence to bromine of mass number 80. Thus Br^{80} *consists of isomeric nuclei emitting electrons of two different periods*. The remaining two of the observed periods, viz; 6.3 mins., and 36 hrs., must be attributed to Br^{78}, a positron emitter, and Br^{82}, an electron emitter, respectively. Confirmatory evidence is had from the fact that the isomeric pair of Br^{30} can be produced by bombarding bromine with deuterons and fast neutrons, as was shown by Snell (1937) or by proton bombardment of selenium, analysed by Da Bridge and others (1938).

Attention has been directed towards two points chiefly : In the cases where the isomers are formed by the "capture" process, nature of the capture level or levels responsible for the phenomenon is investigated. This is usually done either by measuring the "branching ratio" of the two isomeric activities produced under varying conditions of activation or by determining the resonance energies involved in the process by the "boron absorption" method. Thus Amaldi and Fermi by the "branching ratio" experiments (1936) and Von Halben by the "boron absorption" experiments (1937) were able to demonstrate a single capture level in the case of rhodium (Rh^{104}) which exists in two isomeric forms of periods 46 secs. and 4.3 mins.

Separation of isomers has been achieved in several cases using the Szilard and Chalmer method. The γ-rays ordinarily emitted along with like β-activities is utilised for this purpose. The γ-rays may be internally converted and the conversion electrons thus produced may have enough energy to break the chemical bond. Or, if the γ-rays have a high energy, their emission will be accompanied by the recoil of the nucleus, which may be sufficient to liberate the atom from the compound.

In 1939, Segre and his associates have thus been able to separate the two isomers of bromine. Likewise, Seaborg and his associates have chemically separated the three isomeric pairs of tellarium (127, 129, 131). The results are of great importance as they indicate:

(i) That the isomers can be considered as different atoms, though possessing the same mass and the same chemical and physical properties, and

(ii) That the isomeric pairs maybe genetically related, one growing out of the other, since it has been found that in a number of cases the short period isomers grow from the long period ones, *e.g.,* in the case of bromine the 18 mins. period isomer grows from the 4.2 hrs. period isomer; in rhodium the 45 secs isomer from the 4.3 hrs. one

14.8 IMPORTANCE OF ARTIFICIAL RADIOACTIVITY

The study of artificially produced radio elements, gives new ideas in favour of the neutron-proton constitution of the nucleus. The isomeric property of several of the induced radio active elements reveals the existence of low lying metastable excited levels in nuclei.

The process of artificial radioactivity has got much more importance, in many practical applications.

Induced radioactivity provides good knowledge for the study of the processes of nuclear disintegrations.

In other *applied branches of dence,* such as medicine and biology, the importance of artificially produced radioelemants seems not only to be definitely assured but also to be ever on the increase. Their use as "tracera" has already been mentioned. *The potential value of radioactive isotopes of elements which are important in physiological processes, such as phosphorus, sodium, iron, calcium and iodine, is quite apparent.* By choosing a radioactive isotope of a material which the body localises in some definite place, it is possible to apply the desired treatment in that part of the body needing it. For example, it is known that a good part of the iodine taken internally goes to the thyroid gland. Hence for a beneficial treatment of this gland, a radioactive isotope of iodine can be administered, which will for the greater part get localised in the thyroid and thus produce the desired cure without affecting the other parts of the body. In general, the same radioactive isotope serves both the purposes of medical discipline, *viz.,* diagnosis and treatment, being used in small quantities as 'tracers' for diagnosis and in larger doses for treatment. Evidently right choice of isotopes, as regards life, nature of radiation, α, β or γ-rays, etc. is important in the matter of treatment. For example, *radiophosphorus* (P^{32}), period = 14.3 days, β-emitter is found good for surface applications (skin diseases); *radiocobalt* (Co^{60}), period

= 5.3 year, β-emitter, has been utilised in the treatment of deep cancers, in which it is implanted in the form of needles; *radiogold* (Au^{198}), period = 3 days, γ-emitter, is fired right into the cancer in the form of small bullets from a 'gun'; these bullets can be allowed to remain permanently in the system, since gold, after it has lost its activity, ia inert and harmless to the body tissues. It may be noted that the early hopes that these radio-isotopes would prove the 'magic bullets' in the fight of all forms of cancer have not been realised; but in some special forms of cancer they have given relief and even complete cure.

Radiocarbon is another interesting radioactive isotope, which is obtained by slow neutron bombardment of nitrogen according to the reaction

$$_{7}Br^{14} + {}_{0}n^{1} \rightarrow {}^{*}{}_{6}C^{14} + {}_{1}H^{1}$$

The radioactive carbon decays by emitting an electron,

$$^{*}{}_{6}C^{14} \rightarrow {}_{7}N^{14} + e^{-}$$

with a period of about 5,700 years. It has many uses as a "tracer" in biological research and for studies of petroleum products and organic chemicals where radiocarbon w substituted for natural carbon. But an important application known as *radiocarbon dating* is the determination of archaeological and short geological times that are of the order of magnitude of the mean life of radiocarbon. Cosmic rays contain an appreciable amount of secondary thermal neutrons, which on reacting with atmospheric nitrogen produce radiocarbon by reaction. The production of $_{6}C^{14}$ and the subsequent decay are in equilibrium in the earth's atmosphere. Hence C^{14} is present in equilibrium concentration in all living plants. The carbon is absorbed in the form of CO_2, so that dead plants that no longer take in CO_2 from the atmosphere lose their C^{14} by radioactive decay. Consequently, the ages of pieces of wood, grain, peat, charcoal, etc. can be pretty accurately determined by measuring the C^{14} they contain relative to the equilibrium amount present in living plants. The method assumes that the average cosmic ray intensity has remained constant for at least 20,000 years and that atmospheric mixing is rapid compared to the life-time of C^{14}. The archaeologist has only to find some wooden object in the excavations of an ancient settlement and determine its radiocarbon content. Then by a simple calculation he is able to estimate the time of the death of the plant and hence the time of existence of the ancient settlement. Libby using this method and working for a number of years (1949-54) has examined many samples for their C^{14} content and has established dates when life ceased in the samples, which range from 1000 to 16,000 years.

14.9 NUCLEAR FISSION

The nuclear fission was first time done by Fermi, in 1934. The continuous study of artificial radioactivity finally resulted, in the portentous discovery, which is named as nuclear fission.

In 1938, it was discovered that heavy unstable nuclei such as uranium when bombarded with neutrons explode more radically into two more or less equal fragments. This nuclear direction has been named as fission.

During the years 1936 to 1938, the experiments of Fermi were repeated in other laboratories, particularly in Germany by Hahn, Meitner and Strassmann and in France by Curie-Joliot and Savitch. Although Fermi's results were confirmed and extended, it became necessary to assume that the supposed "transuranic" elements in decaying, gave rise to other radioactive elements, which would be isomeric with other heavy atoms. Further, the number of different substances reported became so numerous that it was difficult to fit them into any plausible scheme. It must be said that chemical tests in this region of the periodic table was extremely difficult as the elements wore new and could be obtained only in very small quantities.

The French scientists in one of their chemical separation tests found an "activity" of 3.5 hrs. period, which could be precipitated by lanthanum ($Z = 57$). This result might well have been the key to the right solution of the complicated phenomena observed, thereby allowing Curie and Savitch to be the discoverers of fission; but they missed their chance by sticking to the then commonly accepted view that reactions of the type analysed could not give rise to elements of such low atomic number as lanthanum, although they fully realised the difficulty of finding a place for an element having chemical properties similar to lanthanum among the elements.

The problem concerning "nuclear fission" may be stated us follows. How can the fairly moderate activation of a nucleus resulting from the capture of a neutron lead to such a cataclysmic disruption ? Why does it occur only in a few heavy elements such as U, Th and Pa? The answer to these questions was first given by Meitner and Frisch on the basis of the liquid drop model of the nucleus. Bohr and Wheeler developed it further with definite and quantitative results that could be checked and verified. A nucleus is analogous to a drop of liquid in many ways, such as the close packing of particles, constant density, short-range forces, etc. The resultant effect of these nuclear characteristics is to endow the nucleus with a property analogous to the surface tension of liquids.

Considering a liquid drop, we readily see that the dissipative forces acting between the constituent molecules are counteracted by the surface tension forces. Further, any liquid drop, not subject to outside forces, tends to assume a spherical shape under the influence of surface tension, since a sphere presents the smallest surface for any given volume. Very large drops are impossible, because the surface tension forces are feeble and there is a maximum size of the drop which cannot exist permanently. As this size is approached the drop becomes less and less stable, gets deformed. At this stage, it begins to execute the so-called *surface tension oscillations* with the slightest provocation from outside, which lead to the disruption of the drop into two or more smaller droplets.

Fig. 14.7 : Prof. O.R. Frisch

In atomic nuclei, the short range attractive forces, keeping the constituent particles, *viz.,.,* the charged protons and the uncharged neutrons, stay together in a stable state, play the role of surface tension, while the electrostatic repulsive forces between the protons that of disruptive forces tending to destroy the stability. If the surface tension forces prevail, the nucleus will never break up by itself, and two nuclei coming into contact with each other, will have a tendency to *fuse* just as two ordinary liquid droplets do. If, on the contrary, the electric forces of repulsion have the upper hand, the nucleus will show a tendency to break spontaneously into two or more parts, which will fly apart at high speed; this is the phenomenon of *fission.*

Between the surface tension and electrostatic forces in the nuclei of different elements led to the very important conclusion that while the surface tension forces predominate in the nuclei of all elements in the first half of the periodic table (approximately up to mass 110), the electrostatic repulsive forces prevail for all heavier nuclei. Hence all nuclei of mass greater than 110 are *potentially unstable*, and under the action of a sufficiently strong disturbance from outside would break up into two or more parts with the liberation of a considerable

It was left to the German scientists, Hahn and Strassmann, to clarify the issue in the following manner. In the course of their investigations, they discovered in uranium bombarded with neutrons, a radioactivity of 8.6 mns. period, which could be precipitated by barium ($Z = 56$). They also at first thought, in consonance with the opinion then prevalent, that the observed activity might be a new form of radium, since the chemistry of radium is very much like barium. But radium (Z = 88) could only be formed from uranium by the emission of two α-particles and such a process had never been known to follow neutron capture. Moreover, on further careful tests, such as fractional precipitations and crystallisations of the kind used for separating radium from barium, they found that the activity definitely followed barium, not radium. Hence they proposed, not without hesitation, however,' because of the strange results, a radically new hypothesis that *uranium nucleus after capturing a neutron might, instead of being chipped off by a small fragment as had always been observed previously in all particle disintegrations, break up into two large fragments, each of the size of a moderately heavy atom.* They justified their opinion by showing that among the radioactivities observed was also one with a period the same as that of a known isotope of krypton (Z = 36). On combining the atomic numbers of barium and krypton, (56 + 36) the number 92 could be obtained, which was the atomic number of uranium, so that the process might well be represented by

$$_{92}U^{238} + {}_0n^1 \rightarrow {}_{56}Ba + {}_{36}Kr + {}_0n^1.$$

Hence, they concluded that the α-activities previously ascribed to transuranic elements were probably produced by radioactive isotopes of elements of lower atomic number, as barium and krypton. In this way, for instance, the activity found in lanthanum (Z = 57) by the French scientists could be formed by β-decay of barium (Z = 56). Thus a new type of disintegration, in which a heavy nucleus splits into two other nuclei of comparable size, hence called "nuclear fission" was discovered.

As soon as these results were announced early in 1939, physicists in other laboratories immediately repeated and confirmed these experiments. In the light of the "fission" idea all the facts, which were found disconnected and difficult of synthesis, could be readily coordinated. Many of the fission products could be identified with substances of much lower atomic number than uranium, already known in the study of artificial radioactivity. Thorium (Th) and protoactinium (Pa) were also shown to undergo fission when bombarded with neutrons. Thus the newly discovered process of nuclear fission was definitely established though a theoretical justification was deemed necessary for its formal and final acceptance. But a satisfactory explanation was proposed, almost immediately after the discovery, by Meitner and Frisch, who were also the first to suggest the name of "fission" for the phenomenon.

14.10 SELF-PROPAGATING CHAIN REACTION

Emission of secondary neutrons in the fission process is not an easy task. The increase in fission liberates tremendous amount of energy. Each fission releases a very high energy, in a short time.

It has been calculated that one cubic metre of uranium oxide might develop 10^{12} kilowatt hours in less than 0.01 sec. So it is realised that the phenomenon of fission can be put to practical use as a source of power.

The common sources of power are obtained by combustion processes. Now combustion is always self-propagating : *e.g.*, the lighting of a fire with a match liberates sufficient heat to ignite the nearest fuel, which liberates more heat which ignites more fuel and soon. Although, as a rule, nuclear reactions, unlike chemical reactions, are not self-propagating, yet in the particular case of nuclear fission, on account of the emission of particles of the same kind as those which started the fission, a self-propagating reaction is obtained involving so many nuclei that the whole mass would explode, thus producing enormous energy, many million times that of ordinary fuels.

But to secure an efficient chain reaction is not easy in practice, because there exist several causes that prevent the progressive neutron breeding, essential for the self-sustaining fission reaction. The general condition for the minimising of these causes which militate against the occurrence of the chain reaction is usually expressed by a quantity known as the *multiplication factor* (k) and defined as the ratio of the number of fresh neutrons produced by fission at any place in the material to the number of free neutrons originally present at that place. It is *essential that (k) must be at least equal to 1 for the maintenance of the chain reaction.*

The two chief sources of wastage of neutrons which lead to the collapse of the chain reaction are :

(i) *leakage of neutrons from the system, and*

(ii) *presence of non-fissionable material in the system, which absorb the neutrons.*

The first of these may be reduced by a suitable choice of the size and shape of the fissionable material. The size must be large enough to keep the neutrons within the system so that the probability of their escape, before they could hit nuclei and produce fission is rendered small. *A spherical shape* is also very conducive to obtain the desired result, since the *escape of neutrons is a surface effect* dependent on the surface area and hence, on the square of the radius, while *fission is a volume effect* dependent on the volume and hence on the cube of the radius. Thus the ratio of the rate of escape of neutrons from the system to the rate of their production varies inversely as the radius, decreasing with increasing size. As the radius is increased from a small value, a *critical size* is reached, beyond' which more neutrons are produced than are lost, so that the chain reaction can progress rapidly. This critical size is of great importance, below which the fissionable material is completely stable and perfectly safe, but above which the system becomes unstable and explodes spontaneously. The critical size depends on the nature and shape of the surroundings as well as on the speed of the reacting neutrons.

The second source of loss of neutrons by absorption, non-productive of fission, may be reduced either:

(a) *By carefully purifying the fissionable material from other non-fissionable impurities that absorb neutrons or*

(b) *By neutralising the disturbing action of the non-fissionable materials without actually neutrinos them.*

The probability of non-fission capture of slow neutrons is anishingly small except fora pronounced resonance at neutron energy of about 25eV with an effective cross-section of 1200×10^{-24} cm^2 as estimated by Meitner, Hahn and Strassmann. The light isotope manifests a very high probability of fission capture for neutrons of any energy and the fission cross-section increases with decrease of neutron speed reaching a high value of 400×10^{-24} cm^2 for thermal neutrons. It is to be noted that the resonance non-fission capture by U^{238} produces a new transuranic element called plutonium, ($_{94}Pu^{239}$) which can be chemically separated from uranium; the fission properties of plutonium should closely resemble U^{235} according to the Bohr-Wheeler theory.

14.11 MODERATORS

Fermi suggested a ingenious method to carry the chain reaction, in natural uranium by the use of certain artificial devices. These simple devices were called Moderators.

The essential function of the moderator is to slow down the fast neutrons of fission very rapidly by elastic collisions to thermal energies.

Elements of low atomic weight are used in moderators. It was first thought that hydrogen would be the best moderator, because it could reduce the energy of the neutron by about half at each collision. In about 12 collisions hydrogen could reduce the energy of neutron. As a moderator hydrogen also has many demerits.

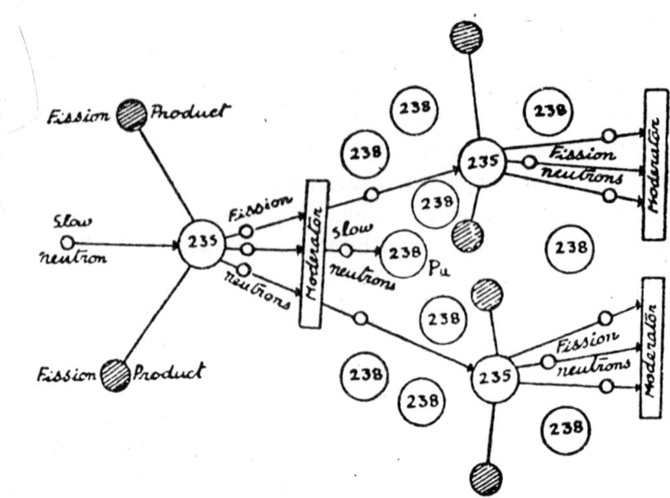

Fig. 14.8 : Principle of Action of the Moderator.

Being a gas, it will take up too much space. Water (H_2O) might be used, but it contains oxygen which complicates matters. Moreover, hydrogen captures neutrons in inelastic collisions to form deuterium, while a good moderator should slow down neutrons rapidly without capturing them. Deuterium, helium, beryllium and carbon are considered as possible good moderators. Carbon in the form of graphite and deuterium in heavy water (D_2O) have been actually used.

This method has the additional advantage of producing at the some time plutonium, which when separated and used pure, is as good, perhaps even better, than U^{235}, as a fissionable material. This new artificial element results from the resonance capture of some of the neutrons in the process of slowing down by U^{238}, as already stated.

Both the methods of developing an efficient chain reaction have been tried with successful results, in connection with the use of atomic energy for practical purposes. It may be noted that the time factor is very important in the production of atomic power for industrial applications. Fission after fission must occur rapidly before the neutrons could escape into the air or be absorbed by the materials without leading to fission.

The enormous energy liberated in nuclear fission coupled with the possibility of practical realisation of rapid chain reaction readily suggests the two most important practical applications of the process, *viz.*,:

(i) *The super-power plant*, and

(ii) *The super-explosive or the atom bomb*. Both of them have been successfully attempted, in spite of complicated techniques involved. The atom bomb project has yielded better results than that of the atomic power plant. The general principle that discriminates the two applications is as follows : For industrial purposes, a graded constant production of power which gives a temperature high enough for the efficient operation of heat engines has to be token into consideration, while for the atom bomb a high-speed cumulative effect is to be realised, where within less than a micro-second the number of fissions is multiplied to such an extent that a considerable portion of the material is affected and the temperature and pressure rise enormously causing a terrific explosion. Accordingly, two different techniques have been developed, one a *slow neutron reacting system* called the *reactor* or *pile* suited for industrial purposes and the other a fast neutron reacting system for the production of atom bombs.

14.12 THE REACTOR

The Reactor or a pile works on the slow neutron chain. A pile consists of a great number of uranium block is, which are disposed in a calculated lattice throughout a huge mass of graphite.

For the chain reaction to take place the size of the must be evidently greater than a certain critical value, when the rate of escape of the neutrons from the surface is not greater than the rate of production. This *critical size* depends upon the precise arrangement of the uranium and graphite and upon whether the surroundings reflect back any of the neutrons which escape from the surface. The size of the pile required to obtain a slow neutron chain reacting system in uranium must be obviously very large, on account of the very small amount of U^{235} in natural uranium and the great quantity of moderator to be used.

The *fact that some of the fission neutrons are delayed in time is utilised to control the operation of the pile*. If the pile is made just above the critical size, the exponential increase in the rate of fission takes place rather slowly due to the delayed neutron emission. Now by inserting into the pile rods of a material like boron or cadmium which absorb stow neutrons strongly, the multiplication of the fissions can be prevented altogether and the pile will not operate. This device is known as arrestor. Slow withdrawal of these rods allows the reaction to begin and there is plenty of time to re-insert them if the multiplication rises too rapidly.

Adjustment of the rods can be carried out automatically by the use of an equipment like a thermostat, which sets the position of the arrestors so that the system rung continuously at any desired energy level.

The fission energy degenerates by collisions into heat and the uranium blocks become hot. This heat may be extracted by cooling the rods, either by blowing, such as air, hydrogen or helium, or by surrounding them with A tube through which water circulates.

The size of the pile may be reduced by replacing the, graphite by a *better moderator* such as heavy water, which is mire effective in slowing down neutrons.

During the building up of the pile, the neutron density was measured as successive layers were added, in order to check the approach of the critical size. With no arresters, the required size was reached before the last layers were in place. Great care was taken lest the chain reaction should get out of hand and cause an explosion. The pile was surrounded by a thick layer of graphite so as to reflect back into it as many neutrons as possible. But even so, in order to make up for the surface loss of neutrons the size of the pile was large, its diameter being 6 to 7 feet. Six tons of uranium and some hundred tons of graphite were used.

When the pile was not in operation several arresters were inserted in a number of slots which brought down the multiplication factor below 1. To operate the pile, all but one of the arresters were removed and the remaining one was slowly pulled out, when the intensity of the neutrons emitted began to increase rapidly. By a careful adjustment of the single arrestor, the pile was made to work at a constant power level. This first pile produced a low power of 200 watts only, but this was enough to check up theoretical expectations.

Nuclear Reactors have various uses. They can be classified in different headings.

1. Uranium Graphite Reactors

These reactors consist efficient air cooling system. These reactors to be operated at a moderately high temperature of 200 to 300°C.

2. Heavy Water Reactors

In this type of reactor, neutrons can be reduced to thermal energies, after about 25 collisions with deuterons. For a heavy water reactor a small volume of moderator is required. Heavy water can be used at the same time as a cooling liquid for the reactor.

3. Homogeneous Nuclear Reactors

In this reactor the fissionable material is concentrated at regularly spaced, intervals throughout the moderator. The first reactor of this type was constructed in 1945 at Los Alamos using enriched uranium salts dissolved in ordinary water as moderator.

14.13 THE ATOM BOMB

The most terrific and devasting explosive-the atom bomb, works on the fast neutron chain. Fissionable materials used are U^{235} or Pu^{239} each of mass somewhat greater, than half the critical value.

The actual production of the atom bomb involves essentially the *preparation of U^{235} or Pu^{239} in sufficiently large quantities*. This is evidently an extremely complicated and costly project, but it has been achieved under the impetus of World War II as a major national effort in the U.S.A. Plutonium is obtained by the use of the uranium pile and of chemical separation plants, as already stated. The technique used for the concentration and separation of U^{235} forming a very small part of natural uranium is still more involved and difficult. But it has been achieved, of course at prohibitive costs, using chiefly the gaseous diffusion method and employing very probably the only known gaseous compound of uranium, UF_6. The gigantic and complicated installations involved in this work may be considered a unique achievement in the history of science. The whole project was highly speculative, until the first test bomb exploded in New Mexico desert according to theoretical predictions.

As pure fissionable materials are used in the atom bomb without any moderator, the size of the bomb is bound to be *far smaller than that of the pile*, which eliminates transport difficulties to a great extent. The critical size depends upon the shape of the bomb and the nature of the material in which the bomb is enclosed.

The two pieces to be detonated are enveloped in a cover of a very dense substance, the tamper, which serves the double purpose of reflecting back into the bomb neutrons which might otherwise escape into air and of delaying the expansion of the bomb until the temperature and pressure of a high explosive is built up.

The atom bomb is the most terrific and devastating explosive so far made by man, although a more powerful competitor, the so-called "hydrogen-bomb", has beaten realised more recently. The ruins of Hiroshima and Nagasaki bear witness to its destructive power, estimated to be about that of 10,000 tons of trinitrotoluene (T.N.T.) In the actual explosion which occurs in a couple of microseconds after the detonation and for which only a portion of the active material is effective the energy liberated is sufficient to raise the temperature to the order of 10,000,000°C and more and to produce a pressure of several millions of atmospheres. Since radiation is proportional to the fourth power of the temperature, the explosion will be accompanied by a violent and intense blast of visible, ultra-violet. X-ray and γ-ray radiations causing a blinding flash. A large quantity of radioactive matter is also released both by the fission products and by the action of the emitted γ-rays and neutrons. Under the conditions of such high temperature and pressure, the active material itself will get quickly vaporised and expand, causing a fall in the density, diminution of neutron absorption and the consequent collapse of the chain reaction. The time for which the explosion lasts is very small and depends on the rate of expansion of the active material.

14.14 THE HYDROGEN BOMB

The hydrogen bomb is the destructive explosive, thousand times more powerful than even the atom bomb. It works just the opposite way to that of the atom bomb.

In the hydrogen bomb very heavy atoms breakdown by fission.

The essential conditions required for the operation of the hydrogen bomb are the extremely high temperatures and pressures that are present in the interior of the stars promoting thermonuclear reactions. To reproduce these in terrestrial matter is not easy. But such conditions have been realised by using the atom bomb as a "*primer*" which by first exploding could provide the very high temperatures and pressures necessary for the successful working of the hydrogen bomb.

The superiority of the hydrogen bomb over the atom bomb is probability due to the fact that it has, not, in principle, the limitation of a *critical size*, unlike the atom bomb. If the active material in the atom bomb exceeds the critical mass, spontaneous explosion results. On the other hand, any amount of active Jaaterial can be used in the hydrogen bomb, which would not start exploding until part of it had been heated to the ignition temperature. The first hydrogen bomb, prepared in America and nicknamed "Mike", was exploded in November 1952 at Eniwetok in the Marshall Islands. The explosive violence of Mike was equivalent to 3 million tons of T.N.T. The fuel used for the thermonuclear reaction was .probably liquid deuterium and was fired with an atomic "bomb "trigger" Mike weighed about 65 tons and was not a transportable bomb.

In August 1953, the Russians exploded a thermonuclear bomb, which had a less explosive power, about 1 million tons of T.N.T. only, but was technically more advanced than Mike, since it was a transportable one and very probably the fuel used was other than liquid deuterium.

A *pure* fusion bomb will not produce any bad after-effects, since the end products are not radioactive; but the radioactive fission products of the 'primer' atom bomb will always be present. In order to produce a *clean bomb, i.e.,* a fusion bomb without radioactive end products, attempts have been made to replace the atom bomb primer by a chemical trigger, *i.e.,* to produce the required temperature of a million degrees by firing a chemical explosive contained in some specially designed container (principle of shaped charges). However the possibility of the large number of high energy neutrons liberated during the explosion inducing radioactivity in the surrounding sir and dust cannot be neglected, but compared to the fission the effect will be very small. The so-called *N-bomb* (neutron bomb) is a clean bomb, produced by tailoring the energy of a fusion explosion (H-bomb), so that, instead of heat and blast, its primary product is a burst of neutrons. Such a burst would operate as a king of death rays. It would do next to no physical damage; it would result in no contamination; but it would immediately destroy all life in the target area. This, of course, would make it an ideal battlefield weapon. A fission atom bomb is inherently *dirty*.

15

LASER

15.1 INTRODUCTION

The word laser is an acronym coin from the words light amplification by stimulated emission of radiation, which is the full decrepitate name of the phenomenon.

The American physicist C.H. Townes, and Russian Physicists N. Basov and A.M. Prokhorov, independently discovered Maser.

In 1960 optical maser or laser was discovered by T.H. Maiman.

The principle of laser is based on the phenomenon of stimulated emission of radiation, the theory of which had been worked out by Einstein in 1917 and was discussed in the previous section. We know that light is emitted from a source when the atoms in the source make transitions from an excited to a lower energy state spontaneously. Normally the atoms exist in the excited state for about 10^{-8} s. If the energy of the excited state is –3 and that of the lower energy state is E_1, then the energy of the emitted photon in this spontaneous transition is

$$hv_{12} = E_2 - E_1.$$

On the other hand, if a photon of energy hv_{12} falls upon an atom in the state of energy E_1, then it will make a transition to the higher energy state E_2 by the absorption of the photon.

Due to spontaneous transition, the light photons are emitted in all possible directions. Besides, there is no definite phase relationship between the different photons so that the emitted light is incoherent in nature.

Besides the above two types, a third type of transition from an upper energy state E_2 to a lower energy state E_1 may be induced by an incident photon of energy $hv_{12} = E_2 - E_1$ giving rise to the stimulated emission of radiation.

The photon emitted during stimulated emission has the same energy as the incident photon. It is emitted in the same direction and has the same phase as the latter. Thus we get two coherent photons in this case. If these two photons are now incident on two other atoms in the state

E_2, then it will result in the induced emission of two more photons so that there will be four coherent photons of the same energy. These four photons may then induce transitions in four other atoms in the energy state E_2, thereby giving rise to the stimulated emission of four fresh coherent photons of the same energy so that the number of coherent photons of energy hv_{12} is now increased to eight. If the process can be made to go on in a chain, we may ultimately be able to increase the intensity of the coherent radiation enormously.

The necessary condition for this type of amplification of the light intensity by the stimulated emission of radiation is that the number of atoms in the upper energy state E_2 must be sufficiently increased.

Normally the number of atoms in the upper energy state is much lower than in the lower energy state ($n_2 << n_1$) due to the Boltzmann factor. As a result the rate of downward induced transitions $B_{12}n_2 u_v$ is much less than that of the upward transitions $B_{12}n_1u_v$ between the same two levels due to absorption of photons. To increase the rate of stimulated emission from E_2 to E_1, it is necessary to make $n_2 > n_1$. This is known as *population-inversion*.

15.2 PRODUCTION OF THE LASER BEAM

The production of laser beam is not an easy task. It requires a careful method.

The lasing medium is enclosed with in a closed vessel. The vessel consists two reflectors, at its two ends.

The first thing is to get population inversion in the atoms of the medium, later amplification of the intensity is done.

Due to repeated reflections from the two reflectors, the beam of light gets sufficient chance to interact with a large number of the excited atoms in the metastable states to cause stimulated emission. There is a small opening at the centre of one of the reflectors through which a fraction of the laser beam comes out. In this way a coherent beam of light, highly collimated and of very high intensity is obtained.

Actually, the lasing medium in the closed vessel acts as a cavity resonator within which stationary waves of definite wavelengths are excited so that there is an integral number of half waves between the two reflectors which act as two nodes. Since the wavelength of the light wave is small compared to the distance d between the reflectors, the number of half waves between the reflectors is very large. The difference in the frequencies between two consequtive modes of vibration is given by $v_{n-1} - v_n = c/2d$ where c is the velocity of light. This shows that there is a spread in the frequency of the laser beam which can be reduced by controlling the quality factor (Q) of the resonator. Usually, the spread in frequency is of the order of a few thousand hertz. Since the frequency of the laser radiation is in the range of $10^{14} - 10^{15}$ hertz, this spread in frequency in negligibly small which accounts for the very high degree of monochromaticity of this radiation.

The energy of the laser beam in the cavity resonator between the two mirrors falls off laterally from the axis. If the lateral distribution is assumed to be a Gaussian function exp $(-\rho^2/w^2)$, ρ being the lateral distance from the axis, then the parameter w may be regarded as the *spot size* which is a function of the longitudinal distance z measured from the *midpoint* between the two spherical mirrors and the wavelength λ and is given by

$$w^2 = w_0^2 + \frac{\lambda^2 z^2}{\pi^2 w_0^2}$$

where w_0 is the spot size at the centre. Its value is determined by the radii of curvature of the mirrors and their separation d. For a symmetrical cavity in which both the mirrors have the same radius of curvature R, w_0 is given by

$$w_0^2 = \frac{\lambda}{\pi}\left\{\frac{d}{4}(2R-d)\right\}^{1/2}$$

For a confocal resonator, using two spherical minors of same R separated by a distance d = R

$$w_0 = \sqrt{\frac{\lambda d}{2\pi}}$$

$$w_0^2 = \frac{\lambda d}{2\pi} + \frac{2\lambda z^2}{\pi d}$$

This gives

So the spot size at the mirrors (z ± d/2) is

$$w_{d/2} = \sqrt{\frac{\lambda d}{\pi}}$$

15.3 MAIMAN'S FIRST LASER

In 1960 T.H. Maiman discovered the first laser. It was a ruby laser. The 8 cm lengthy (5 mm in diameter) cylindrical rod served as the lasing medium.

Ruby is actually a crystal of aluminium oxide. In place of some of the aluminium atoms there are chromium atoms.

The two plane faces A and B of the rod were highly polished. The face A was silvered so that the laser rays produced within the ruby cylinder were fully reflected from it and proceeded backward towards the other face. The face B was half-silvered so that even though the major portion of the light falling on it was reflected back into the crystal, a small fraction (about 1%) could come out through it as the laser beam.

The ruby crystal was enclosed within an electronic flash lamp L in the shape of a coil. Light of wavelength 5500 Å was emitted from it. The ruby laser is a pulsed laser, usually operated by high power pulses of short duration.

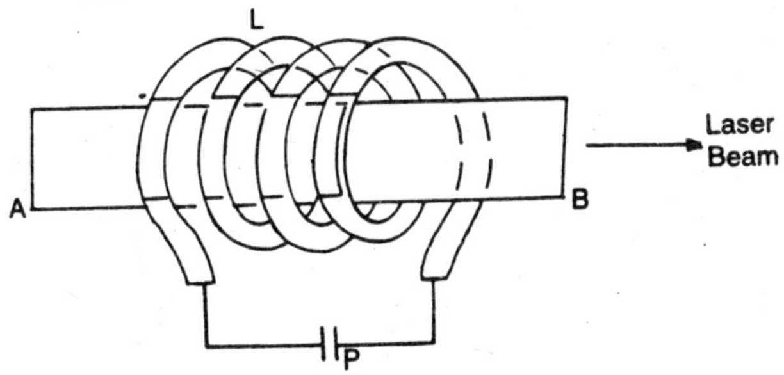

Fig. 15.1 : Ruby Laser.

The first excited state E_1 is a metastable state having a mean life of about 3×10^{-3}s which is about 105 times longer than the normal mean life of an excited atomic state. The higher excited state E_2 1s a normal excited state having a mean life of the order of 10^{-8} s.

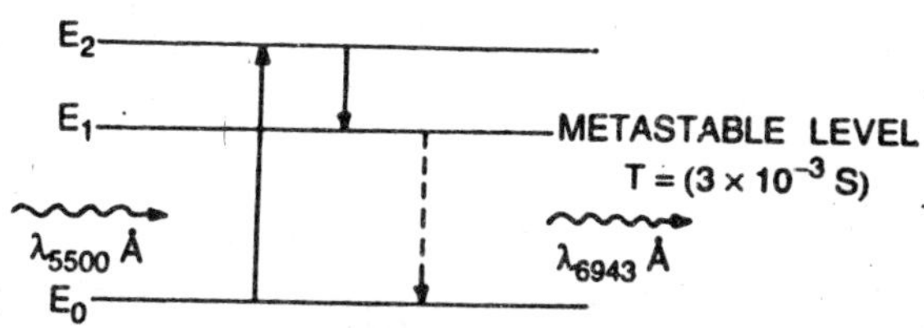

Fig. 15.2 : Energy Levels of the Chromium Atoms.

When the lamp L is switched on, a large fraction of the chromium atoms in the ground state Eg are raised to the excited state E_2 by the absorption of light of wavelength 5500 Å. These make transitions to the lower excited state E_1 within about 10^{-8} s and are trapped in this metastable state. Thus the number of atoms in the excited state E_1, becomes ultimately greater than the number in the ground state so that there is population inversion between them by the method of optical pumping.

When the chromium atoms in the metastable state E_1 make spontaneous transition to the ground state, red light of wavelength 6943Å is emitted. A photon of this wavelength emitted from one of the chromium atoms can then induce transition in another chromium atom in the excited state E_1 so that two coherent photons of the above wavelength are obtained. These two coherent photons then induce further transitions to give rise to a large number of coherent photons of the same wavelength by the chain process discussed earlier. Thus, there is light amplification by stimulation emission.

In the ruby laser, due to repeated reflections from the two silvered faces or the ruby rods the light of wavelength 6943Å travels a considerable distance within the rod. As a result, the probability of these light photons to interact with other chromium atoms raised to the metastable level E_1 increases appreciably. Obviously only the beam of light incident normally upon the two end faces can undergo such repeated reflections so that the intensity of this beam which travels parallel to the axis of the ruby cylinder is considerably-amplified. Out of this, about 1% comes out through the half silvered face as the laser beam.

Various other types of solid state lasers were invented after Maiman's original ruby laser, *e.g.*, by using calcium tungstate ($CaWO_4$) or glass with neodymium as impurity which gives laser beam in the intra-red region of wavelength 1.06 microns. A high efficiency solid state laser uses neodymium impurity in yttrium-aluminium garnet (YAG) giving radiation of the same wavelength (1.06 microns).

15.4 POPULATION INVERSION

Population inversion is a process which is employed in the production of a laser beam.

Most of the atoms in a medium are in the ground state of Energy (E_0). To raise them do an excited state four different methods can be employed, they are as follows:

1. Excitation by photons.
2. Excitation by electrons.
3. Excitation made by the collision between atoms.
4. Excitation by chemical method.

Let us study all these methods one after another.

1. Excitation by Photons

In this method atoms are exposed to electromagnetic radiation. Then selective absorption of radiation takes place by atoms then they are raised to excited state.

2. Excitation by Electrons

In the production of gas lasers electrons are used to excite atoms. The electrons are accelerated to high velocities due to the strong electric field inside the discharge tube. When the electrons collide with the neutral gas atoms the atoms are raised to excited state.

3. Excitation Made by the Collision Between Atoms

When the electrons are passed through a gas which consists two types of atoms, some of these atoms are raised to excited state.

Then the number of the excited atoms goes on increasing. This type of collision is known as collision of second kind.

This method is employed in helium-neon laser. Helium has an excited state which is metastable. As the excited He atoms in the metastable state collide with the Ne atoms in the ground state, the latter are raised to the excited state. The number of such excited Ne* atoms goes on increasing continually because of the metastability of the generating He* state. Methods (i) and (ii) are known as *electrical pumping.* (iii) Excitation by chemical method : Sometimes an atom or a molecule which is a product of a chemical reaction is produced in an excited state. As an example hydrogen and fluorine combine to produce HF molecule in the excited state. Thus, the number of such excited atoms or molecules may be considerably greater than those in the ground state giving rise to the population inversion.

Optical pumping is more suitable for solid state lasers or liquid dye lasers. Light from a powerful source is absorbed by the active material and the atoms are thereby pumped into the pump-level (or ralfier pump-band, since the line broadening mechanisms in these materials produce very large broadening). As a result these bands can absorb a fairly large fraction of the light emitted from the exciting lamp.

The exciting lamp is usually in the form of a helix surrounding the laser rod or a cylinder placed alongside the cylindrical laser rod enclosed in an elliptical or circular reflecting cylinders. The energy stored in a capacitor bank is discharged into the exciting flash lamp.

The pumping efficiency can be split up into three factors :

(i) transfer efficiency (η_t);

(ii) lamp radiative efficiency (η_r);

(iii) pump quantum efficiency (η_q).

The transfer efficiency η_t, is the ratio of the energy pumped into the laser rod to the power emitted by the exciting lamp. η_r is the efficiency of conversion of the electrical power into light entering the laser rod in the wavelength range of the effective pumping band λ_1 to λ_2 . η_q is the ratio of the number of atoms decaying from the upper laser level to the number of atoms actually raised to the upper laser level by the incident exciting radiation.

η_t for a helical exciting lamp is usually much smaller than that of a linear lamp in an elliptical enclosure. However, the former provides more uniform pumping.

Gas lasers usually employ electrical pumping which is also used in the case of semiconductor lasers.

In this method, a current in passed through the gas producing ions and free electrons. They acquire additional kinetic energy due to the electric field and excite some neutral atoms by collision. This type of impact excitation is caused more readily by the electrons which have much higher average energy than the ions. After a while an equilibrium condition is established.

Both elastic and inelastic collisions are involved in electron impact excitation. The *pumping rate* depends on the total collision cross section ground state population of the atoms (N_g) and the electron flux which depends on the electron velocity depends on the electron beam energy and has a maximum at an electron energy of few eV higher than the threshold value to cause the transition to the laser level from the ground state.

Actually, since the electrons have an energy distribution within the discharge, one has to integrate the expression for the pumping rate for monoenergetic electrons assuming a suitable energy distribution for the electrons. The pumping rate is found to depend on the ratio X/p where X is the electric field and p is the gas pressure. A change in the pumping rate can be achieved by changing the current density in the gas discharge.

Electrical pumping of gas lasers is a very complicated process and a closed expression for the pumping rate cannot be obtained unlike in the case of optical pumping.

15.5 APPLICATIONS OF LASER RADIATION

We know that the radiation from the laser tube is monochromatic, and the laser beam is spatially coherent.

Because of these properties of laser it is possible to perform different types of interference experiments with it.

The very high intensity off the laser beam implies that the amplitude of the corresponding electromagnetic wax is very large. So it is possible to investigate the nonlinear optical properties of different materials with the help of laser rays.

The production of three dimensional pictures by the method of holography, is based on the coherence property of the laser rays.

Laser beams have been used for very accurate alignment of objects. The two mile long linear accelerator of the Stanford University (SLAC) was aligned with the help of laser beam. Since the rays in the laser beam are almost perfectly parallel, it is possible to focus them within an extremely small area. If a laser beam with a cross section of radius a is focused by a lens of focal length f, then the lateral spread of the beam is $f\lambda/a$ where λ, is the wavelength. Since f and a are of the same order of magnitude, the lateral spread is of the order of λ. Hence the area of cross section of the focused beam is of the order of λ^2 which is $\sim 10^{-14}$ m^2. This is much smaller than the cross section of the focused beam in the case of ordinary light.

Since the laser beam can be focused into such a small spot, a large amount of light energy can be concentrated in an extremely small area. For this reason, it is possible to cut minute holes in different materials as also to cut sheets of metals of high melting point with the help of laser beam. It can also be used for medical purposes, *e.g.*, in repairing retinal detachment in the eye.

Because of the highly directional property of the laser rays, it is possible to measure distances very accurately with the help of laser beam. The American astronauts landing on the moon had placed reflectors on the surface of the moon. Laser beams sent from the earth were reflected back by these. By measuring the time required for the light signal to travel from the earth to the moon and back, the earth to moon distance has been measured with an error of only 0.3 m. This is much less than the error of about 80 km in the measurements made by other methods.

Application of laser beams in diverse fields of science and technology is now quite common. These include determination and control of the motion of moving objects like aircrafts or rockets. The method thus makes possible the control of the motion of missiles to hit distant targets during warfare. The field of communication technology has been revolutionized by the use of laser in conjunction with optical fibres use of the very high frequency of the carrier light waves it is possible to transmit a very large number of speech over the same channel.

16

PROBLEMS OF RADIOACTIVITY

16.1 INTRODUCTION

We know well that, radioactivity is a nuclear phenomenon. The rays emitted by radioactive substances have their origin in the nucleus.

While studying about radioactivity one must know about the emission of γ, β and γ particles.

Certain radioactive elements emit α-particles. Since outside the nucleus nothing is accepted except electrons. So the α-particles must be supposed to proceed out of the nucleus itself.

Radioactive elements which emit β-particles, identified with electrons. The β-particles which cause the disappearance of their very parent atoms can not originate from the extranuclear group of electrons.

The nuclear origin of γ rays is identical with electromagnetic radiation. The γ-rays emitted by some radioactive substances are of extremely short wavelength.

Any radiation of shorter wavelength as atleast some of the γ-rays must come from the nucleus.

To day Researches were made to determine more precisely the nature and circumstances of the origin of the three radiations of radioactivity. They form one of the most interesting problems of the active phenomenon.

The activity is spontaneous in the sense that it arises solely from intrinsic natural causes unaffected by any external agent physical or chemical. As a matter of fact heat enough to melt any element, cold enough to freeze any substance, electric fields high enough to tear electrons out intense magnetic fields, all these have been tried on radio active elements, but in every instance the activity remains unaltered.

16.2 ALPHA-RAY SPECTRA

All the α-particles emitted by a given radioactive substance came forth from their parent atoms with a single speed and hence with the same kinetic energy except for slight variations due to the straggling effect caused by statistical fluctuations of the collisions in the absorbing medium. This belief was continued for quite a long time and the Geiger and Nuttall empirical law was based upon it.

When S. Rosenblum discovered a real fine structure in the α-rays belonging to the main group, situations changed.

Fresh and more delicate researches were undertaken, which resulted in the definite establishment and satisfactory interpretation of the long range and the fine stricture characteristics of α-rays. This revision of idea was extremely fruitful, as it led to important discoveries concorning the internal structure and activity of the nucleus, especially to the existence of quantum energy levels in it. We shall give here a brief summary of the experimental study made on a-ray spectra and of the theoretical explanation of the results thus obtained, leading up to the construction of nuclear energy levels.

The *long α-range particles* were *first discovered by the scintillation method.* Rutherford and Wood, in 1916, measuring the ranges of the α-particles from ThC' by the use of absorbing screens and a spinthariscope found a very few particles with a high range of 11.6 cms. among the main group of range 8.6 cms. A few years later (1919) similar long range particles were observed by Rutherford among the α-rays emitted by RaC' using the same arrangement. This method, in spite of its singular delicacy and value, is wearisome and taxing, since all the observations are ocular and there is no permanent record left behind except in the observer's memory or notes.

In more recent years, *three other more efficient methods* have been used in the study of these long range particles.

The *first* of these is the *Wilson's Cloud Chamber* by means of which the tracks of the α-particles are photographed.

Study of such photos gives information regarding the number of particles involved and their ranges. Now and then, on the same photograph, one or more long range particles are recorded among an enormous number of normal particles,.as seen in the one reproduced here. But it is a long, tedious and costly work to obtain a sufficient number of such photographs required for plotting really good distribution in range curves.

The *second,* known as the *differential ionisation chamber method,* was devised by Rutherford, Ward, Liewis and Wynn Williams in 1931. The differential ionisation chamber is a very effective and sensitive device for detecting single charged particles and hence well suited to the study of the extremely small number of the long range α-particles, almost lost in the overwhelming majority of the normal group. It consists of a pair of very shallow ionisation chambers separated by a metal foil, of negligible thickness, which acts at the same time as the negative electrode of one of the chambers and the positive electrode of the other. The charge collected by this special electrode is then the difference between the ionisations in the two chambers which can be measured by connecting the electrode to an electrometer through an amplifier.

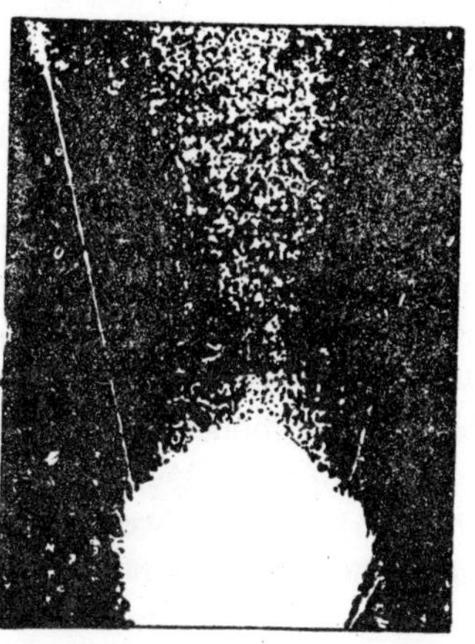

Fig. 16.1 : Long Range α-Particles from RaC'.

If the chambers are traversed by a particle which is yet far from the end of its range, the difference in ionisation will be small and even imperceptible; if by α-particle which is approaching its maximum ionising power (the peak of Bragg's curve), the difference will be appreciable and of one sign; if by a particle which is coming to the end of its range, the difference will again be appreciable but of the opposite sign. Such an arrangement is therefore sensitive above all to particles which "are nearing the ends of their ranges. With this delicate detector of single particles almost at the ends of their ranges very accurate data for the plotting of the distribution-in-range curve can be easily obtained by recording the readings of the electrometer corresponding to different thicknesses of the absorbing screen interposed between the source of the a-particles and the differential chamber.

A *third* method of investigating the long range particles is the *magnetic spectrograph with annular magnet*, devised also by Rutherford and his colleagues. As this apparatus is very effective in the analysis of the fine structure of the normal particles as well, we shall speak of it presently.

The *fine structure of α-rays has been studied by the magnetic spectrograph semi-circular focusing method.* But the method has to be adopted to obtain a perceptible resolution of the fine structure whose order of magnitude is not much greater than the variations due to the straggling effect. This is achieved by the following two devices :

(i) Use of a very strong and extensive magnetic field; with the fields actually employed it is possible to separate particles whose velocities differ by only 0.02 per cent.

(ii) High vacuum along the path of the a-rays to minimise the straggling effect.

Rosenblum was the first to obtain these conditions and thereby succeeded to place in clear evidence the fine structure of α-particles. He used the huge magnet of the Academy of Sciences at Bellevue, Paris, with solid circular pole-pieces 75 cms. in diameter and 6 cms apart, capable of producing an essentially uniform field of 25,000 gauss over a region of some 35cms. in diameter. After passing through collimating slits, the α-particles were deflected through a semi-circle by the magnetic field and made to fall on a photographic plate. High vacuum was maintained in the semi-circular box which contained the source at one end of its diameter and the plate at the other. In this arrangement, all the groups of particles of different speeds, deviated simultaneously in circular paths, each of its own particular radius, fell upon the plate producing what looked like an optical line spectrum.

Fig. 16.2 : Fine Structure of α-Particles of ThC. (Rosenblum).

16.3 THE TECHNIQUE OF THE MAGNETIC SPECTROGRAPH

In 1933 Rutherford and his associates modified the technique of the magnetic spectrograph.

They used narrow, circular rings instead of solid disc shaped pole pieces. The source of α-particles was placed in the narrow annular space between the faces of the rings at one side, while on the diametrically opposite side the photographic plate was replaced by a simple ionisation chamber connected through an amplifier to a scaling system.

Arrangement was made to vary the magnetisation of the rings over sufficiently wide limits, so that the field strength could be adjusted step by step in order to bring group after group of the α-particles to the ionisation chamber. Although this entailed the accurate determination of the field strength at the different stages, which was no easy work, yet the very possibility of choosing any desired field strength combined with the very sensitive electronic detector increased the efficiency of the apparatus in the investigation of not only extremely weak groups but also the rare long range particles, (d) The annular space and everything within it was evacuated, being walled in by means of a ring band.

The apparatus constructed with these modifications, *viz.,* a fixed radius of curvature, a variable magnetic field and an electronic adjacent photo. The rings were 40 cms. in radius, 5 cms. broad and 1 cm. apart. The field strength was adjustable up to 12,000 gauss, which was sufficient for α-particles of range up to and even beyond 11.5 cms.

Fig. 16.3 : Annular Magnetic Spectrograph Used by Butherford for the Analysis of α-Ray Spectra.

Rutherford, Lewis and Bowden with this annular magnetic spectrograph were able to analyse the fine structure of the α-particles from several radioactive substances with great accuracy. Plotting the different values of the field strength against the corresponding readings of the detector, they obtained curves which snowed a peak for every group. From the value of the field strength corresponding to a peak and the radius of the ring pole-pieces, the energy of the group responsible for the peak could be readily computed. With ThC the results of Rosenblum were confirmed and thus the fine structure of the α-ray spectrum was definitely established.

As regards the *long range* particles, working with RaC', all the peaks indicated by the differential ionisation chamber were clearly separated, a hump which had suggested two groups was resolved into three maxima, and an extra group was discovered. Thus it was shown that there were in all 12 groups of long range particles from BaC'. Only two groups of long range particles were found with ThC', one at 9.7 cms. and the other at 11.54 cms. which agreed well with the results obtained by the measurement of more than 600 cloud chamber tracks of long range particles of ThC". This fact indicates also the advantage of the magnetic spectrograph method over the cloud chamber method from die point of view of economy of cost and labour.

More recently, in 1940, Boy Bingo, in America, has constructed a magnetic spectrograph for α-ray analysis, which claims to have overcome certain inherent defects of the methods of Rosenblum and Rutherford, such as the immobility of die Bellovue magnet, which prevents its use in conjunction with reactions produced by artificially accelerated particles, the background effect in the detector of the annular magnet method caused by contamination and cosmic rays, which necessarily limits the accuracy of results, the impossibility of repetition of experiment in quick succession, arising from the fact that the source and detector are placed in a highly evacuated region inside the magnetic field in both the methods, etc.

16.4 ALPHA DISINTEGRATION

The current idea of nuclear constitution says that the a-particles made up of 2-protons and 2, neutrons. Further, the 2-particles inside the nucleus are not lying in any confused and idle manner, but are arranged in distinct energy levels, in active motion.

A serious difficulty arises against the emission of α-particles by radioactive nuclei due to the existence of an electrostatic *potential barrier* surrounding the nucleus. For, the energy of any α-particle inside the nucleus cannot be greater than the height of the potential barrier, or it would not stay there at all, which means that a activity could have never been detected, contrary to observation. On the other hand, if the energy of the α-particle is less than the height of the potential barrier, the α-particle should never be able to get out.

Classical mechanics, which considers material particles as solely corpuscular, is quite incapable of explaining this difficulty. But in wave mechanics, which attributes to every material particle a wave aspect also, the difficulty disappears, since an electrostatic potential barrier, though very "high, cannot completely hinder the passage of a wave through it. For, there is a certain probability of penetrating through the barrier, however small that probability may be, for all positive energies of the material particle.

More precisely, the Schröedinger wave function ψ whose square measures the density of probability of the particle in a certain region, does not vanish in the regions where the potential energy V is higher than the total energy E and where the particle in the classical model would have a negative kinetic energy; instead, although decreasing exponentially with the distance, the wave function ψ maintains a finite Value provided E is greater than zero. This statement can be established by the following considerations:

The apparently paradoxical behaviour of the particle in this case has a complete analogy in certain phenomena connected with the reflection of light. If a beam of light falls on the boundary between two media at an angle of incidence greater than the critical angle, then according to geometrical optics, total reflection will occur, *i.e.,* all the light will be reflected at the surface of separation and no disturbance will enter the second medium. According to the wave theory, however, the process of total reflection is much more complicated. On this theory, the disturbance in the second medium is not everywhere zero, but within the space of few wavelengths decreases exponentially to become entirely negligible at greater distances. This 'forbidden' penetration of the disturbance cannot be described in terms of rays of light at all, the lines representing the directions of energy-flow being curved and returning to the surface again. If now, the second medium be confined to a thin sheet, of thickness less than the range of penetration of the disturbance, and if it is immediately followed by the first medium, a small fraction of the disturbance which has penetrated the sheet will emerge from it and enter the first medium. This transmission of energy is obviously in contradiction to the prediction of geometrical optics; it is, however, established by experiment. The change from classical mechanics to wave mechanics introduces an exactly similar possibility for the transmission of α-particle through a potential barrier which would otherwise be insurmountable.

16.5 BETA DISINTEGRATION

The emission of β-particles by radioactive nuclei was easily understood. In the actually accepted opinion of the composition of the nucleus with protons and neutrons, the β particles do not pre-exist in the nucleus but are created just at the moment of their emission.

It is believed that the β-radioactivity is due to the transformation of neutrons into protons or vice versa inside the nucleus. It is more correct to say that the proton and neutron represent two quantum states of one and the same fundamental heavy particle which is subject to transformation from one state to the other.

The explanation of the origin of β-rays was evolved only after laborious and ingenious researches which have not yet reached completion. There were two *main difficulties* to be solved. The first one arose from the very nature of the β-particles which are nothing but electrons, pure and simple, like those surrounding the nucleus. Now, since the radioactive atoms, like other atoms, can emit electrons from the peripheral electronic structure under the influence of radiation by a process, such as the photoelectric effect, there is a possibility of confusing the two types of electrons, the one arising from the nuclear change and the other from the extranuclear family as a photoelectric emission, chiefly given the fact that electrons, whatever be their genesis, are the same in mass and charge. Moreover, in the case under study, the admixture of the two kinds of electrons is real, as the radioactive elements which exhibit β-ray disintegration usually emit also γ-rays. A portion of this high frequency radiation gets absorbed in the outer electronic structure of the disintegrating atom itself and as a result of this, electrons from the different shells, K, L, M, etc., are ejected. This process, known as the *internal conversion of γ-rays*, has been subjected to direct lost and proved to be true. Further if the γ-rays are of different frequencies, since each of these frequencies can give rise to as many different speed electrons as there are shells and sub-shells in the electronic structure of the atom, complex groups of electrons of discrete and characteristic energies will be ejected from the radioactive element along with the real disintegration electrons, and it is not easy to distinguish between the two. As a matter of fact, *internal conversion electrons or secondary β-rays*, as they are now called, *were initially mistaken for the primary β-rays*, until carefully devised researches clearly differentiated the two phenomena.

Hahn and Meitner using the magnetic spectrograph method were able to show that *the internal conversion electrons always consist of groups, whereas the disintegration electrons have a continuous energy spectrum.* The photographic plate which records the magnetic spectrum of the β-rays emitted by a radioactive substance clearly demonstrates a continuous background within certain limits, produced by the disintegration electrons, on which, is superposed the line spectrum formed by the groups of secondary electrons. Chadwick using a Geiger counter instead of the photographic, plate in the magnetic spectrograph and counting the number of electrons of different energies produced by a radioactive substance was able to prove that the *'groups' formed only a small portion of the total emission. The experimental data gave curves of the type.* The integrated intensity of the continuous part is much greater than the sum total of the Intensities of the individual peaks. In the case of RaB, for instance, the total intensity of the individual lines is found to be only about 30% of the total emission.

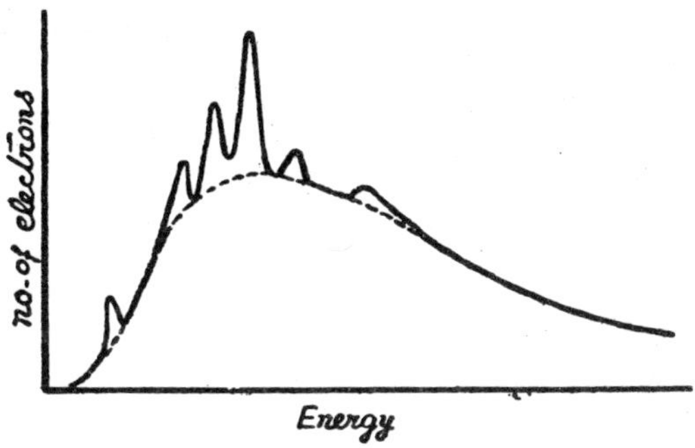

Fig. 16.4

Further, researches were made to confirm this interpretation which distinguished the two kinds of electrons met with in β-ray emission. Thus, for Instance, Rutherford, Robinson and Rawlinson replaced the fine wire source of radon of the magnetic spectrograph method by a small glass tube containing radon, around which could be wrapped thin metal foils. The γ-rays emitted by the source liberated secondary electrons from the metal sheath. The foils were of sufficient thickness to stop most of the primary electrons and to straggle widely even the most penetrating ones among them so that no traces due to the primary β-rays could appear on the plate. In spite of this, a definite line spectrum was obtained.

This clearly labelled the electrons producing the observed line spectrum as purely secondary, since there was no chance for any primary β-rays to interfere.

Gurney, replacing the Geiger counter by a Faraday cylinder in the magnetic spectrograph, was able to measure the number of β-particles given off per second by different radioactive substances and established that *in a β-disintegration one disintegration electron is ejected per atom in consonance with theory, while the number of the accompanying internal conversion electrons is much smaller*. Thus, it is found that RaB, where γ-rays also are emitted, gives off 1.25 electrons per atom approximately, while RaE which has no γ-emission, ejects only one electron per atom. The excess (0.25) in the former case is evidently due to the secondary electrons emitted by the γ-rays.

16.6 GAMMA-RAY EMISSION

The Gamma rays are electromagnetic in nature like other radiations. We know well that, a body heated above the temperature of its surroundings emits heat radiation. An atom when excited emits also radiations comprised between infra-red and X-rays. When the excited body returns to its normal state the excess of energy is given out in the form of electromagnetic radiations.

The γ-ray emission usually accompanies the α- or β disintegration of a radio active nucleus. The transition can take place from the excited state to another of lower energy with the consequent emission of the excess of energy as a γ-rays. γ-rays are not emitted in every α or β disintegration; certain atoms are known to disintegrate without emitting γ-rays.

As regards the γ-rays associated with α-disintegration, there are reasons to believe that the γ-rays may be emitted *either before* or *after* the departure of the α-particle, although the latter is the more common occurrence.

It may be noted that as the emission of γ-rays involves no change in nuclear charge or mass, being analogous in this respect to the emission of radiation by atoms, there is a tendency nowadays to restrict the term "radioactivity" to the emission of α- or β-particles alone. But, it might be argued against this opinion that although light quanta are never considered as a constituent part of an atom, their mass in the form of the energy of the electromagnetic field makes an appreciable contribution to the total mass of the system them. Consequently, indirect methods based on measurements of the secondary electrons produced by the γ-rays have been developed. There exist three such methods, *viz.*,

(i) The natural secondary β-ray or the conversion electron method,

(ii) The excited secondary electron or the photodectran method, and

(iii) The Compton electron method.

The first two employ a magnetic spectrograph, while the third a Wilson's cloud chamber for the experimental technique. As the first method is the most important of the three and has provided almost all the available information about y-ray spectra it will be described somewhat in detail, while the other two will be more briefly stated.

The natural secondary β-ray method. This method, first worked out independently by Ellis and Meitner, in 1921, is based on the phenomenon of the internal conversion of γ-rays, which, as we have already noted, can be regarded as a special case of' the photoelectric effect, in so far as the γ-ray quantum issuing from the nucleus is absorbed in the extranuclear electronic structure of the disintegrating atom itself and thereby liberates the electrons found there. But according to the wave-mechanical thee interpretation is shown to be simpel there we also another possibility of mechanical transmission of energy from the excited nucleus to one of the peripheral electrons of the same atom. Whatever be the theoretical explanation of the process, the effect produced is the same as in the external photoelectric effect, but with two additional advantages, *viz.*,

Fig. 16.5 : Lise Meitner.

(i) The probability of the yield of electrons being greater on account of the internal nature of the phenomenon, and

(ii) The spectral lines produced by the internal conversion electrons being better defined and sharper than those obtained in the external photoelectric effect, on account of the better elimination of absorption effects.

16.7 INTERPRETATION OF THE β-RAY CONTINUUM

We know well that β-particles are actually ejected from the nucleus with a continuous distribution of energies. This happens because of the following reasons.

Every radioactive β emitter gives rise to continuous spectrum whether it emits or not a line spectrum. In all cases, where reliable measurements have been made, the number of continuous β-particles ejected per unit time is equal to the number of atoms disintegrating during the same interval.

All the experimental findings, considered together, prove definitely that the β-particles constituting the continuous spectrum are really of nuclear origin.

Having thus established that the β-particles are emitted by the nucleus itself with continuously variable energy, an extremely embarrassing situation is created, which may be expressed as follows:

While all the nuclear phenomena, such as the α- and the γ-ray spectra, show the existence of nuclei in perfectly definite quantum energy states, how is it that the β-ray spectra alone, manifesting a continuous distribution of energies, appear to deny the existence of such discrete energy states in the nucleus? The difficulty may be stated in yet another way. As regards the origin of the β-rays, it is now generally admitted that the disintegration electron arises as a result of the transition of the nucleus from the neutron quantum state to the proton quantum state. This implies that the energy of β-disintegration must have a definite and constant value equal to the difference of the energies of the initial neutron and the final proton quantum states involved.

The energy of the initial state minus the energy of the final state is not equal to the release of energy during β-disintegration. *But according to the principle of conservation of energy it is to be expected that the β-rays should be expelled either all with identical speed or at least a-discontinuous group*, as in the case of the fine structure of α-rays. How, then, to reconcile the fact of a continuous β-spectrum with the existence of discrete quantum states in nuclei without violation of the law of conservation of energy.

This *difficulty* concerning energy conservation is *supplemented by another equally serious one arising from the point of view of conservation of momentum*. Experimental researches show that every nucleus has an angular momentum (nuclear spin), which is a half integral value of h/2π for nuclei of odd mass number and a whole integral value of h/2π for nuclei of even mass number. This is just what would be expected if all nuclei were constituted with neutrons and protons, each with spin ½. The β-particles, like all electrons, has also a spin ½; but when it leaves a nucleus of odd mass number, the remaining nucleus still retains a half-integral spin, since its mass number remains odd, the β-disintegration effecting no change in the mass number. Evidently this goes against the law of conservation of momentum, since a half cannot be removed from a half integral quantity, the balance still remaining half integral. The same is true if the parent nucleus has an integral spin.

Several solutions have been suggested, but only one of them is now generally accepted as the most satisfactory. Some of the attempted but improbable solutions, which can be easily dismissed are:

The pre- and post-disintegration energies may not be the same but different for different nuclei in a given element, so that the energy available for disintegration need not show a rigorous constancy but may vary continuously from zero to an arbitrary maximum. But this would mean, side by side, that the attendant nuclear process, *viz.,* α-ray, γ-ray emission, etc., must be different for different nuclei, which, however, is contrary to experimental facts.

Granting that during disintegration, the total energy (equal to the observed maximum energy) is transferred to the primary or disintegration electron, part of it may be subsequently lost through collisions with the outer electrons of the atom in a statistical fashion giving rise to the continuous Mcwellian-like velocity spectrum. But experiments rule out such an assumption. For, over and above the indirect experimental evidences in favour of the nuclear origin of the very continuous nature of β-ray spectrum, already stated, experiments performed by Ellis and Wooster and by Meitner and Orthmahn directly disprove this hypothesis. These authors determined the average energy of disintegration per nucleus in the case of RaE, which is almost a pure β-emitter, by measuring the total heat energy produced by a known amount of the element with a microcalorimeter whose walls were thick enough to absorb all the emitted β-particles. If the continuous β-spectrum were due to secondary scattering caused later by the external electrons of the atom, then the total energy should remain inside the calorimeter and one should expect the energy per nucleus measured by the heating effect to be equal to the observed maximum given by the upper limit of the continuous β-spectrum curve, which is 1.05 MeV for RaE. The actually measured energy was, however, only 0.34 MeV, which closely agreed with the average value deduced from the distribution curve. Thus it appears certain that the original energy of the β-particles is, on the average, the same as that which one finds these particles possess when they get out of the atom. They have not, therefore, all been ejected with an energy corresponding to the upper limit and then lost varying amounts in collisions in the manner supposed. In other words, the continuity of the β-spectrum is a property of the disintegrating nucleus itself.

The discrepancy between the constant amount of energy released and the continuously varying energy observed might be explained by assuming that two electrons which are simultaneously emitted share the total constant disintegration energy in the observed continuous fashion. But Gurney's experiment, clearly demonstrating that the number of electrons in the spectrum of a β-emitter is exactly one per disintegration, excludes such a possibility.

17

MAGNETISM IN SOLIDS

17.1 INTRODUCTION

All the solids may not exhibit magnetism. In most of them the magnetization, induced by the external magnetic field is feeble. In some it may be quite high.

Let us consider 'M' is the magnetization induced by the magnetic field H.

Then we can write the relation as follow:

$$M = x . H$$

where x = magnet susceptibility.

One thing we have to bear in mind that, in an isotropic substance, M is in the direction of H and X is constant.

According to their magnetic properties, all solids can be grouped in to three groups:

1. Diamagnetic Solids
2. Paramagnetic Solids
3. Ferromagnetic Solids

In Diamagnetics magnetic susceptibility is less than 'O' and x is a constant independent of field.

In paramagnetics magnetic susceptibility will be positive. In these substances also, x is a constant and independent of H.

For ferromagnetic substances magnetic susceptibility is positive and has very high value. Iron, nickel, cobalt, and rare earth elements, gadolinium, holmium, erbium, exhibit ferromagnetism.

For ferromagnetics the magnetization increases rapidly with the increase of magnetic field, and ultimately becomes saturated at higher fields. Hysteresis is the another important characteristic feature of ferromagnetics. The ferromagnetics for which the coercive force is low are known as magnetically soft.

17.2 ORIGIN OF MAGNETISM

Many substances exhibit permanent magnetic dipole moments. Paramagnetism owes its origin to the existence of permanent dipole moments.

Now let us study about the, classical theory of paramagnetism developed by Langevin.

Let,

μ = magnetic dipole moment of each atom

μ = External magnetic field

E_m = magnetic energy.

then $\quad Em = \vec{\mu} \cdot B_0 = -\mu B_0 \cos\theta$

{θ = the angle between $\vec{\mu}$ and H.}

The atomic dipoles which are aligned at random in the absence of the magnetic field tend to align in the direction of the applied field so that their magnetic energies are lowered.

According t o Maxwell-Boltzmann classical statistics, the probability of a dipole to align at an angle between θ and $\theta + d\theta$ will be

$$W(\theta)\, d\theta \propto \exp(-\varepsilon_m/kT)\, d\Omega$$

where $d\Omega = 2\pi \sin\theta\, d\theta$ is the solid angle within θ and $\theta + d\theta$ and k is Boltzmann constant.

Using Eq. we can write

$$W(\theta)\, d\theta = C \exp(\mu B_0 \cos\theta/kT) \sin\theta\, d\theta$$

where C is a constant. The magnetization induced in the substance by the applied field H is

$$M_H = n < \mu \cos\theta >$$

where n is the number of atoms per unit volume and $< \mu \cos\theta >$ is the mean value of the projection of the magnetic moment in the field direction.

$$< \mu \cos\theta > \int_0^{\pi} \mu \cos\theta . W(\theta)\, d\theta / \int_0^{\pi} W(\theta)\, d\theta$$

$$\frac{\mu \int_0^{\pi} \exp(\mu B_0 \cos\theta/kT) . \cos\theta \sin\theta\, d\theta}{\int_0^{\pi} \exp(\mu B_0 \cos\theta/kT) . \sin\theta\, d\theta}$$

If we substitute

$$x = \cos\theta \text{ and } \alpha = \mu B_0/kT\ \mu_0 \mu H/kT.$$

we get $< \mu \cos\theta > = \mu \int_{-1}^{1} x \exp(\alpha x)\, dx / \int_{-1}^{1} \exp(\alpha x)\, dx$

$$= \mu \left\{ \frac{\exp(\alpha) + \exp(-\alpha)}{\exp(\alpha) - \exp(-\alpha)} - \frac{1}{\alpha} \right\} = \mu \left(\coth \alpha - \frac{1}{\alpha} \right)$$

Hence from Eq. we get

$$M_H = n\mu L(\alpha) = n\mu \left(\coth \alpha - \frac{1}{\alpha} \right)$$

where $L(\alpha) = \coth \alpha - \frac{1}{\alpha}$

is known as Langevin function.

The random thermal motion of the molecules opposes this type of tendency. Each atomic magnetic dipole is turned to some extent in the direction of the applied field and assumes an equilibrium position. The magnetic moment induced in the substance by the external field will be equal to the algebraic sum of the projections of the moments of all the dipoles in a unit volume.

As we know the paramagnetic susceptibility is inversely proportional to the absolute temperature. There are some metals which exhibit feeble paramagnetism independent of temperature.

In metals the free occupy successive energy levels starting from the bottom of the potential well upto the Fermi level. Each level is occupied by two electrons with their spins aligned in opposite directions.

Langvin's theory says, when a magnetic field is applied the permanent dipole moments to the atoms in a paramagnetic material tend to align parallel to the field direction, the probability for which exceeds that of antiparallel alignment by a factor about $\mu B_0/KT$.

17.3 STRUCTURE OF FERROMAGNETIC MATERIALS

In a ferromagnetic substance, the electrons are aligned parallel to one another. Due to the influence of the exchange interaction the spins of electrons in the partially filled subshells of all neighbouring atoms are aligned parallel to one another.

A ferromagnetic material may be regarded as made up of a large number of regions of very small size. Within each region the magnetization is saturated. They are of macroscopic dimensions and are known as *ferromagnetic domains*. Between two neighbouring domains there is small region within which the directions of orientation of the atomic spin moments gradually change from the direction of magnetization of the one domain to that of the next. This transition region is called the *Bloch wall*. The width of the Bloch wall is of the order of 1000 Å.

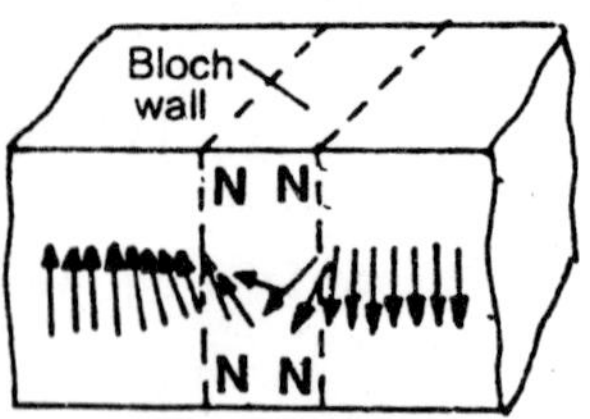

Fig. 17.1 : 'Bloch Wall' Region in a Ferromagnetic Substance.

The theoretically expected pattern of the ferromagnetic domains is shown diagrammatically which have been confirmed by photographs of the domain structure. In the figure the arrows indicate the directions of spontaneous magnetization within the neighbouring domains.

In the absence of any external magnetic field the different domains are oriented at random so that the resultant magnetization is zero. In this case the *free energy* of the specimen is the minimum. By the application of a magnetic field, the specimen becomes magnetized. The process or magnetization may be divided into three steps:

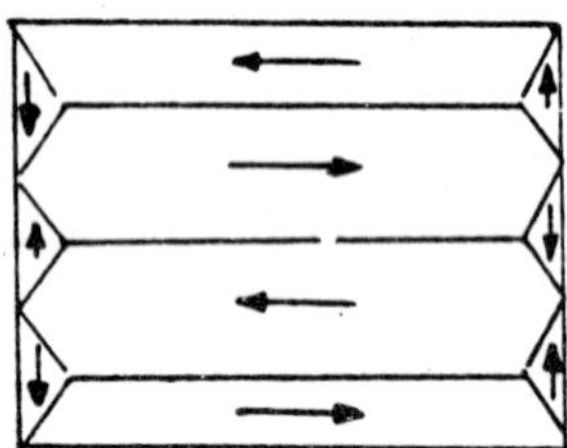

Fig. 17.2 : Pattern of Ferromagnetic Domains (Theoretical).

Magnetization by Displacement of the Domain Boundaries

The magnetic moments of the different Swains are oriented at different angles with respect to the applied magnetic field H. The size of the domain, the magnetic moment of which attends the minimum angle with respect to H increases with the increase of H while the sizes of the domains, the magnetic moments of which are oriented at larger angles with respect to H decrease with increasing H. Such changes in the sizes of the domains cause a diminution in the magnetic energy of the system, which therefore produces a more stable configuration. Since these changes in the domain size are produced by the displacement of domain boundaries, the process is known as the *displacement process of magnetization*. This process actually occurs in the initial stages of magnetization The process ends with the domain a pervading the whole crystal.

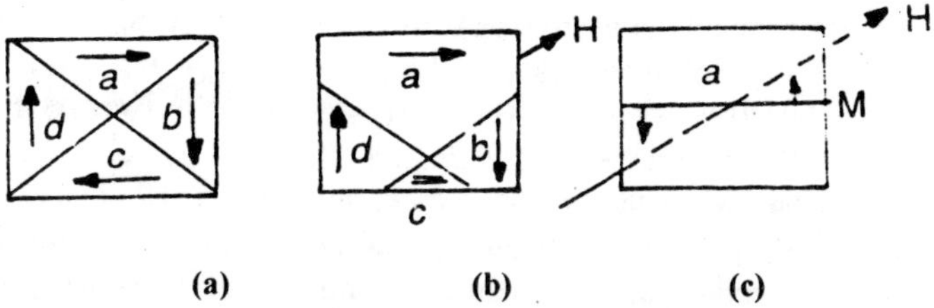

(a) (b) (c)

Fig. 17.3 : Effect of External Magnetic Field (H) on Domains.

At low magnetic fields this process of magnetization takes place continuously and is reversible. If the magnetic field is increased by a small amount and then reduced by the same amount, then the increase and decrease of magnetization take place along the same path, so that at the end of the complete cyclic change, the specimen comes back to the initial state of magnetization.

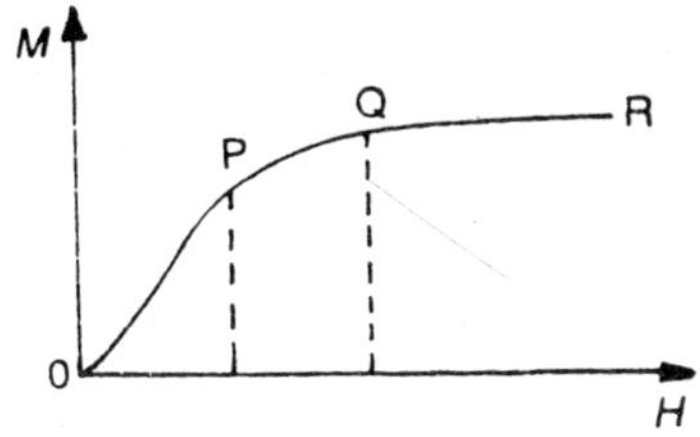

Fig. 17.4 : Magnetization Curve (M vs H) in a Ferromagnetic.

At higher magnetic field, the magnetization occurs in discontinuous steps and is irreversible.

Magnetisation by Rotation of the Magnetic Moment Vectors

After the domain a has expanded to spread over the whole crystal, further increase of H causes its magnetic moment to turn towards the direction of H Due to this, the magnetization increases further. When the magnetic moment of the domain turns fully in the field direction, the specimen shows what is known as *technical saturation.*

Para Process

With further increase of H, the magnetization M increases slowly by a small amount. We have assumed that in the spontaneously magnetized domains, spin moments of all the atoms are alined parallel. Strictly speaking, this happens only at 0 K. At higher temperature, due to thermal collision, complete saturation of magnetization within a domain is not achieved. Some of the spin moments may be aligned and parallel. At higher magnetic fields, these are realigned parallel to the other spin moments which causes slight increase of magnetization beyond technical saturation. This is known as the para process.

For n atoms gives a net magnetic moment $nB_0\mu^2kT$. However in the case of the free electrons in the metals the probability of turning over to the direction of the magnetic field from antiparallel alignment is zero since the states of parallel spin orientation are already occupied. Only the electrons near the top of the Fermi distribution have a chance to turn over to the field direction from antiparallel orientation. As we have seen that only a fraction (T/T_f) of the total number of electrons which occupy the energy levels in the energy interval kT below the Fermi level $\varepsilon_f = kT_f$ can make such transitions. Hence $\chi \sim (\mu_0\ n\mu^2/kT) \times (T/T_f) = n\mu_0\ \mu^2/kT_f$ which is independent of temperature and has a value 10^{-6} which is much lower than χ predicted from Langevin's theory. The above result holds for the case $\mu B << kT$ which is true at the room temperature for fields as high as 10^2 T.

The presence of a small temperature-independent paramagnetism in some metals is an additional confirmation of the hypothesis of the existence of free electrons in metals.

The theory of ferromagnetism was first developed by the French physicist P. Weiss in 1907. It is known as the *theory of molecular field.* Two main assumptions of the theory are:

There are some *macroscopic domains* within a ferromagnetic material which are spontaneously magnetized. The magnetization of the whole body is the resultant of the magnetic moments of these domains.

The spontaneous magnetization of the macroscopic domains arises due to the existence of molecular fields within the material. Due to this field all the magnetic dipoles within a domain are aligned parallel to one another.

We have seen that paramagnetic substances attain saturation magnetization by the application of very high magnetic fields (> 10^3 T). On the other hand, one important characteristic of the ferromagnetic materials is that they attain saturation magnetization by very weak magnetic fields. For example in the case of silicon-steel, saturation is attained by a field as low as 10 A/m. Under this condition magnetization

$M \sim 10^5$. For a paramagnetic material the magnetization is about 10^{-4} for the same field which is only 10^{-9} that of a ferromagnetic substance.

The high value of magnetization produced in a ferromagnetic material by a very weak external magnetic field shows that within the mascopic domains in these materials, all the permanent magnetic dipoles are aligned parallel to one another. Thus, there is some kind of mutual interaction between the dipoles within the domain which is responsible for their spontaneous magnetization.

Weiss assumed that the total magnetic field acting on the magnetic dipoles within a ferromagnetic material is

$$H_m = H + \gamma M$$

where H is the external field, M the magnetization of the material and γ is a constant, known as the *molecular field constant* or *Weiss constant* γM predominates over that of the external field H.

Weiss original theory was based on Langevin's classical theory of paramagnetism. In our discussion below we shall however use the results of the quantum theory of paramagnetism. According to this, the magnetization of a paramagnetic material is given by

$$M = ngJ\ \mu_B B_j\ (\alpha)$$

where $$\alpha = J\ g\ \mu_B\ B_0/kT$$

Here $B_0 = \mu_0 H$, H being the external field.

For the ferromagnetic material, we have to use the field H_m defined above in place of H so that we have

$$\alpha = Jg\mu_B\mu_0 + (H + \gamma M)/kT$$

For spontaneous magnetisation, H = 0 so that we can write

$$\alpha = Jg\mu_B\mu_0\ \gamma M/kT$$

$$M = \frac{\alpha kT}{\gamma g \mu_0 \mu_B J}$$

17.4 EINSTEIN AND DE HAAS EXPERIMENT

In 1915, Einstein and de Hass performed the gyromagnetic experiment, to produce magnetization by rotation in a ferromagnetic material.

The specimen in the form of an iron rod R is suspended vertically from an elastic fibre S. A coil of wire C is wound over the rod R. When an electric current is passed through the coil an axial magnetic field is produced which magnetizes the specimen in the field direction. We have seen that this magnetization is due to the regular alignment of the magnetic moments of the atoms of the material in the field direction. Since these magnetic moments are antiparallel to the angular momentum vectors of the electrons in the atoms, these are also aligned in a regular manner along with the magnetic moments which gives rise to a resultant angular momentum for the whole rod. Due to this, the specimen rod turns about its own axis and the suspension fibre is twisted which can be measured with the help of a lamp and scale arrangement. From this the angular momentum P of the rod can be estimated. By measuring the magnetization M of the rod at the same time, the ratio MIP can be found.

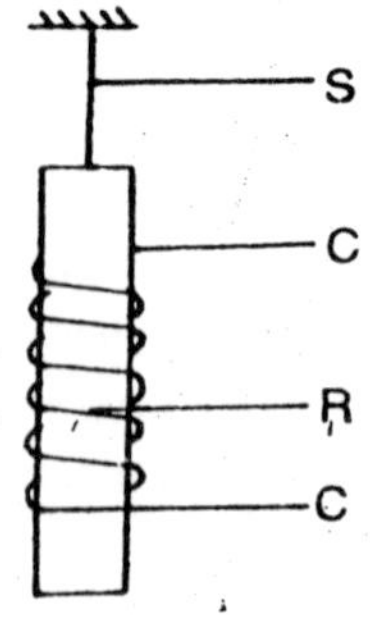

Fig. 17.5 : Gyromagnetic Experimental Set-up.

Similar experiments were performed by others later. From these and other similar experiments, the gyromagnetic ratio is found to be 2:

i.e.,
$$\frac{M}{P} - 2 \times \frac{e}{2m} = -\frac{e}{m_e}$$

An experiment which is just the opposite of Einstein-de Haas experiment in principle was performed by S.J. Barnett in which the magnetization produced in a rapidly rotating iron rod was measured. This and similar other experiments on *magnetization by rotation* confirmed the results of Einstein-de Hass experiment.

From all these measurements it is definitely established that the magnetization of a ferromagnetic substance is due to the spin motion of the atomic electrons. Their orbital motion does not contribute to the magnetization.

18

COSMIC RAYS

18.1 INTRODUCTION

As early as 1900 C.T.R Wilson, Elster, and Geitel discovered a new kind of rays, later called as cosmic rays. It was considered that there must be some external agency other than air which continually made good the loss of ions due to the residual discharge. In 1903 this surmise was confirmed by Rutherford and cooke.

As regards the real existence of highly penetrating radiation coming from outer space however there is no doubt, and the name cosmic rays first employed by Millikan to designate them, has come tube universally used. Thus the discovery of cosmic rays was established.

It was thought for a time that radioactive matter occurring in small quantities everywhere, in the earth, surrounding atmosphere, etc., might be responsible for this penetrating radiation. If this were so, the rate of discharge of an electroscope should diminish considerably at very great altitudes, far removed from such influences. This inference was tested by balloon observations by three German scientists. In 1909, Gockel measuring the rate of discharge with a Wulf electroscope at different altitudes up to a maximum height of 4,500 metres, found that the rate of discharge diminished much more slowly with height than could be explained by the hypothesis that the penetrating radiation was of terrestrial or atmospheric origin.

In 1911-14, Hess and Kolhorster, using sealed ionisation chambers, extended the balloon observations to much higher altitudes than Gockel had done and found that above a few hundred metres the intensity of radiation increased with height continuously up to 9,000 metres, as much as five to ten times the value at ground level, which, of course, could not happen if the radiation was due to radioactive contamination of the earth, since it would then be almost completely absorbed in going up through a few hundred metres of atmosphere. Hence it appeared certain that the radiation had its origin outside the earth and its atmosphere.

Furthermore, the fact that the radiation produced effects at the earth's surface and was able therefore to penetrate the entire atmosphere showed that it was far more penetrating than even the γ-rays from radioactive substances. Kolhorster's experimental data proved, that the new radiation was absorbed only to about 1/10 the amount of the hardest γ-rays. Hence these authors concluded that an *extremely penetrating type of radiation whose origin was entirely beyond our atmosphere*, fell upon the earth almost uniformly in all directions. They called it by the special name *"hohenstrahlung* or *"ultrastrahlung"*, (*i.e.*, radiation from high above or beyond our atmosphere). All later investigations have supported this interpretation of the phenomenon, which may therefore be considered as the discovery of cosmic rays.

The first World War (1914-18) interfered with the continuation of the researches on the newly found radiation and it was only in 1922 that further investigations were restarted by Kolhorster and his co-workers in Germany and Millikan and his associates in America. In 1923-24 Kolhorster worked at the mountain station, Jungfraujoch, in Switzerland, in order to obtain a reasonable intensity of ionisation, and found that the intensity of the incoming radiation was the same during day and night, even during a solar eclipse. This was disproved later. But Kolhorster concluded from his observations that the *origin of the radiation could not be the sun, nor even the stars, but should be somewhere far beyond our galaxy.*

Millikan and his associates, in 1925-26, studying the absorption of these rays in the snow-fed (and hence non-radioactive) water of high mountain lakes and finding that the absorption in water agreed with that in air confirmed the opinion that the origin of the rays was outside the atmosphere. Further, from the data obtained from these absorption experiments, Millikan made bold to affirm that the *main part of these penetrating rays might consist of energetic photons of a complex nature, (i.e., electromagnetic radiation)*, emitted in the process of the building of helium nuclei from hydrogen in interstellar space. However, as the evidence has become overwhelming that most of these rays are electrically charged and of a higher order of energy then was supposed.

RESEARCHES ON COSMIC RAYS

From the time of their definite discovery, extensive investigations have been carried out on these radiations, chiefly with a view to know more about their nature, composition and properties. The existing atomic tools have been modified and adapted to suit the needs of the study of cosmic rays which, are characterised by the smallness of the effects to be observed. The experimental data gathered are vast and often overwhelming.

18.2 STUDY OF COSMIC RAYS

For the study of cosmic rays, the ionisation chamber has been improved chiefly in two directions, in order to measure as accurately as possible the small ionisation current produced by them.

Now, let us study about the compton's model 'C' cosmic ray meter. This instrument was used by compton and Turner.

The ionisation chamber is very larger made up of spherical steel bomb of capacity 20 litres filled with pure argon at a high pressure of 50 atmosphere. The use of argon at high pressure provides an ionisation nearly 70 times that which is produced in air at atmospheric pressure.

The cathode of the chamber is a copper sheet lining inside the steel bomb, well insulated from the latter with wax, while the anode is a fork at the centre of the sphere. The chamber is shielded from local radiations by surrounding the steel bomb with an outer shell filled with lead shots to an equivalent thickness of about 11 cms. solid lead. A *method of balancing* is used in the measurement of cosmicray intensity by the following device. The current to compensate the ionisation current due to cosmic rays is supplied by the ionisation produced within a small auxiliary chamber (*balance chamber*) by the β-rays from a uranium source U. The compensating current is adjusted to approximate equality with the mean cosmic ray ionisation current, so that small changes in cosmic ray intensity rather than the total intensity is recorded. A great advantage of this method is that it automatically compensates for pressure or temperature changes of the gas. *A Lindemann electrometer* connected to the central electrode records the changes in the net compensated intensity, the displacement of the shadow of the electrometer needle being projected through a compound microscope on to a continuously moving strip of photographic film. The readings of a barometer and a thermometer are also recorded on the same film.

In operation, the central electrode is grounded for a period of 3 mins. at the beginning of every hour by a clockwork mechanism; it is then left insulated for 57 mins to gather the net charge in the system and communicate it to the electrometer needle. At the end of every four-hour period an automatic sensitivity calibration is performed by the application of a known voltage to the central electrode.

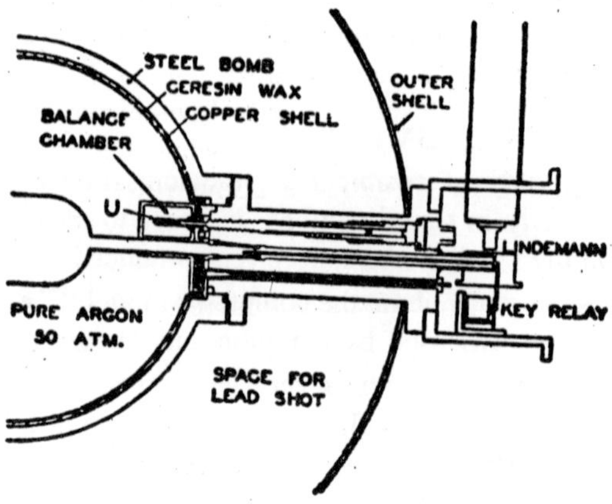

Fig. 18.1 : Model "C" Cosmic Ray Meter.

The Geiger-Müller counter with its associated electronic circuit is a very efficient instrument for detecting the passage of single ionising particles, and should therefore be a very be useful tool in the investigation of cosmic rays, where one has to deal with relatively small numbers of ionising particles and in many eases even with single particles. But one groat drawback of this apparatus is that it is equally sensitive to all type of radiations and in consequence gives rise to an ever present *spurious background effect* due to radioactive contamination of the surroundings, which masks to a large measure the small effects due to cosmic rays. This limitation has been removed to a great extent by a simple and ingenious device known as the *coincidence counters*, first conceived and used by Bothe and Kolhorster, in 1928, and then improved by Rossi and others so that the G.M. counter has now become a very powerful weapon in researches on cosmic rays.

The device consists in arranging two or more counters in such a way that a record is made only when all the counters discharge almost simultaneously. This type of discharge is known as *coincidence*, which may be due either to a single particle passing through all the counters at the same time practically, or to two or more separate particles coming by chance one to each counter at the same instant. The former is considered *true coincidence* while the latter *chance coincidence*. In cosmic ray work, the two kinds of coincidences can be differentiated, since the number of chance coincidences can be readily estimated from the separate counting rate of each counter.

Rossi, in 1930, improved the coincidence counter technique in the following two ways: (i) *Use of three counters in a row and recording only when all three counters respond simultaneously.* In this way the possibility of chance coincidences, which can hardly traverse all the three counters, is much more effectively eliminated than in the Bothe-Kolhorster two-counter arrangement, while the genuine coincidences due to penetrating particles passing through all the three counters remain unchanged. (ii) *Use of an electromechanical device, instead of photographic film, for recording the discharges of the counters.* This is accomplished by means of modern *electronic selecting circuits.*

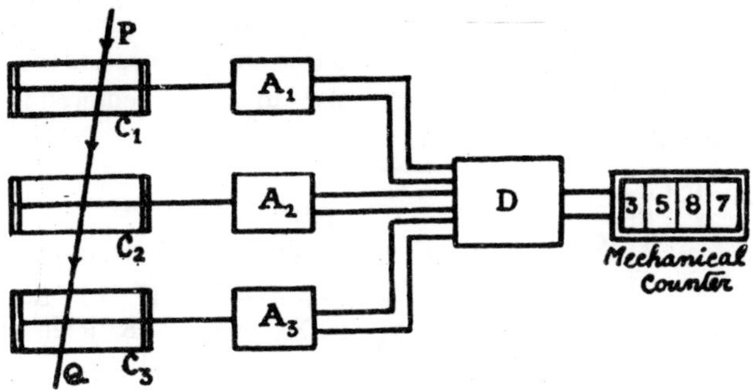

Fig. 18.2 : Coincidence Counter Set. (Rossi).

Three counters C_1, C_2 and C_3, arranged one below the other in a line, are connected to the grids of three triodes A_1, A_2 and A_3 in such a way that the discharge of a counter momentarily lowers the potential of the grid of the corresponding triode and stops the flow of thermionic current in it. The plates of these three valves are connected among themselves and communicate with the grid of a detector D through the intermediary of a grid bias and a resistance. These elements of the circuit of the plates are so regulated that the grid potential of the detector is sufficiently negative so that no current passes in its anode circuit. Now, if an energetic cosmic ray, such as PQ, traverses all the three counters almost at the same time a discharge is produced simultaneously in all of them and the flow of current in all the three corresponding valves is completely cut-off. This causes a momentary decent rise in the grid potential of the detector and a sufficiently strong pulse of current flows in its plate circuit actuating a telephone which produces an audible which or a mechanical counter which records the coincidence. If, on the other hand, only one or two of the counters are affected, the currents in the corresponding valves are cut, but as nothing happens in the third valve, the grid potential of the detector valve is not sufficiently altered to permit a pulse of current in Its output circuit.

The arrangement of valves acts as a selector, which chooses, among the various discharges in the counters, only those that occur simultaneously in all the counters, thus eliminating the non-coincident effects in the counters. The circuit is quite *symmetric* with respect to the counters allows the use of any number of them. It is to be remarked that the selector circuit has the following teaks to accomplish, *viz.,* (a) to make the amplitude of the final poises to be as constant as possible and their period as short as possible in order to reduce the number of chance coincidences, without however musing true coincidence and (b) to modify the period of the final coincidence pulses to a value required to function a mechanical counter.

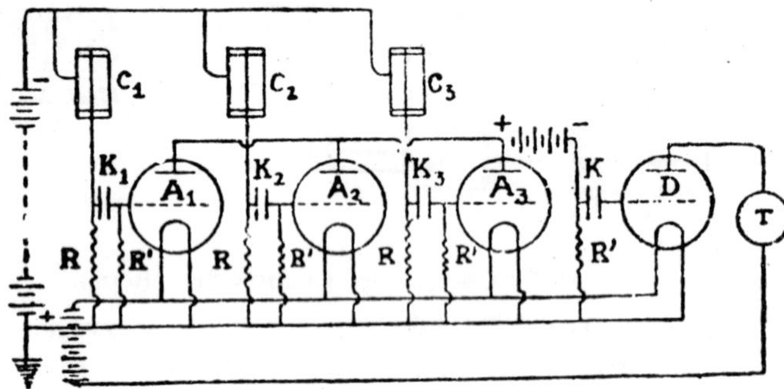

Fig. 18.3 : Coincidence Circuit for G.M. Counters.

This land of grouping of counters is sometimes called a *cosmic ray counter telescope*, for it can be pointed in Buy given direction and made to record the passage of the cosmic rays in that direction. Bernardini and Johnson, using such an arrangement of counters, were able to establish an outstanding characteristic of cosmic rays, *viz.,* their incidence on the earth with uniform intensity from all directions.

18.3 THE BUBBLE CHAMBER TECHNIQUE

At the university of California Dr. L Alvarez and his group have built a bubble chamber circular in shape and 2.5 inches in diameter.

Using liquid hydrogen at about 400° below zero Fahrenheit as the super heated liquid and exposing the chamber to neutrons, they have been able to take photos of protons in the liquid hydrogen recoiling from collisions with the incoming neutrons.

The bubble chamber technique is still in its infancy, but it promises well to become a standard detection instrument for work with the high-energy particle accelerators in laboratories as well as with cosmic rays. Its special advantages over the cloud chamber and the emulsion are :

(i) *The greater frequency with which interesting events are recorded.* In the Brookhaven Laboratory, it has been shown that the *six-inch* bubble chamber catches as many events as would an ordinary cloud chamber 140 feet long. What is still more interesting is that the first 22 photos taken in 11 minutes with the bubble chamber contained eight events similar to the one stated above,

(ii) *Its adaptability to the different conditions of research.* The chamber may be filled with a light liquid which does not deflect particles much and therefore will permit magnetic field experiments as the cloud chamber, or with a highly dense liquid which will produce a great amount of scattering are the emulfum.

(iii) *The tracks are not much distorted*, unlike in the cloud chamber will to convection currents,

(iv) *The density of the bubbles weight prove a better index of particle energy than the droplet density in a cloud chamber.* Above all these, the bubble chamber will certainly speed up the rate at winch one can gain information about strange particles and still stranger nuclear forces.

It may lie noted that quite recently Alvarez has been able to build a very big bubble chamber, 6 ft. long containing 150 gallons of liquid hydrogen. He intends to shoot into that chamber antiprotons from the Berkeley 6 BeV bevatron and photograph their tracks that will reveal the innermost secrets of matter. In particular, he has been able to trace the birth, death and after-effects of an *anti-lambda* particle, the counterpart of lambda particle which is produced when high energy protons hit protons at rest.

Fig. 18.4 : Dr. Alvarez.

In another series of experiments, bombarding the chamber with extremely powerful particles from the bevatron, he has been able to obtain photos of tracks made by μ-mesons knocked out of smashed atoms. A few of these tracks are found to have gaps in them that are quite puzzling at first sight. One of the photon which such gaps is reproduced on the opposite page Alvarez has offered the following ingenious explanation, which seems to point to a new and better method of obtaining nuclear energy.

The μ-mesons, liberated from the bombarded atoms and having negative charges, get attached to positive hydrogen nuclei and revolve around them as electrons normally do. Since, however, the μ-mesons are about 200 times heavier than electrons, the radii of their orbits will be about 200 times smaller, than those of the electronic orbits; hence they will revolve close to the nucleus. The atoms formed in this way are known as *"mesic" atoms*, which are somewhat heavier than the ordinary hydrogen atoms, but still extremely small. They can therefore pass through the electron defences of ordinary atoms and fuse With their nuclei. This is what has happened in the photo under study Mesons form mesic atoms with the nuclei of heavy hydrogen (deuterium) present in the bubble chamber. Mesons so occupied make no 'bubbles, which accounts for the gaps in the meson tracks. But when a mesic deuterium atom hits an-atom of ordinary hydrogen atom, the nuclei fuse together forming an atom of helium 3 and releasing 5-4 MeV of energy. The meson ejected from the helium atom thus formed, carries away the energy as velocity. It may eventually stop and decay into an electron. In the photo, the μ-meson enters the chamber at upper right and after travelling a good distance (the long thick track slightly curved downwards), forms a mesic atom with deuterium, which drifts slightly to left forming a gap in the track. At the left side of the gap the mesic deuterium atom has fused with an atom of ordinary hydrogen. The meson ejected from the resulting helium (3) atom moves towards the left, stops after a while and decays into an electron, which produces thinner track curving upwards.

The meson that shoots off from the helium (3) atom may also form another mesic deuterium atom and cause it to fuso with another ordinary hydrogen atom. When mesons act in this way; they behave like catalysts, which induce chemical reactions without themselves being changed. If such catalysed fusion could be made practical, it would have advantages over known methods of releasing nuclear energy. It would not require expensive fuel, as uranium fission does, and it would not create harmful radioactive fission products. It would not require exceedingly high temperature as thermonuclear fusion reactions do. It might burn peacefully, releasing a considerable amount of nuclear energy.

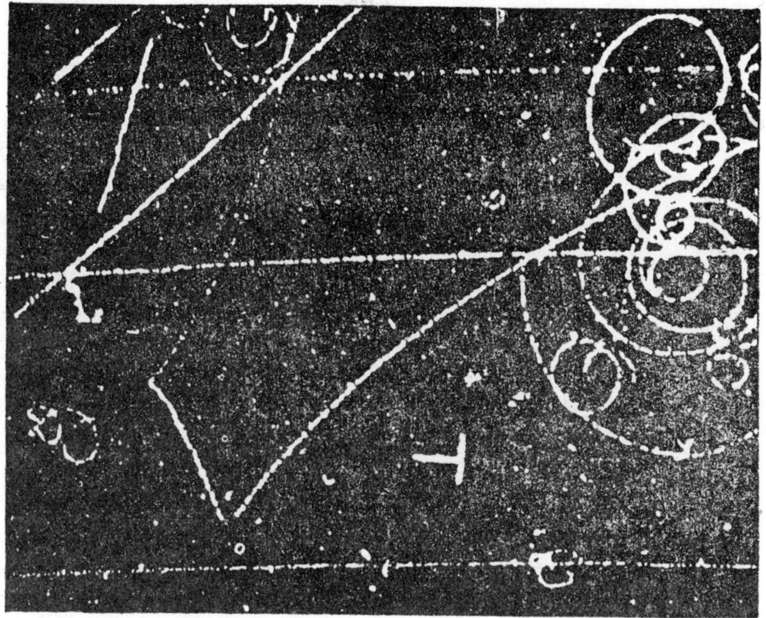

Fig. 18.5 : μ-Meson Acting as Catalyst—Bubble Chamber Photo.

18.4 CHANGES IN COSMIC RAY INTENSITY

In 1937 compton and Turner studied thoroughly about the changes, in the intensity of cosmic rays.

They fund that the shape of the intensity latitude curve depended slightly upon the season in which the data were taken. They were able to resolve the observed intensity latitude variation into two components.

1. The atmospheric effect.
2. Pure magnetic effect.

Forbush in America has made a continuous record of cosmic ray intensity over a period of more than 18 months at four different places covering the latitudes from 48°S to 50°N. Duperier in London, working with a triple coincidence counter telescope made up of several large counters at each row, thus presenting a big receiving area, has been able to record continuously the cosmic ray intensity over a much longer period of four years. From these data, the following results concerning seasonal variations have been obtained: (i) The effect has an amplitude ranging from zero at the equator to 1 or 2% of the total intensity at a latitude of 50°; (ii) it is opposite in the two hemispheres; (iii) it is definitely correlated with atmospheric temperature, being a maximum always during the colder part of the years—in January in the northern hemisphere and in July in the southern.

The cause of the seasonal variation, according to the explanation now generally accepted and first proposed by Blackett, in 1938, is the temperature variation in the mesotron-producing layer of the upper atmosphere. The mesotrons, which constitute the highly penetrating component of cosmic rays arid are produced in the upper atmosphere, are radioactive with a very short period of a few microseconds so that some of them decay before reaching the surface of the earth. Now, during summer, the air-layers in which mesotrons are produced are shifted to higher altitudes and are less compact than in winter, so that the mesotrons have a larger chance of decaying before they reach the recording apparatus in summer than in winter.

There exists also a *diurnal change, i.e.,* a variation of cosmic ray intensity with *solar time* of the day, as clearly established from the data gathered over long periods of many years by different workers such as Mesaerschmidt (1932) in Germany, Compton and Turner and Forbush in America and Duperier in England. Experiments were conducted without and with lead screens surrounding the measuring instrument in order to study separately the soft and hard components of cosmic rays. The results obtained are: (i) There is a small diurnal variation in the *softer* component, the intensity reaching a maximum at about midday and a minimum at about midnight. On the other hand, the intensify of the penetrating component is practically constant. (ii) This variation of intensity with solar time appears to be independent of both latitude and altitude.

The actual cause of the variation of cosmic ray intensity with solar time is not known. But there are several evidences to show that the sun might be responsible for these changes. Over and above the periodic variation per 24 hours with maximum intensity at noon and minimum at midnight, a 27-day cycle, corresponding to the moon's revolution, and chiefly a marked change in intensity, as much as 10%,. during the solar bursts associated with sun-spot activity, which cause major magnetic storms in the earth, clearly indicate the influence of the sun on cosmic ray intensity. But there appears to be no mechanism correlating these phenomena, though it has beeofloggested that the magnetic field other sun might be responsible for the observed small diurnal changes of intensity in cosmic rays.

The study of the variation of cosmic ray intensity with altitude, by means of which the discovery of cosmic rays was first made was undertaken anew, in greater detail, in order to arrive at a proper knowledge of the nature and origin of these rays. The chief aim was to measure very carefully the ionisation produced by the cosmic rays at higher and higher altitudes, reaching up practically to the top of the atmosphere. For this purpose, *mountain stations* at altitudes of 2 to 3 *miles*, such as Jungfrau in Switzerland, Mount Blane in the Italian Alps, Pie du Midi in the Pyrenees, Bolivian Andes in South America, Mount Evans in the U.S.A., have been used. These permit the use of massive equipments and long periods of continuous observation. *Airplanes up to altitudes of 6 or 7 miles*, and *stratosphere manned balloons* (first devised by Piccard) *reaching up to 10 to 11 miles* have also been used. This technique has the advantage of allowing the observer to keep watch on what is happening. Still higher altitudes have been attained by *unmanned sounding balloons* (initiated by Regener) that can *ascend to about 20 miles* which is almost the top of the atmosphere, and record for hours with very delicate apparatus the cosmic ray ionisation in that region. Recently, *rockets* carrying ionisation chambers, counters, photographic emulsions and even cloud chambers, have pushed the inquiry to altitudes of *more than 100 miles*, although their flight-time is very short, three to tour minutes. With the introduction of *space rockets* it is now possible to reach out to several thousands of miles and more.

18.5 ABSORPTION OF COSMIC RAYS IN AIR, WATER ROCKS

We can say cosmic rays are complex, powerful and mysterious radiations. By following different methods it is possible to measure the rates of absorption of cosmic rays, in media other than the atmosphere.

Now let us start with the absorption of cosmic rays in water.

Millikan and Cameron were the first, in 1926, to make a systematic study of the absorption of cosmic rays in water. They mounted electroscopes in water-tight containers and immersed them metre by metre in mountain lakes, such as Muir Lake (altitude 3,505 metres) down to 20 metres, which was later (1931) extended to 72 metres. From the electroscopic records of cosmic ray intensity at various depths they were able to arrive at the following conclusions: (a) The effect was still appreciable at the greatest depths they were able to reach, (b) there was a gradual hardening of tile radiation as it passed through more and more matter and (a) a considerable portion of the rays reaching the earth could be classified into two groups, with different absorption coefficients, one of them much more penetrating than the other.

Very accurate and extensive observation of cosmic ray intensity under water was made by Begener in 1928. Using a large, strongly built ionisation chamber of 39 litres capacity, filled with CO_2 at a pressure of 30 atmospheres, in order to increase the intensity of ionisation, he lowered it in Lake Constance (altitude 395 metres) down to a depth of 280 metres, from the data obtained he was able to plot a curve showing the manner in which the intensity continuously falls off with increase in depth. The observed intensity at the bottom of the lake was still one per cent that at the surface. From the observed rate of absorption, he found three different absorption coefficients indicating three different components of the cosmic ray. The less penetrating components were filtered first, leaving mainly the hardest component after great depths of water had been traversed.

More recently (1938) Clay and his associates have extended these underwater studies to a depth of 440 metres. The residual ionisation at such great depths of water indicates the presence of very highly penetrating particles among the cosmic rays.

V. C. Wilson, in 1938, employing a four-tube coincidence counter cosmic ray telescope, made measurements down to a depth of about 384 metres in a, mine. Measurable radiation penetrated this nearly one quarter of a mile of rock, equivalent to about 1,100 metres of water or 100 metres of lead. The intensity was of the order 1/10,000 that incident upon the rock. By tilting the apparatus so that it recorded particles that came obliquely passing through a greater thickness of the rock, he could still detect the radiation at a depth equivalent to about 1,400 metres of water. Gemert (1938) was able to detect radiation at a depth of 610 metres in a coal mine, equivalent to 1,600 metres of water. Some of the cosmic rays are thus extremely penetrating although only 1/20,000 of the radiation incident on the earth actually penetrates to these depths.

Bothe and Kolhorster were the first, in 1929, to attempt a direct measurement of the absorption of cosmic ray particles in metallic screens using two coincidence counters with an interposed block of gold about 4 cms. thick. They were able to prove the existence of ionising particles of mass absorption coefficient of 3×10^{-3} cms.^{3} gm.$^{-1}$, which was of the same order of magnitude as the absorption coefficient of cosmic rays in air at sea level, deduced, from Millikan and Cameron's experimental data.

Rossi, in 1932, using three counters in coincidence arranged in a vertical line studied the absorption in lead of cosmic ray particles which had already been filtered through 7 oms. of lead. He was able to establish the existence of particles capable of traversing one metre of lead placed between the counters. Assuming the absorption to be exponential through each successive layer of lead, the following values were found for the mass absorption coefficient : in the first 10 cms., $\mu/\rho = 1.8 \times 10^{-3}$; in the next 15 cms., $\mu/\rho = 0.5 \times 10^{-3}$; and in the last 76 cms., $\mu/\rho = 0.55 \times 10^{-3}$. This indicates the existence of a soft and a hard component in the cosmic rays. Since the absorption coefficient is practically constant for rays that have traversed 16 cms. of lead, whatever passes through this thickness of lead may be considered as the *hard* component.

18.6 THE POSITRON

The positron is identical with the electron except for the positive sign of its charge this is also a fundamental particle, which is the exact counterpart of the electron.

Anderson and Blackett are the two scientists who worked hard to discover the positron.

Anderson found the characteristics of positron, which bent in a direction opposite to that of a negatively charged particle. Later Anderson carried series of experiments on these particles.

The first alternative was very improbable in the actual arrangement of the apparatus, chiefly since cosmic rays of great energy, came from above. In any case the sign of the charge of the particle could be definitely fixed only when the direction of motion of the particle was known.

In order to decide this point, Anderson took photographs of tracks with a lead sheet placed across the centre of the chamber. One of these historic and highly important photographs, which definitely showed that the sign of the charge of the particles which made positive curvature in a magnetic field was undoubtedly positive and thereby led to the discovery of the positron, is reproduced here. The track obviously represents the path of a single particle which traverses the lead plate of 6 mm. thickness. Its general appearance closely resembles that ordinarily produced by an electron. It is less curved below the lead sheet than above. The difference in curvature must be evidently due to a loss of energy by the particle in the lead plate, the direction of motion being assumed to be from the side where the track is lose curved to that where it is more curved indicating a diminished speed. Measurement of the radius of curvature of the track below and above the plate gives the energy of the particle as 63 MeV and 28 MeV respectively. The-loss of energy an the absorber is therefore 40 MeV, which is approximately the amount of energy one would expect such a particle to lose in a lead plate, 6 mm. thick. Hence, the particle must have been moving upward. If this conclusion is correct, then the sense of the curvature of the track and the known direction of the magnetic field show definitely that the particle carries a *positive charge.*

The fixing up of the sign of the charge of the particle as positive does not, however, prove at once that the particle is a positive electron. It might very, well be a proton, even if the general appearance of the track resembles that ordinarily caused by an electron. For, at the high energies involved the relativistic mass of an electron approaches that of a proton and in consequence there exists very little difference in the specific ionisation (directly proportional to the actual mass of the ionising particle) produced by them so that their tracks are no more easily distinguishable.

Fig. 18.6 : Discovery of the Positron (Anderson).

But, Anderson, continuing his analysis of the track, was able to show that *it was definitely not produced by a proton.* A proton leaving a track having a curvature of the amount seen above the plate would have an energy of only 300,000 eV and a proton of such low energy is known to ionise much more heavily than what is indicated by the track. Moreover, the range of such a proton can be only about 5 mm., whereas the actual track shows a range of 5 cms. and more. Hence, he concluded that the track was can fed by a positively charged particle which must have a mass much less than that of a proton, but about the same as that of the electron. Thus a hitherto unknown particle was discovered, which Anderson called the *positive electron or positron.*

The discovery of such a new particle would have remained doubtful—for, after all, a few photos where the tracks-are curved inversely might be attributed to some parasitical complications—had not other observations made by other workers soon followed which placed beyond doubt the existence of positrons.

One such very important confirmatory evidence was the phenomenon met with in cosmic rays themselves and known as cosmic ray showers. Skobelzyn was the first to indicate the existence, in his cloud chamber photos, of several tracks of cosmic rays arising simultaneously from about the same point. Anderson himself had noted groups of such rays,, bent in opposite directions and radiating from a common place. But it was Blackett and Occhialini who, with their ingenious counter controlled cloud chamber, were the first to place in clear evidence this interesting cosmic ray phenomenon of "showers". One of the earliest photos of such showers, obtained by Blackett, is reproduced here. It consists of two groups of tracks bent in opposite directions, originating from a common point, usually in the massive metal parts of the apparatus, such as the mass of copper wire surrounding the chamber. The two groups consist of an approximately equal number of tracks, thereby indicating that a shower contains positively and negatively charged particles in about equal numbers. The mean aspect is the same for the negative

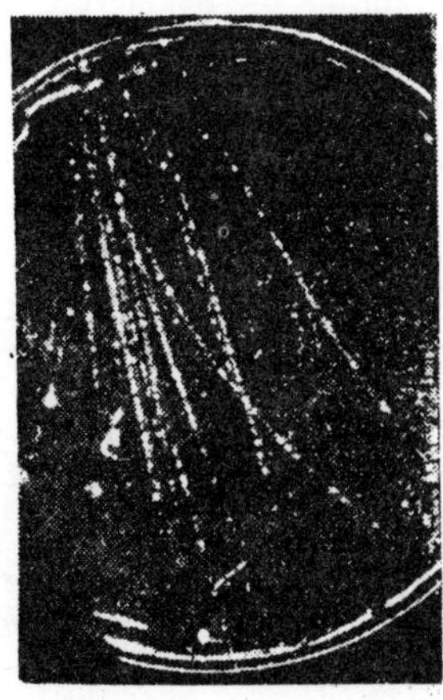

Fig. 18.7 : Cosmic Ray Shower (Bluckott).

as well as the positive tracks, suggesting a perfect symmetry between the two kinds of particles which would thus have the same intrinsic mass with equal charge but opposite in sign. From the measurement of energies and amount of ionisation, it can be shown that *all the tracks in a shower are due to particles which are electronic in nature.* The common spot from which a shower is found to originate decides the direction of motion of the shower particles, *i.e.,* outward and not inward. Under these conditions, the sense of curvature of the tracks unambiguously fixes the sign of the charge. A more detailed study of the cosmic ray showers will be considered later. For the present, it is enough to remark that the "shower" phenomenon proves the existence of positrons in a very convincing manner.

18.7 ABSORPTION OF POSITRONS IN MATTER

Thibaud, after continuous study was able to show that positron behaved in the same way as electrons of the same speed.

The absorption of positron in matter is of much greater interest and importance from the point of view *viz.,...* the possibility of checking the theoretical prediction of the phenomenon of annihilation of matter.

The positrons, penetrating into the interatomic regions of matter which swarm with negative electrons cannot live there free for long. For, it ia an established experimental fact that the positive electrons play no part, unlike the negative electrons, in the phenomenon of conductivity, gaseous, metallic or electrolytic. On the other hand, positrons and electrons, existing side by side, would necessarily rush together and destroy each other, unless a *negative proton* be postulated, which might bind the positron into a new type of hydrogen atom, in a manner analogous to the positive proton binding the electron into the ordinary hydrogen atom. For, in all atoms, including that of hydrogen, the mutual destruction of positive and negative electricity is avoided through the usual quantum conditions holding within the atom.

According to Einstein's relativistic law, such a fusion must give rise to radiant energy, since the disappearing potential energy of two separated attracting systems, corresponding to m in the relation $W = wc^2$, must of necessity appear in radiant form. Further, as the rest-masses of the two particles that have disappeared are equivalent to 1 MeV approximately, the total energy content of the resulting radiation must be about 1 MeV. It is assumed that the kinetic energies of the particles at fusion are negligibly small. Thus, we arrive at the phenomenon which is usually referred to as the annihilation of matter or *dematerialisation.*

Applying the laws of conservation of energy and of momentum to the process, two possible cases arise concerning the nature and energy content of the radiation that results from the union of positron with electron:

If the combination takes place in free space, removed from any third particle, i.e., if the positron unites with a free or loosely bound electron, *two photons of equal energy, about* 1/2 MeV each, will be emitted in opposite directions, since only thus can both energy and momentum to be conserved.

If, however, the combination takes place near a third particle, i.e., it the positron units with electron tightly bound to the nucleus, *a single photon of about 1 MeV will be emitted,* since the third particle can take up the momentum and energy required for their conservation, and, being sufficiently heavy, as would be the nucleus of any atom, it would take up only a negligible amount of energy.

Theoretically then, one would expect both types of photons, some of 1/2 MeV and some of 1 MeV. But, since positrons, traversing matter would lose practically all of their kinetic energy before being annihilated and hence would be unable to penetrate the interior of atoms, the first type of annihilation where two 'photons, each of 1/2 MeV energy, are emitted would be the most probable process that could be observed, while the other single photon emission process would be much less probable.

These theoretical predictions have been well confirmed in the positron-absorption experiments carried out by several experimeters. Chaor was the first, in 1930, to observe the expected emission of 1/2 MeV photons in his experiments on the absorption of the 2.62 MeV γ-rays of ThC" in lead. Gray and Tarrant, in 1932, performing similar experiments, were able to detect two secondary y-radiations of energies nearly 1/2 MeV and 1 MeV among the scattered rays. They also placed in evidence the following special features of the observed γ-rays: (i) their energy did not depend upon either the energy of the primary y-rays or the material used as absorber; (ii) they could not be identified with any of the characteristic γ-rays emitted by the absorber; (iii) their wavelengths were independent of the angle of scattering and hence they could not be attributed to Compton scattering; (iv) they were observed only when the energy of the incident γ-ray was greater than 1.5 to 2 MeV. All these characteristics of the observed secondary y-radiations can be easily explained, as Blackett and Occhialini first pointed out, only if it be admitted that these rays rise from the annihilation in lead of positrons which are produced in the lead itself by the primary γ-rays. The chief drawback of these initial experiments was that the estimate of the energies of tile secondary radiations could not be made with great accuracy.

More direct and convincing proofs of the existence of the annihilation radiation were soon obtained. Thibaud and Joliot, in 1933, working independently with the trochoid method and using aluminium bombarded by the a-rays of polonium are source of positron-electron pairs and a photographic plate or G.M. counter as detector, were able to observe the γ-rays emitted in the annihilation process when a stream of positrons was directed on a metal absorber such as Pt or Pb. By a suitable change of direction of the magnetic field, either the positrons alone or the electrons alone were made to strike the absorber. When the positrons fell on the metal plate, there emerged from the point of concentration γ-rays of much greater intensity than when, by reversing the field, electrons were made to fall on the absorber. By interposing absorption sheets between the emitted radiation and the detector, the energy of the radiation was found to be about 1/2 MeV. Joliot measured also the number of photons corresponding to the annihilation of a single positron and found it close to two in agreement with theory.

18.8 THE MESOTRON

The mesotron is a fundamental particle, which carries the mass intermediate between that of the electron and that of the proton. Mesotron was discovered in 1937. The credit of the discovery is given to Neddermeyer and Anderson.

Now let us study about the experimentally observed facts, which constituted the starting point in the discovery of the mesotron.

The existence of the two main components is cosmic radiation *soft* and *hard*, the latter forming about 75% of the total radiation at sea level. The hard component consisted of high speed particles positively and negatively charged in approximately equal numbers. These particles often left isolated single tracks in a cloud chamber, the appearance of which was very much like those produced by electrons as illustrated in the adjacent photo containing the tracks of two highly penetrating cosmic ray particles which traverse in one case two lead plates each 1 cm. thick without any appreciable change of curvature, thereby indicating that they lose very little energy in the absorber. The most important characteristic of these particles was, therefore, that they were unusually penetrating, nearly 50% of them passing through 100 cms. of lead, while the particles of the softer component, presumably electronic in nature, were completely absorbed in about 10 to 15 cms. of lead. A statistical study of the loss of energy of the two types of particles in going through solid absorbers showed that different mechanisms were followed in the two cages. For the particles of the soft component, the loss of energy was found to be proportional to the incident energy of the particle as well as to the square of the atomic number of the absorber, while with the particles of the hard component the loss of energy exhibited no such simple relation with the incident energy and was found to be approximately proportional to the atomic number of the absorber.

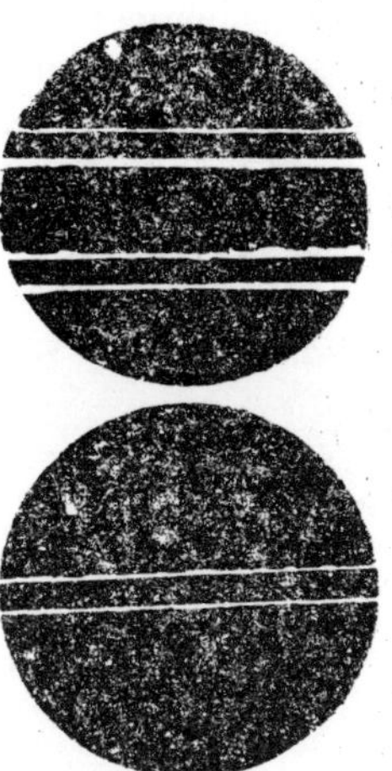

Fig. 18.8 : Tracks of Highly Penetrating Particles of Cosmic Rays.

An attempt to interpret adequately these experimental data in the light of the *existing theories* concerning the loss of energy 'by fast charged particles passing through matter gave a strong though indirect evidence to the effect that the highly penetrating particles of cosmic rays were neither electrons nor protons but were of a new type possessing mass intermediate between those of the electron and proton, though carrying a single elementary charge, either plus or minus.

The theories relevant to the present problem are those which deal with the loss of energy by fast charged particles due to two types of interaction with matter, known as ionising collision and radiative collision, whose main and distinctive features may be briefly stated as follows:

Ionisation takes place, as we know, due to inelastic collision between a fast charged particle and the atoms of the material through which it passes, with a real transfer of energy from the former to the latter. The mephanism of the phenomenon may be visualised as follows: As the fast moving charged particle passes near an atom, an electric force develops between it and the orbital electrons of the atom, resulting in a transfer of energy to the electrons which in consequence free themselves from the parent atom and ionisation follows. The ejected electrons are known as secondary electrons, and may sometimes be sufficiently energetic to ionise other atoms. Such energetic electrons are sometimes called *"knoci-on" electrons*, as illustrated in the adjacent photo, where a very high energy cosmic ray particle of about 10^{10} eV produces in the gas of a cloud chamber a "knock-on" electron curved into a circular are under the influence of a magnetic field of 10.000 gauss, with an estimated energy of 6 MeV.

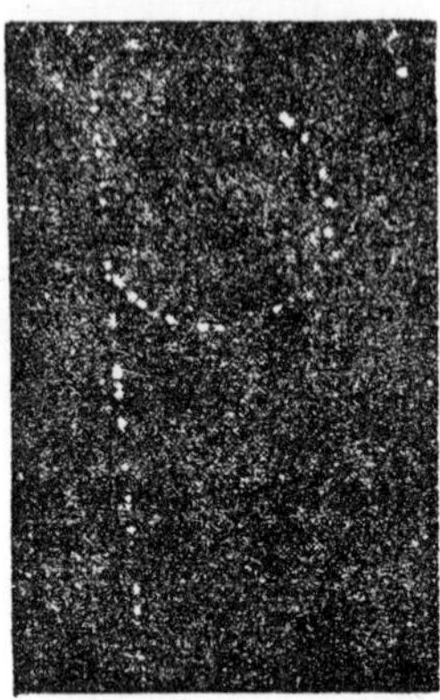

Fig. 18.9 : Production of a "Knock-on" Electron by a Very Energetic Cosmic Ray Particle (10^{10} eV) (J.G. Wilson)

In 1934 Bethe and Heitler on the theory, of interaction of fast charged particles with matter on the basis of quantum electrodynamics. They came to the following conclusions:

1. The energy loss by radiation per unit length of path is inversely proportional to the square of the mass of the particles.
2. The rate of energy loss by radiation is nearly proportional to the energy of the particle.
3. The energy loss by radiation is roughly proportional to the square of the atomic number of the material traversed.

Further studies indicate that the mesotrons are neither electrons nor protons.

The general appearance of their tracks and the existence of positives and negatives among them might, at first sight appear to suggest that they should be electronic in nature. But there were several serious objections against such an assumption. First of all, the approximate energy that an electron should possess in order to have the high penetration of the particle under study, when theoretically calculated, led to values very much larger than any of the energies associated with cosmic ray particles. Secondly the high penetrating power of the particles suggested that Bathe and Heitler's theory of collision radiation involving rapid .absorption as expected by the Z^2 law failed in their case; this, however, could not be due to their high energies, since theory was found valid right up to very high energies. Hence the true cause for the failure must be sought elsewhere.

The *low* probability of radiative collision process for heavy particles might well account for the-breakdown of the theory. In fact, if the particles were heavy enough not to be suddenly stopped in their interaction with atomic nuclei, their energy would not be largely transformed into impulse radiation and their loss of energy would not be as important as demanded by the radiative collision, but could be explained on the basis of ionising collisions alone. Hence it followed that the *highly penetrating cosmic ray particles must be heavier than the electrons.* Since, however, when last, they showed an ionisation much like that of the electron, the theory of ionising collisions, according to which the amount of ionisation depends upon the square of the charge carried by the ionising particle, demanded that *they should have not multiple but single charge, like electrons.*

Not Protons : For, in the first place, the small number of slow protons actually observed in cosmic 'rays at sea level and below (about 10%) could hardly represent all the fast particles constituting the hard component. Secondly, the specific ionisation of the particles was found too low to admit their identification with protons and suggested that *they were definitely lighter than protons.*

Thus, cosmic ray studies, both from the experimental and theoretical points of view, led to the conclusion that the highly penetrating cosmic ray particles forming the hard component were most probably a new type of particles with a single charge as the electron and proton and a mass intermediate between the masses of the electron and proton.

Anderson and Neddenneyer, in 1936-37, by further careful investigations made on the cloud chamber tracks of the penetrating cosmic ray particles were able to confirm the above conclusion and place in clear evidence the existence of the new particle. Their general method of procedure in the analysis of trade, which were varmed to have been made by the new particle, was to show that the observed range combined with the observed curvature as well as the actual amount 'of ionisation could not be adequately accounted for, if the proton or the electron-had been responsible for the tracks, whereas they fitted well a singly charged particle of intermediate mass. Several among the cloud chamber photographs taken, in 1936, at Pike's Peak, subjected to such teets, gave *direct evidence* to the existence of the new particle.

As an illustration, the first photograph, in which a track of such a particle was recognised, is reproduced here. Out of the six ionising particles ejected from the same point of the lead plate, probably produced in a nuclear disintegration by a noniodising cosmic ray, one is far more strongly ionising than all the others. This one, ejected nearly vertically upwards, has a range of 4 cms. nearly; its measured radius of curvature is 7 cms. about; the applied magnetic field being 7,900 gauss, its Hp is approximately 5.5×10^4 gauss-cm. Now the energy of a proton having the observed range of 4 cms. would be 1.5 MeV. But a proton of this energy would have a value of $H\rho = 1.7 \times 10^5$ gauss-cm., or a radius of curvature of 20 cms. in the magnetic field used. This is about three times the actual value. Hence, if the observed curvature were produced entirely by the applied magnetic field it would be necessary to conclude that this track must have been left by a massive particle with and e/m value much greater than that of a proton,—which meant a particle having a mass smaller than that of the proton, but still far larger than that of the electron. Since the track analysed had a positive curvature, the particle must be positively charged.

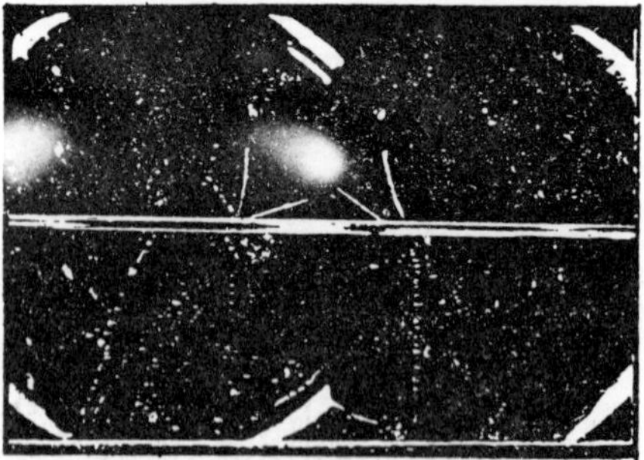

Fig. 18.10 : The First Photograph Which Gave Evidence to the Existence of the Mesotron, (Anderson Ss Neddenneyer).

Street and Stevenson, in 1937, in order to make sure that such a particle of intermediate mass really existed, devised a cloud chamber which would record only those penetrating particles which were nearing the ends of their ranges and thereby produced-heavy ionisation.

18.9 THE STUDY OF SHOWERS

In 1932 Rossi and in 1933 Occhialini and Blackett were the first to place in clear and direct evidence the phenomenon of cosmic ray showers.

Rossi discovered the simultaneous emission of groups of particles from matter as secondary products of cosmic rays.

The different characteristics of the shower phenomenon have been studied with three different techniques.

1. The cloud-chamber.
2. Shower counting array of coincidence counters.
3. Bursts in cosmic ray ionisation chamber.

One must bear in mind that, the chance of shower production depends upon the amount of matter.

But heavy matter alone is not enough for producing showers; a radiation that can effectively interact with matter is required. Cloud chamber photographs make it clear that *showers may be induced either by an ionising particle or a non-ionising photon*, as illustrated in the two

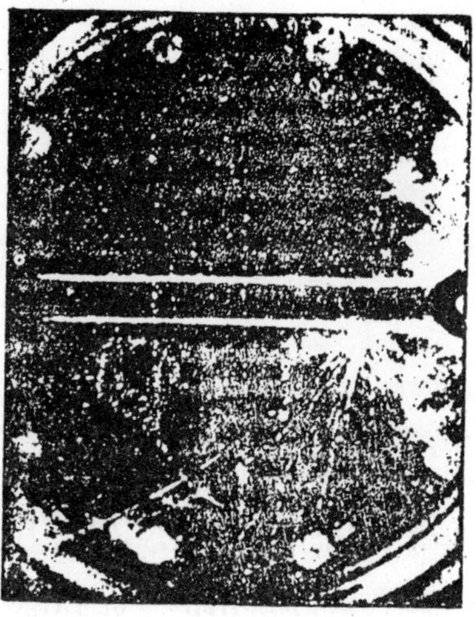

Fig. 18.11 : Shower Produced by an Ionising Particle (Anderson).

photos on this page. In the first one an ionising particle passing through the lead plate ends at the point of origin of the shower below the plate, while in the second a shower emerges from the plate with no apparent track of a particle entering- into the plate. *There is some evidence for the two types of radiation taking roughly equal shares in the production of showers.*

Even within a single shower, sometimes the particles originate not from one point but rather from a narrow region. Furthermore, a shower passing through the lead plate produces other showers radiating, from different foci. These observations indicate that *showers do not happen suddenly in a single elementary process, but they grow as a result of many successive elementary processes* taking place within a short distance in the substance.

Fig. 18.12 : Shower Produced by in Non-Ionising Photon (Anderson).

It can be easily established from the shower tracks that a shower is made up of an approximately equal number of positrons and electrons. There are also strong evidences to admit that a *shower contains photons as well* though these leave no tracks in the chamber. Thus, for instance, when a shower reaches a metal plate and more particles are formed in the plate, small showers appear to come out where nothing visible went in. Granting that something connected with the shower must have gone into the plate to produce this effect, one can conclude that there must be non-ionising particles, *i.e.*, photons associated with the shower which can do this. It belongs to theory to suggest how these non-ionising links are able to transfer energy from shower to shower. As far as the shower photographs are concerned, one has no way of deciding this point or of judging how many photons may have been present in the shower. There are indications, however, from other types of measurements, that *the number of photons present in a large shower is comparable to that of ionising particles, perhaps even slightly in excess.*

Anderson and Neddermeyer in 1936, using a counter-controlled cloud chamber, with a lead plate 35 mm. thick across it, placed in a magnetic field of 7,900 gauss, were able to obtain very interesting data about the *number of ionising particles contained in showers*. At Pasadena (sea level), out of 2,684 photographs of cosmic ray tracks, 383 or 14% contained showers. On Pike's Peak (4,300 metres altitude), out of 1775 photographs, 752 or 42%, contained showers. In both places *two-particle showers were most frequent*; next came those containing 6 to 10 particles; among the photographs taken at Pike's Peak, some half a dozen showed about 100 particles each as the one in the photo below; one particular photo showed more than 300 tracks of electrons and positrons. The number of particles in a shower often increased as the shower traversed the lead sheet.

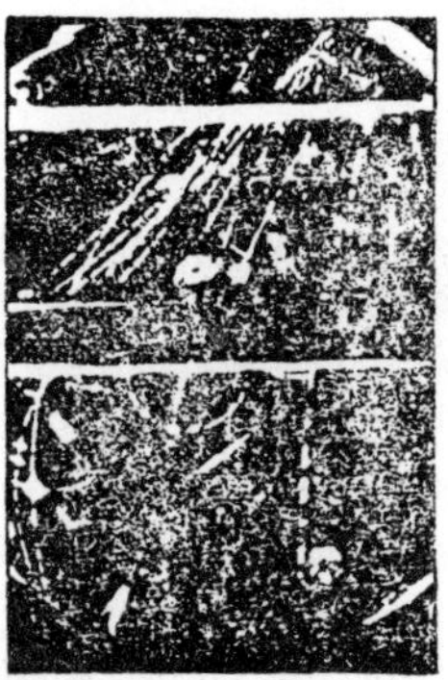

Fig. 18.13 : Two Associated Showers, One Originating from Above the Chamber and the Other in the Lead Plate. (Anderson & Neddermeyer).

The *increased frequency of occurrence of showers at higher altitudes* is a point worthy of note. At Pasadena, only 34 photographs per hour contained showers, while at Pike's Peak, 120 per hour. Chiefly the larger showers were noted to increase at a higher altitude. This observation points out that *the shower particles form the chief constituents of the soft component of cosmic rays.*

From a statistical study of the curvature of the shower tracks, it was found that the energies of the shower particles (both positive and negative) ranged from 1 MeV to 500 MeV, 5 to 20 MeV being the most frequently occurring energy. The total energy in the biggest shower of 300 particles was estimated to be greater than 15,000 MeV.

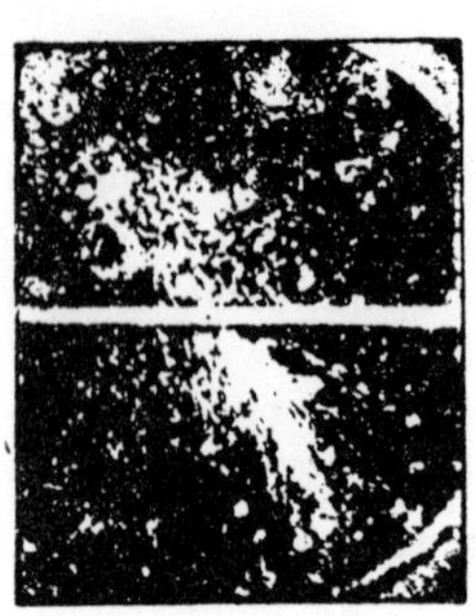

Fig. 18.14 : A Complex Shower with More Than 100 Tracks Obtained at Pike's Peak.

When shower particles passing through a plate produce other showers, the particles of the newly produced ones are found to be less energetic, as seen from the greater curvature of their tracks in the magnetic field. This suggests that *in shower formation, we are dealing with a process of continuous dissipation of the energy available at the start, without am/replenishment in the intermediate stages*; hence the more particles there are in a showers the less energy each will have and the quicker the shower will die out.

Most of the particles constituting a shower appear to leave the point of origin within a cone of 40° and less. The maximum number of particles diverge at still less angles, about 20°, which, however, increase slightly with the thickness of the metal plate used for producing the shower. This indicates that a *large fraction of the momentum of any given shower particle lies along the direction of the shower*—produc*ing particle*. This would be expected if the shower particles received most of their energy directly from the primary shower-producing ray, which fact is bound to throw light on the mechanism of shower formation.

Shower particles seem to give rise to further showers far more readily in a metal plate than the normal single particles, which might mean that the two kinds of particles are different in nature. As a matter of fact they are; the shower particles are mostly electrons belonging to the soft component of cosmic rays, while the bulk of single particles are mesons constituting the hard component.

18.10 NATURE OF COSMIC RADIATION

The primary cosmic radiation arriving at the top of the atmosphere from outerspace is constituted largely with protons. Measurements on the charge spectrum of the primary radiation show that the intensity decrease with increasing charge and no certain cases of nuclei with charge greater than 26 have been observed.

The secondary cosmic radiation reaching sea level through the atmosphere sea level through the atmosphere there are evidences to show that it also contains more massive components such as protons neutrons and so called star particles.

The neutrons from a more important component in cosmic rays than protons.

Let us study in more detail about primary and secondary cosmic rays.

Primary cosmic rays, the vast majority of the incident particles are protons. Recent experiments have shown that there is a very small percentage of *heavy nuclei.* The velocity range of all types of nuclei is approximately the same. At any given velocity, there are 85 helium nuclei and 6 heavier nuclei for every 1,000 protons. There might be also a small number of electrons. Even if these are not really, primaries, they certainly come into existence in the first few layers of the atmosphere. These primaries as they reach the top of the atmosphere, have energies ranging between 3×10^9 and 3×10^{10} eV roughly, the others being prevented from reaching the earth by the action of the magnetic field of the sun and the earth. But some of the primaries have possibly much higher energies going up to 1017 eV, as deduced from the study of extensive air showers. Within these energy limits, the number of particles N (E) crossing unit area per sec. at a given energy E varies roughly as the inverse square of the energy, $N(E) \approx K.E^{-2}$ (K a constant). This means that there is a preponderance of particles of low energy.

Secondary cosmic rays found in the atmosphere, which have been the subject of more direct research, are highly complex in composition and nature. They contain *heavy nuclear fragments, α-particles, protons, neutrons, positive and negative mesons* and *electrons* as well as *photons*, in different proportions, that vary considerably with-the different points of observation, .due to the different complicated processes by which they are produced. The sequence of events from the top to the bottom of the atmosphere, as far as our present knowledge permits, appears to be somewhat as follows:

As the primary protons, having energies of the order of 10^{10} eV and a consequent relativistic mass ten times as great as their restmass penetrate the atmosphere, they begin to interact with the atoms of oxygen, nitrogen and other gases of the air in a complex series of collisions. These disrupt the nuclei of the atoms that are hit, causing them to explode into heavy fragments, protons, neutrons and mesons—in the so called *"star" effect.* Each of the members of disruption, in its turn, demolishes other nuclei, creating other stars, and the process goes on until the energy of the primary ray is dissipated. At 60,000 feet, less than a third of primary protons survive. But at this altitude an enormous number of secondary particles has already been released, including secondary protons, neutrons, mesons and electrons. As the surviving primaries plunge earthward into the relatively dense lower atmosphere, the frequency of collisions increases and the particles rapidly lose energy. At 14,000 feet, only a comparatively small numbers is sufficiently energetic to be detected, and at sea level the number is still less. Thus, ends the life-history of the primary protons.

As regards the other two probable components of the primary radiation, the heavy nuclei undergo a fate similar to that of the protons. But the primary electrons, coming down into the atmosphere, immediately lead to cascade showers, thereby causing a sharp increase in the total number of electrons, quite at the top of the atmosphere, as indicated by the maximum in the altitude intensity curves. The increase, however, does not extend very far, since the majority of the primary electrons have Only energies 40 to 100 times the critical energy (10^8 eV) in air, and beginning at almost one-seventh of the atmosphere from the top, the number of electrons steadily falls. Some of the higher energy primary electrons give rise to the *extensive air showers* which however will not reach sea level, unless the energy of the initiating particle is of the order of 10^{14} or 10^{15}eV.

The main result of the action of the primary rays on the top layers of the atmosphere is the *production of mesons.* Each primary particle loses only a fraction of its energy in each meson-producing collision and is therefore able to make many collisions and to produce many mesons. It is also possible that a single primary particle may give rise to a shower of mesons, consisting of five to ten mesons, in a single collision. In both oases, the mesons formed .are of comparatively low energy and therefore will come to the end of their career relatively soon. Furthermore, there is evidence that the process of meson production may be a two-stage affair, consisting of generation of a heavy π-meson which decays quickly into a lighter meson. Both types are positively or negatively charged. These mesons lose their energy-chiefly by ionisation, in which they occasionally produce fast electrons through the so-called "knock-on" process. The final behaviour of the meson depends on what kind of charge it carries; if positively charged, it will spontaneously decay producing a positive electron, while if negatively charged it will he captured by a nucleus and eventually absorbed by it, thereby releasing an electron. Sometimes a slow meson can give rise to a star disintegration, thereby producing fresh supply of protons, neutrons, α-particles and heavier nuclei. By these different processes of collision, decay and capture, electrons both positive and negative are created.

These electrons are capable of radiating photons when they decelerate in the electrostatic field of the air nuclei. The generated photons are absorbed quickly by pair production at energies above 20 MeV in air or 5 MeV in lead and by Compton effect below these values. The combination of radiative collision and pair production often results in .the building of *cascade showers*, a self-multiplying process, in which a single election may set off a train of events to create hundreds and even tens of thousands of pairs of particles in a fraction of a second. At the level in the atmosphere where the cosmic ray intensity is greatest, such showers make up approximately 5/6 of the total radiation.

Fast protons and mesons lose energy in small amounts chiefly by ionisation when their energies aft below 1,000 MeV. On the other hand, fast electrons lose energy in *large* amounts by radiative collision and the consequent shower production, when their energies are above 100 MeV in air or 10 MeV in lead. On account of tills fact, the total cosmic radiation divides itself into a penetrating or *hard component*, consisting of mesons, a very small number of protons and possibly some very energetic electrons or photons, and a readily absorbed or *soft component*, consisting of electrons, about an equal number of photons and possibly some very slow mesons or protons. At the position of the maximum for the total intensity, the hard component is only about a fifth, of the soft component. The soft component after passing through its' maximum decreases uniformly till at sea level its intensity is less than one per cent of the maximum value. The hard component, on the other hand, decreases continually from the top of the atmosphere until at sea level its intensity is about 6% of the value at great altitudes. The total radiation at sea level is made up of about 25% soft and 75% hard component and has an intensity of a little more than one ionising particle per minute per sq. cm.

Neutrons also are present among the cosmic rays in the atmosphere, and its intensity increases with altitude. The source of these neutrons are the nuclear interactions between primary protons and air atoms; secondary neutrons originate from cosmic ray "stars" which are nuclear disintegrations caused by neutrons in most cases or by the capture of slow mesons.

When we study more and more about cosmic rays, the question arises in out mind, where do the primary cosmic rays originate and how are they produced?

About the place of origin it is now well established that the cosmic rays originate outside the earth's atmosphere.

About the origin of primary cosmic rays the different suggestions given are as follows:

Remote interstellar space for beyond our galaxy at an effective distance of about 10^{10} light years. The galaxies are distributed at large distances, and the intergalatic space is very empty and contains very few stars. Cosmic rays are supposed to have originated in this space.

This opinion was the first to be suggested by several authors, such as Compton, Blackett, etc., in an attempt to account for the isotropic distribution of cosmic radiation. A suppofled slight variation of cosmic ray intensity with sidereal time was adduced as an additional support for the interstellar origin. If the source of cosmic rays were outside our own galaxy and at rest relative; to its central of gravity, then the motion of our galaxy as a whole, combined with earth's rotation would lead to a small diurnal variation with sidereal time. But, more recently, it has' been definitely proved that this sidereal time effect is fictitious. Furthermore, it is not easy to account for all the protons and especially the heavy nuclei contained fit cosmic rays if their sources placed in the extremely attenuated interstellar space.

Initial Explosion of the Expanding Universe

In order to account for the existence of protons and heavy nuclei in the primary radiation as well as the high energies involved, Lemaitre, Begener and others have suggested that the cosmic radiations are the dust flow the original cosmic explosion that occurred some 3×10^9 years ago, when the whole mass of the universe was concentrated in a single nucleus. There can be no doubt that during such an explosion a fantastically great amount of radiation was formed, with the highest individual energies now known And possibly much more. Certainly protons and other nuclei were shot out in all directions with all energies and in sufficient numbers. These particles would then have been circulating in the universe for the requisite time (a few times 10^9 years) during which some of them would have disappeared by absorption. But the earth would find itself embedded in this cosmic particle radiation, isotropically distributed. This hypothesis though intriguing is not an impossible one and cosmic rays may indeed be residual radiation left over from the largest nuclear explosion of all time.

Within Our Own Galaxy

Considerations of the main protonir constitution of the primary radiation and an adequate mechanism which could accelerate the protons to the requisite high energies have led Fermi to suggest that the cosmic rays originate in our own galaxy. The diffuse material in our galaxy, photo-ionised by the star nearby, is the source of cosmic rays. Clouds of ionised diffuse matter in rotational motion along with the galaxy can set up fields which am impart to the charged particles in them the required high energies. This theory appears to be an attractive one, which may contain the right solution to the problem, but the chief difficulty is that it fails to explain this presence of heavy nuclei in the primary rays.

Within the War System Itself

In order to account for the presence of heavy nuclei in the primaries, the apparent large intensity of cosmic rays and the observed influence of the sunspot activity on them, E. Teller and B. D. Richtmyer, in 1948, have suggested that cosmic rays originate in the solar system itself, *on or near the sun.* The heavy nuclei would be more readily found along with the protons in the solar system than in interstellar space. These particles (both protons and heavy nuclei) could be accelerated to the high energies of cosmic rays by an *extended (wandering) magnetic* field of solar origin, reaching oat to die outermost planet. The same field could distribute the particles throughout the solar system with a high degree of uniformity. This theory, which seems to be a revival of Swarm's idea proposed already in 1933, though in a different form, has the chief merit of assuming that a negligibly small fraction of the sun's energy would be sufficient to generate the accelerating field of the necessary intensity. Such a field has not been actually observed so far, obviously due to the fact that the intensity is too low to be easily detected.